The
Laser
Guidebook

Optical and Electro-Optical Engineering Series
Robert E. Fischer and Warren J. Smith, Series Editors

Other Books of Interest

The
Laser
Guidebook

Jeff Hecht

Second Edition

McGraw-Hill, Inc.
New York   San Francisco   Washington, D.C.   Auckland   Bogotá
Caracas   Lisbon   London   Madrid   Mexico City   Milan
Montreal   New Delhi   San Juan   Singapore
Sydney   Tokyo   Toronto

Library of Congress Cataloging-in-Publication Data

Hecht, Jeff.
    The laser guidebook / Jeff Hecht. — 2nd ed.
        p.   cm
    Includes bibliographical references and index.
    ISBN 0-07-027737-0   ISBN 0-8306-4274-9 (pbk.)
    1. Lasers—Handbooks, manuals, etc. I. Title.
TA1683.H43 1992
621.36'6—dc20   91-19406
                                                    CIP

ISBN 0-07-027738-9

First McGraw-Hill paperback edition published by TAB Books, 1992.
TAB Books is a division of McGraw-Hill, Inc.

*The sponsoring editor for this book was Daniel A. Gonneau, the editing
supervisor was David E. Fogarty, and the production supervisor was Pamela
A. Pelton. This book was set in Century Schoolbook. It was composed by
McGraw-Hill's Professional Book Group composition unit.*

*The paperbound cover photograph is courtesy of Coherent, Inc., Palo Alto,
California.*

8 9 10 DOC/DOC 9 9 8 7

Contents

ABOUT THE AUTHOR

Jeff Hecht, an electronics engineer, is cofounder and senior contributing editor of *Lasers & Optronics* magazine. As a full-time writer and consultant on lasers and their applications, he is a frequent contributor to *Fiberoptic Product News, New Scientist, Optics and Photonic News,* the German magazine *Laser und Optoelektronik,* and *Electronic Times.* Mr. Hecht obtained his B.S. from the California Institute of Technology in 1969. He is a member of the American Physical Society, the Institute of Electrical and Electronics Engineers, the Optical Society of America, and the American Geophysical Union.

Preface

I wrote this book because it is something I long wanted to have on my bookshelf. Several excellent textbooks describe the physics of lasers. Handbooks tabulate laser lines reported in the scientific literature. But I have never found any other book devoted to the functional characteristics of commercial lasers—information vital to those of us who work with lasers.

This book is intended both as a reference and a tutorial introduction to the practical aspects of lasers. It cannot duplicate the in-depth coverage of laser physics found in the best textbooks. However, the first six chapters give an overview of lasers and the accessories used with them that should help the reader new to the field make sense of the chapters on specific lasers. The overall treatment is qualitative, intended to give the reader the intuitive understanding of lasers needed to put them to work.

Most chapters share a common structure to make reference use easier. Laser types in wide use are described in the most detail, including the characteristics of commercial devices. Readers looking for a quick tabulation of commercial lasers should turn to the appendix, which lists lasers by wavelength, and cites the relevant chapter where more details can be found.

The impetus for this new edition was the impressive advances over the past few years in many areas of laser technology, notably in semiconductor, solid-state, free-electron and x-ray lasers. This edition greatly expands the coverage of semiconductor lasers, reflecting their growing importance in the field. Where progress is most rapid, the chapters look at research developments which are likely to have an impact on commercial laser technology during the lifetime of this edition. It isn't easy to keep up, and it sometimes seemed that something new appeared each time I thought I had finished a chapter—but that's one thing that makes laser technology so stimulating. An inevitable tradeoff of expanding the new coverage has been reducing the space devoted to some older lasers, such as ruby, which are being replaced for many applications.

Earlier, shorter versions of many chapters originally appeared in *Lasers & Applications* magazine, now *Lasers & Optronics*, which also has published excerpts. I am grateful to that magazine's founding publisher, Carole Black, for sharing my belief that this material would be useful to the laser community. I also received much valuable help from members of the magazine's editorial staff, particularly Richard Cunningham, Robert Clark, Breck Hitz, Jim Cavuoto, and invaluable encouragement from some of the magazine's readers. Comments on the first edition have helped me fine-tune this version.

Many people have given graciously of their time to check for errors and misunderstandings in earlier versions of these chapters, both in the first edition and this one. Thanks particularly to Bob Anderson, Steve Anderson, Stephanie Banks, Tony Bernhardt, Mark Boehm, Dan Botez, Gerry Bricks, Joan Bromberg, Carl Burns, Evan Chicklis, Paul Crosby, Brian Davis, Mark Dowley, Gary Forrest, Horace Furumoto, Bob Goldstein, John Grace, Tim Grey, Hans-Peter Geieneisen, Ken German, Don Heller, Randy Heyler, Chuck Higgins, Jim Higgins, Dan Hogan, Bill Hug, Steve Jarrett, Bill Jeffers, Tony Johnson, Joanne LaCourse, Steve Lewellan, Andrew Kearsley, Paul Kenrick, Gary Klauminzer, Tom Kugler, Phil LadenLa, Kurt Linden, Arlan Mantz, Peter Moulton, Ed Neister, Ross Payne, Stephen Picarello, Bob Pitlak, Dick Roemer, Bob Rudko, Ken Sample, Roger Sandwell, Mike Sasnett, Dick Steppel, Jim Stimson, Tom Stockton, Tim VanSlambrouck, Bill Vaughan, David Wall, Colin Webb, Sicco Westra, Dave Whitehouse, Ben Woodward, Bill Young, and Peter Zory. Any mistakes that slipped through are my fault, not theirs.

I also owe thanks to my editors at McGraw-Hill, Harry Helms, Roy Mogilanski, Rich Krajewski, and Dennis Gleason for the first edition, and Dan Gonneau and David Fogarty for the second. Finally, thanks to my wife Lois for running innumerable errands and applying encouragement when needed.

Jeff Hecht

Introduction

The word *laser* is an acronym for "light amplification by the stimulated emission of radiation," a phrase which covers most, though not all, of the key physical processes inside a laser. Unfortunately, that concise definition may not be very enlightening to the nonspecialist who wants to *use* a laser, and cares less about its internal physics than its external characteristics. The laser user is in a position analogous to the electronic circuit designer. A general knowledge of laser physics is as helpful to the laser user as a general understanding of semiconductor physics is to the circuit designer. However, their jobs require them to understand the operating characteristics of complete devices, not to assemble lasers or fabricate integrated circuits.

This book is written and organized with the needs of the laser user in mind. The first few chapters describe the basic ideas behind lasers, how they work, and the characteristics that are most important from a user's standpoint. They give an overview of the field, but cannot replace a good textbook or tutorial introduction to lasers. The following chapters focus on individual laser types, first outlining operating principles, then describing specific characteristics such as output power, wavelength, and input power requirements. The Appendix tabulates the most important characteristics of major lasers. The chapters on individual types of lasers follow the same basic outline and are structured both to be read as chapters and to be used for looking up specific data. Inevitably, such a structure brings with it some redundancy; wavelength and power range, for example, may be mentioned a few times in the same chapter. However, what may seem repetitive when reading the chapter also makes the information easier to find when searching out specific data in the book.

As a guide to practical laser technology, this book concentrates on lasers which are available commercially or are otherwise important to current and potential laser users. Details on performance are given for lasers where the technology is reasonably well established, such as helium-neon; newer types such as free-electron lasers are described in more general terms. Development of high-energy laser weapons is covered only briefly because the technology is classified and of limited interest. In a few cases—notably chemical, free-electron, and x-ray lasers—far more effort has gone into military weapons programs than into development of lower-power lasers for civilian use.

An annotated bibliography and reference list follows each chapter. The listing includes selected review articles as well as references cited in the text so that readers can search out more details if they desire. Some references may appear dated, but some fields of laser technology are not as fast-moving as they may appear from a distance. For example, the fundamentals of rare gas ion lasers have changed little in the past few years, although commercial models have pushed further into the ultraviolet. In addition, few good review articles on more mature lasers—in particular, many gas types—have come to my attention in recent years.

No book can hope to keep up with the continual changes in the product lines of laser manufacturers. Specific models are mentioned in a few places, but only as examples of the characteristics to expect in commercial products. Companies also come and go in the field. The best way to follow the laser industry is by reading one or more of its monthly trade magazines: *Lasers & Optronics, Laser Focus World*, and *Photonics Spectra*. (To make my biases perfectly clear, I will identify myself as a cofounder and contributing editor of *Lasers & Optronics*. The other magazines provide the sort of competition that keeps all three publications working hard to serve their readers.) Many other magazines and scholarly journals publish useful material on lasers and related technology.

All three major trade magazines publish industry directories, and those listings are the best way to locate manufacturers of specific lasers. They also tabulate specifications for selected lasers and advertisements from manufacturers. I find the *Lasers & Optronics Buying Guide* the easiest to use, but that may be because I helped organize it. The bibliography gives information on all three directories.

Definition and Description of a Laser

From a practical standpoint, a laser can be considered as a source of a narrow beam of monochromatic, coherent light in the visible, infrared, or ultraviolet parts of the spectrum. The power in a continuous beam can range from a fraction of a milliwatt to around 25 kilowatts (kW) in commercial lasers, and up to more than a megawatt in special military lasers. Pulsed lasers can deliver much higher peak powers during a pulse, although the power averaged over intervals while the laser is off and on is comparable to that of continuous lasers.

The range of laser devices is broad. The laser medium, or material emitting the laser beam, can be a gas, liquid, glass, crystalline solid, or semiconductor crystal and can range in size from a grain of salt to filling the inside of a moderate-sized building. Not every laser produces a narrow beam of monochromatic, coherent light. Semiconductor diode lasers, for example, produce beams that spread out over an angle of 20 to 40°, hardly a pencil-thin beam. Liquid dye lasers emit at a broad or narrow range of wavelengths, depending on the optics used with them. Other types emit at a number of spectral lines, producing light that is neither truly monochromatic nor coherent.

Practically speaking, lasers contain three key elements. One is the laser medium itself, which generates the laser light. A second is the power supply, which delivers energy to the laser medium in the form needed to excite it to emit light. The third is the optical cavity or *resonator*, which concentrates the light to stimulate the emission of laser radiation. All three elements can take various forms, and although they are not always immediately evident in all types of lasers, their functions are essential. Figure 1.1 shows these elements in a ruby and a helium-neon laser; the internal workings of lasers are described in more detail in Chap. 3.

There are several general characteristics which are common to most lasers which new users may not expect. Like most other light sources, lasers are inefficient in converting input energy into light. Efficiencies range from under 0.01 to over 30 percent, but few types are much above 1 percent efficient. These low efficiencies can lead to special cooling requirements and duty-cycle limitations, particularly for high-power lasers. In some cases, special equipment may be needed to produce the right conditions for laser operation, such as cryogenic temperatures for the lead salt semiconductor lasers described in Chap. 21. Operating characteristics of individual lasers depend strongly on structural components such as cavity optics, and in many cases a wide range is possible. Packaging can also have a strong impact on laser characteristics and the use of lasers for certain applications. Thus wide ranges of possible characteristics are specified in many chapters, although single devices will have much more limited ranges of operation.

Differences from Other Light Sources

The basic differences between lasers and other light sources are the characteristics often used to describe a laser: the output beam is nar-

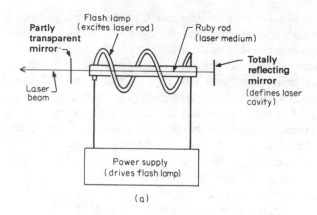

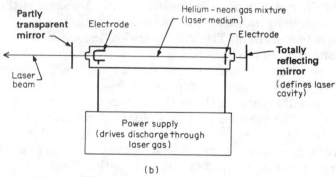

Figure 1.1 Simplified views of two common lasers, (*a*) ruby and (*b*) helium-neon, showing the basic components that make a laser.

row, the light is monochromatic, and the emission is coherent. Each of these features is important for certain applications and deserves more explanation.

Most gas or solid-state lasers emit beams with a divergence angle of about a milliradian, meaning that they spread to about one meter in diameter after traveling a kilometer. (Semiconductor lasers have much larger beam divergence, but suitable optics can reshape the beam to make it much narrower.) The actual beam divergence depends on the type of laser and the optics used with it. The fact that laser light is contained in a beam serves to concentrate the output power onto a small area. Thus a modest laser power can produce a high intensity inside the small area of the laser beam; the intensity of light in a 1-milliwatt (mW) helium-neon laser beam is comparable to that of sunlight on a clear day, for example. The beams from high-power lasers, delivering tens of watts or more of continuous power or higher peak powers in pulses, can be concentrated to high enough intensities that they can weld, drill, or cut many materials.

The laser beam's concentrated light delivers energy only where it is focused. For example, a tightly focused laser beam can write a spot on a light-sensitive material without exposing the adjacent area, allowing high-resolution printing. Similarly, the beam from a surgical laser can be focused onto a tiny spot for microsurgery, without heating or damaging surrounding tissue. Lenses can focus the parallel rays in a laser beam to a much smaller spot than they can the diverging rays from a point source, a factor which helps compensate for the limited light-production efficiency of lasers.

Most lasers deliver a beam that contains only a narrow range of wavelengths, and thus the beam can be considered monochromatic for all practical purposes. Conventional light sources, in contrast, emit light over much of the visible and infrared spectrum. For most applications, the range of wavelengths emitted by lasers is narrow enough to make life easier for designers by avoiding the need for achromatic optics and simplifying the task of understanding the interaction between laser beam and target. However, for some applications in spectroscopy and communications, that range of wavelengths is not narrow enough, and special line-narrowing options may be required.

One of the laser beam's most unique properties is its *coherence*, the property that the light waves it contains are in phase with one another. Strictly speaking, all light sources have a finite coherence length, or distance over which the light they produce is in phase. However, for conventional light sources that distance is essentially zero. For many common lasers, it is a fraction of a meter or more, allowing their use for applications requiring coherent light. The most important of these applications is probably holography, although coherence is useful in some types of spectroscopy, and there is growing interest in communications using coherent light.

Some types of lasers have two other advantages over other light sources: higher power and longer lifetime. For some high-power semiconductor lasers, lifetime must be traded off against higher power, but for most others the life-vs.-power trade-off is minimal. The combination of high power and strong directionality makes certain lasers the logical choice to deliver high light intensities to small areas. For some applications, lasers offer longer lifetimes than do other light sources of comparable brightness and cost. In addition, despite their low efficiency, some lasers may be more efficient in converting energy to light than other light sources.

The Laser Industry

Commercial Lasers. There is a big difference between the world of laser research and the world of the commercial laser industry. Unfortunately, many text and reference books fail to differentiate between types of lasers that can be built in the laboratory and those which are readily available commercially. That distinction is a crucial one for laser users.

Laser emission has been obtained from hundreds of materials at many thousands of emission lines in laboratories around the world. Extensive tabulations of these laser lines are available (Weber, 1982), and even today researchers are adding more lines to the list. However, most of these laser lines are of purely academic interest. Many are weak lines close to much stronger lines which dominate the emission in practical lasers. Most of the lasers that have been demonstrated in the laboratory have proved to be cumbersome to operate, low in power, inefficient, and/or simply less practical to use than other types.

Only a couple of dozen types of lasers have proved to be commercially viable on any significant scale; these are described in the rest of this book and summarized in the Appendix. Some of these types, notably the ruby and helium-neon lasers, have been around since the beginning of the laser era. Others, such as vibronic solid-state, are promising newcomers. The family of commercial lasers is expanding slowly, as new types such as titanium-sapphire come on the market, but with the economics of production a factor to be considered, the number of commercially viable lasers will always be limited.

There are many possible reasons why certain lasers do not find their way onto the market. Some require exotic operating conditions or laser media, such as high temperatures or highly reactive metal vapors. Some emit only feeble powers. Others have only limited applications, particularly lasers emitting low powers in the far-infrared, or in parts of the infrared where the atmosphere is opaque. And some simply cannot compete with materials already on the market.

Market Size. Sales of lasers per se—including power supply and optics—was about $750 million in the noncommunist world in 1990 (*Lasers & Optronics*, 1990). Total sales of systems containing lasers was much higher, probably several billion dollars. The difference reflects an important fact: for many applications, the cost of the laser represents only a small part of the system cost.

Because some lasers are much more expensive than others, there are large dichotomies between rankings of laser sales based on number and those based on dollars. Over 25 million semiconductor diode lasers were sold in 1990, representing 98 percent of the total number of lasers. However, diode lasers accounted for less than 30 percent of the dollar volume of sales. Conversely, some 4000 carbon dioxide gas lasers accounted for 21 percent of 1990 laser sales dollars (*Lasers & Optronics*, 1990).

Types of Organizations. Many companies in the laser industry are young, and some are not far removed from humble origins in a garage or basement. Some large companies started their own laser divisions, but most laser groups began as independent start-ups. The first wave of laser start-ups were formed in the early to mid-1960s, in the wake of the demonstration of the first laser in 1960. That wave included the two companies generally considered the largest laser manufacturers, Spectra-Physics Inc. and Coherent Inc. Many more laser companies have been formed since, with start-ups continuing to play an important role in developing new technology through the 1980s.

The last half of the 1980s saw many mergers and much restructuring of the laser industry. Two of the largest independent laser companies, Spectra-Physics and Lumonics Inc., were acquired by much larger corporations, Ciba-Geigy and Sumitomo, respectively. (However, in mid-1990, Ciba-Geigy sold Spectra-Physics to Pharos AB, a Swedish company affiliated with the Nobel group, but comparable in size to Spectra-Physics.) Some smaller companies also were sold, reflecting a general consolidation of the industry, but that was offset by formation of new companies and spin-offs of laser divisions of some larger firms.

No single company offers every kind of laser, but many of the larger firms do produce several different kinds. In contrast, most smaller companies specialize in just one or a few types of lasers. While some companies produce related products, others stick to lasers. Likewise, many other companies concentrate on related products such as optics or power supplies and stay away from producing lasers per se. Some companies build systems around lasers they purchase from outside vendors; others produce their own lasers for use in systems. The field is a dynamic one, and readers who want to keep up with who's making what should follow the trade magazines which cover the field: *Lasers & Optronics, Laser Focus World,* and *Photonics Spectra.*

Market Structure. There are few clear-and-fast patterns in the structure of the laser market. The market has become an international one, with major activity in North America, western Europe, and Japan. The two largest laser makers, Spectra-Physics and Coherent Inc., for many years concentrated on gas and dye lasers, but both now offer broader product lines. Both are nominally based in the United States but have strong ties to Europe; Spectra-Physics is a subsidiary of a Swedish company, while a principal Coherent subsidiary is based in Germany. A third large laser manufacturer, Lumonics Inc., is a Canadian subsidiary of a Japanese company, and in turn has a major division based in the United States. Japanese electronics giants dominate the world market for low-priced semiconductor lasers, but they have some competition in other semiconductor lasers, and so far have not become major suppliers of high-power semiconductor laser arrays.

Lasers have long been among the military sensitive technologies subject to export controls by the United States and many of its allies. The severity of restrictions depends on the political climate, and eased with the decreased international tensions of 1990. However, as of 1990, there is little trade in laser or related technologies between the Soviet Union, eastern Europe, and China on one hand and North America, Japan, and western Europe on the other. In general, developing countries must import laser equipment because they lack domestic industries.

The biggest laser makers also offer optical components and accessories for single-stop shopping, but many laser makers offer accessories only if they can be used with their own products. There are many small companies (and some large ones) which make laser optics, often custom-made for specific applications. There are several companies which function as catalog supply houses for optics and accessories. All told, about a hundred companies make lasers, and several hundred more provide laser accessories and offer services to the laser industry. The cast of companies changes; for current information consult an industry directory.

An Introduction to Laser Safety

Laser safety has been controversial since lasers began appearing in laboratories. The two major concerns are exposure to the beam (which presents much more danger to the eyes than to the rest of the body) and high voltages within the laser and power supply. Many standards have been developed covering either the performance of laser equipment or the safe use of lasers; some developed by government agencies have legal status, while others are recommendations by voluntary organizations. Tabulations of these standards are available (Weiner, 1990).

High-power laser beams can burn the skin, but the most important hazards of laser beams are to the eyes, which are the part of the body most sensitive to light. Like sunlight, laser light arrives in parallel rays, which the eye focuses to a point on the retina, the layer of cells that responds to light. Just as staring at the sun can damage vision, exposure to a laser beam of sufficient power can cause permanent eye damage.

Eye hazards have attracted considerable attention from standards writers and regulators. They depend on parameters such as laser wavelength, average power over a long interval, peak power in a pulse, beam intensity, and distance from the laser. Wavelength is important because only certain wavelengths—between about 0.4 and 1.5 micrometers (μm)—can penetrate the eye well enough to damage the retina. Ultraviolet light can damage surface layers of the eye (and some can penetrate to the retina, especially in people whose natural ocular lens has been removed). Infrared light also can damage the surface of the eye, although the damage threshold is higher than that for ultraviolet light. Eye response also differs within the range that penetrates the eyeball because the eye has a natural aversion response, which makes it turn away from a bright visible light which is not triggered by infrared wavelengths longer than 0.7 μm.

High-power laser pulses pose dangers different from those due to lower-power continuous beams. A single high-power pulse lasting less than a microsecond can cause permanent damage if it enters the eye. A lower-power beam presents danger only for longer-term exposure. Distance reduces laser power density, thus decreasing the potential for hazards to the eye.

Many countries have safety standards which must be met by laser products sold within their borders. The National Center for Devices and Radiological Health, part of the U.S. Food and Drug Administration, has established standards in the United States (CDRH, 1985). Many other countries have individual standards based largely on recommendations of the International Electrotechnical Commission (IEC, 1984). Laser product standards include provisions for warning labels indicating hazard class. (In the United States, the hazards increase with class number, with Class IV covering the most powerful lasers.) Depending on hazard classification, lasers sold in the United States may require beam shutters to block the beam when not in use, key interlocks, and other safety features.

The United States does not have federal standards for the safe use of lasers, but several states have set their own standards, and many other countries have standards for laser use. The IEC recommended standard covers the safe use of lasers. The American National Standards Institute (ANSI) has developed a voluntary standard for laser use. These standards concentrate on avoiding eye exposure.

Systems incorporating lasers also must meet safety standards. In many of these systems the beam is contained so that it cannot leave the box or reach the user's eye, thus avoiding the need for warning labels. Laser printers and compact-disk players are the most common examples. Some scan a low-power beam so rapidly that the beam does not stay in one place long enough to pose a hazard; laser scanners at supermarket checkout counters are an example. High-power lasers used in manufacturing may operate behind beam-absorbing shields, which separate the operator from the workpiece and laser beam. Such shields also may be used in some laboratories.

People working where they cannot avoid exposure to laser beams that pose eye hazards should wear safety glasses or goggles. Special filters in the safety goggles block light at certain laser wavelengths, while transmitting enough other light to enable the worker to see. The proper goggles should be used for work with each type of laser; it does no good to block the blue-green light from an argon laser when working with a red helium-neon laser. Users can select general-purpose goggles that fit over the eyes and spectacles, or can have prescription spectacles made from laser-blocking filter glasses. Goggles should block all possible paths for laser light to reach the eye, because hazardous pulses can be reflected from any shiny surface and arrive from unexpected angles.

High-voltage hazards have attracted less attention, but in practice are far more deadly. There is no public record of fatal injury due to a laser beam, but several people have been electrocuted by high voltages in a laser or power supply. Except for semiconductor lasers, virtually all lasers require high voltages and sometimes high currents to generate a beam. The high voltage may be applied directly to the laser medium or to a pump lamp, but it is present in the system. Capacitors and some other components may retain dangerous voltages for a considerable time after the laser is used, especially in pulsed lasers. Be wary of these electrical hazards.

If you plan to work directly with lasers, you should read a book that discusses laser safety in more detail. The most exhaustive survey is by David Sliney and Myron Wolbarsht (1980). A briefer guide to laser safety with more emphasis on practical measures has been written by D. C. Winburn (1990). The Lawrence Livermore National Laboratory has issued a booklet on safety eyewear (1987). Documents also are available from the National Center for Devices and Radiological Health (1390 Piccard Dr., Rockville, MD. 20850), and the American National Standards Institute (1430 Broadway, New York, NY 10018).

Overview of Laser Applications

Laser applications are far too broad and diverse to cover in any detail here. Individual chapters cover the most important applications of each major type. The industrial and commercial applications fall into several basic categories:

- Materials working
- Measurement and inspection
- Reading, writing, and recording of information
- Displays
- Communications
- Holography
- Spectroscopy and analytical chemistry
- Remote sensing
- Surveying, marking, and alignment
- Surgery and medical treatment

There are also several important research applications pursued on various scales by government and industrial organizations and in universities:

- Laser weaponry
- Laser-induced nuclear fusion (sometimes called inertial confinement fusion)
- Isotope enrichment (particularly of uranium and plutonium)
- Spectroscopy and atomic physics
- Measurement
- Plasma diagnostics

Historically, much support for laser research has come from the Department of Defense or quasi-military organizations. Indeed, the principal justification for research in laser fusion has been military interest in simulating the effects of nuclear weapons. The largest military research program now is in efforts to develop laser weapons for use against satellites or for defense against nuclear attack. Laser systems are now in production which serve as range finders, target designators for smart bombs, and scorekeepers for battle-simulation systems.

Bibliography

ANSI Z-136.1-1980, ANSI Standard for the Safe Use of Lasers, American National Standards Institute, New York, 1980 (standard).

Jeff Hecht: *Understanding Lasers*, Sams, Indianapolis, 1988 (tutorial introduction to lasers).

Laser Focus World Buyers' Guide, published annually by PennWell Publishing, P.O. Box 989, Westford, MA 01886. Write publisher for information.

Lasers & Optronics (staff report), "The laser marketplace—forecast 1990," *Lasers & Optronics 9* (1):39–57 (1990) (annual market overview).

Lasers & Optronics Buying Guide, published annually by Elsevier Communications, P.O. Box 650, Morris Plains, NJ 07950-0650. Write publisher for information.

Lawrence Livermore National Laboratory: *A Guide To Eyewear for Protection from Laser Light*, Lawrence Livermore National Laboratory, Livermore, Calif., 1987.

Performance Standards for Laser Products, National Center for Devices and Radiological Health, Publication No. HFX-430 (federal standard). To obtain a copy, write NCDRH, 1390 Piccard Dr., Rockville, MD 20850 (standard).

Photonics Directory, published annually by Laurin Publishing Co., Berkshire Common, P.O. Box 1146, Pittsfield, MA 01202. Write publisher for information.

David Sliney and Myron Wolbarsht: *Safety with Lasers and Other Optical Sources*. Plenum, New York, 1980 (exhaustive review of laser safety, totaling over 1000 pages).

Marvin J. Weber (ed.): *CRC Handbook of Laser Science and Technology*, 2 vols., CRC Press, Boca Raton, Fla. 1982; also Marvin J. Weber (ed.): *CRC Handbook of Laser Science and Technology Supplement 1*, CRC Press, 1989.

Robert Weiner: "Status of laser safety requirements," *Lasers & Optronics 1990 Buying Guide* 327–329, 1990 (tabulation of standards documents, updated in each annual edition).

D. C. Winburn, *Practical Laser Safety*, 2d ed., Marcel Dekker, New York, 1990 (practical guide to laser safety, by former laser safety officer at the Los Alamos National Laboratory).

A Brief History
of the Laser

The key concepts of laser physics are so far removed from the normal realm of classical physics that it took many years to put the conceptual pieces together into a working laser. The basic idea of the stimulated emission of radiation dates back to the early part of this century, but it wasn't until the 1950s that physicists began to work on putting it to practical use. Once the conceptual logjam was broken, the early 1960s saw an amazing and rapid proliferation of laser types, followed by years of experimentation with the new playthings. Although important research advances have also been made in the past two decades, the most important advances in that period have been in developing laser applications ranging from fiber-optic communications and information handling to laser fusion and high-energy laser weapons.

This chapter can only touch on some of the highpoints in the tangled history of the laser. There are really two parallel and largely intertwined stories to be told: the intellectual evolution of the laser concept and the competition among a number of strong personalities to build the first laser. Readers interested in the intellectual evolution of the field would do well to study *Masers and Lasers, An Historical Approach*, by Mario Bertolotti (1983) or an objective early account by Bela Lengyel (1966). The interplay of individuals has not been treated in as much detail, but I have written historical chapters in books on laser applications (Hecht and Teresi, 1982) and laser weapons (Hecht, 1984). Two more books on laser history are due to appear at about the same time as this book. One is a collection of interviews with laser pioneers, including an overview of laser history (Hecht, 1991). The other is a scholarly analysis by science historian Joan Bromberg (1991), written as part of the Laser History Project sponsored by the American Institute of Physics.

Stimulated Emission

The idea of stimulated emission of radiation originated with Albert Einstein (1916). Until that time, physicists had believed that a photon could interact with an atom in only two ways: it could be absorbed and raise the atom to a higher energy level or be emitted as the atom dropped to a lower energy level. Einstein proposed a third possibility—that a photon with energy corresponding to that of an energy-level transition could stimulate an atom in the upper level to drop to the lower level, in the process stimulating the emission of another photon with the same energy as the first.

In the normal world, stimulated emission is unlikely because at thermodynamic equilibrium more atoms are in lower energy levels than in higher ones. Thus a photon is much more likely to encounter an atom in a lower level and be absorbed than to encounter one in a higher level and stimulate emission. The first evidence for stimulated emission was not reported until more than a decade after Einstein's prediction (Ladenburg, 1928). For more than two decades thereafter, stimulated emission seemed little more than a laboratory curiosity.

The Coming of the Maser

The first efforts to use stimulated emission were in the microwave region and led to the invention of the maser (from "microwave amplification by the stimulated emission of radiation"). The maser concept evolved nearly simultaneously in the United States and the Soviet Union, but the name usually linked with the invention is that of Charles H. Townes, a physicist then on the faculty of Columbia University and now professor emeritus at the University of California at Berkeley.

Townes says that the maser idea struck him early one morning in the spring of 1951, while he was sitting on a park bench in Washington before going to a meeting of the American Physical Society (Townes, 1978). He had been puzzling over ways to build a source of millimeter waves when he realized that molecules had the right characteristics. He envisioned starting with a molecular beam, which could be split into excited and unexcited portions. Molecules in the excited part could be stimulated to emit microwaves when placed in a special resonant cavity designed to enhance the emission. He outlined the idea to postdoctoral fellow Herbert Zeiger and graduate student James P. Gordon, and by 1953 they had a working maser. Meanwhile, Aleksander M. Prokhorov and Nikolai Basov (1954) of the Lebedev Physics Institute in Moscow had completed detailed calculations of the conditions required for maser action, which were published shortly after Townes's results. The contributions of all three men were recognized in 1964 when they shared the Nobel prize in physics.

Theoretical Groundwork for the Laser

Masers proliferated rapidly in the 1950s. Meanwhile, Townes and other physicists began looking beyond the microwave region to shorter wavelengths. They recognized that at those shorter wavelengths the physical conditions required to produce stimulated emission would be very different than in a maser. Many of the key principles were worked out by Townes and Arthur L. Schawlow, then a Bell Laboratories researcher and now on the faculty at Stanford University. Their results were published in a major paper (Schawlow and Townes, 1958), which pointed out important differences between requirements in the microwave and visible regions, including cavity structure, spontaneous emission ratios, energy differences between energy levels, and excitation mechanisms. The two also filed a patent application before the paper was published, leading to a now-expired U.S. patent (2,929,922).

Meanwhile, a Columbia graduate student, Gordon Gould, was working out his own analysis of the conditions required for stimulated emission at visible wavelengths. Gould wrote his proposals in a set of notebooks dated 1957 but did not try to publish his results promptly. Considering himself more an inventor than a researcher, Gould wanted to patent his work. He had his notebooks notarized promptly, but because of bad legal advice he did not file a patent application until April 1959, about 9 months after the Schawlow-Townes patent application was submitted, and after publication of their paper.

The Schawlow-Townes patent was granted promptly, but Gould's application ran into a lengthy set of interferences. Gould pursued the patent himself for many years, losing some rounds and winning others, but in the process running up massive legal bills. Finally, he turned the litigation and part interest in his patent position over to the Refac Technology Development Corp. of New York. Eventually he was granted four patents (Gould, 1977, 1979, 1987, 1988) based on divisions of his original application. The 1977 patent covers optical techniques for pumping or energizing the laser medium, such as using a flashlamp to drive a dye or neodymium laser. The 1979 patent covers a range of laser applications. The 1987 patent covers lasers pumped by electric discharges and the 1988 patent covers the Brewster angle windows used in many lasers. The Gould patents have withstood legal challenge, and most of the U.S. laser industry now licenses rights through the Patlex Corporation of Las Cruces, New Mexico.

Schawlow and Townes have received many scientific honors for their work, but Gould received little recognition until his patents were issued. Ironically, Gould benefited financially from the long delay in issuing the patents, because laser sales, on which royalties are calculated, are much larger today than in the 1960s and 1970s. The question of who really deserves credit for conceiving the laser may never be definitely resolved. However, while Gould did lose the prestige race, it was he who first coined the word *laser* in his notebooks. Schawlow and Townes described their idea as an "optical maser."

The Race to Build a Laser

Publication of the Schawlow-Townes paper stimulated many efforts to build lasers, and interest spread beyond the narrow scientific community. Gould and executives from TRG Inc., a small Long Island company, got an enthusiastic reception when they presented the Pentagon with a proposal to build a laser. Indeed, the Department of Defense's Advanced Research Projects Agency was so excited it gave TRG a million dollars for the project, rather than the $300,000 the company had requested. Then the military agency proceeded to deny Gould a security clearance to work on the project because of his youthful but long-dormant interest in Marxism (Hecht and Teresi, 1982).

Schawlow, Gould, and most other researchers thought that the best materials for building lasers were gases. Theodore H. Maiman, a young physicist at Hughes Research Laboratories in Malibu, California, quietly disagreed, preferring synthetic ruby crystals, although some theorists insisted that ruby would not work. Maiman, who had studied energy levels in ruby extensively while working on ruby masers, proved that the theorists were wrong. In mid-1960 he proudly demonstrated the world's first laser: a rod of synthetic ruby with reflecting coatings on the ends, surrounded by a helical flashlamp. When the lamp was pulsed, a pulse of red light emerged from one end of the rod, which had a partially transparent coating. The laser era was born.

Maiman's course was not an easy one. Hughes management told him to stop working on the ruby laser, although once it worked the company's public relations officials were delighted to show off the results. The prestigious journal *Physical Review Letters* rejected his report of the ruby laser as "just another maser," a blunder few scientific journals can claim to have matched. A shorter version was published in the British journal *Nature* (Maiman, 1960). Today, however, Maiman is universally recognized as the person who built the first laser and has received a number of honors, including the Japan Prize. The ruby laser remains in commercial use, although its popularity has declined in recent years.

The Great Laser Explosion

Maiman's demonstration of the ruby laser opened the floodgates. Before the year was out, a second type of solid-state laser was reported, trivalent uranium ions in calcium fluoride, by Peter P. Sorokin and M. J. Stevenson (1960) at the IBM Corp. That laser has never found significant practical use. However, it was followed soon afterward by the demonstration of the helium-neon laser by Ali Javan, W. R. Bennett Jr., and Donald R. Herriott at Bell Telephone Laboratories in Murray Hill, New Jersey (Javan et al., 1961). Their first helium-neon laser (Fig. 2.1) operated at 1.15 micrometers (μm) in the near-infrared; later other researchers found the 632.8-nanometer (nm) red line which has made the helium-neon laser one of the most widespread types (White and Rigden, 1962).

The laser boom really got going in 1961. L. F. Johnson and K. Nassau (1961) demonstrated the first solid-state neodymium laser, in which the neodymium ion was a dopant in calcium tungstate ($CaWO_4$). Elias Snitzer demonstrated the first neodymium-glass laser at American Optical that same year. However, it would be three more years before today's best choice of neodymium host for most commercial applications—yttrium aluminum garnet (YAG)—was demonstrated as a laser material by J. E. Geusic, H. M. Marcos, and L. G. Van Uitert (1964).

The first semiconductor diode lasers were demonstrated nearly simultaneously by three separate groups in fall 1962 (Hall, 1976). All three teams—at General Electric Research Laboratories in Schenectady, New York; the IBM Watson Research Center in Yorktown Heights, New York; and MIT's Lincoln Laboratories in Lexington, Massachusetts—demonstrated similar gallium arsenide diodes cooled to the 77 K temperature of liquid nitrogen and pulsed with high-current pulses lasting a few microseconds.

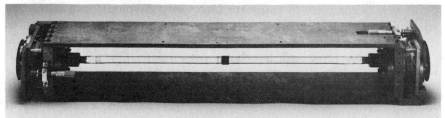

Figure 2.1 The first gas laser, a helium-neon type conceived and developed by Ali Javan, which was demonstrated for the first time on December 12, 1960, at Bell Telephone Laboratories in Murray Hill, New Jersey. The helium-neon gas mixture was contained in an 80-cm-long quartz tube with inner diameter of 1.5 cm: its output was in the near-infrared. (*Photo by Lee Nadel: courtesy of Ali Javan and Laser Science, Inc.*)

The next few years saw the births of several more of today's most important lasers. William B. Bridges (1964) observed 10 laser transitions in the blue and green parts of the spectrum from singly ionized argon, the basis of today's argon ion laser. C. Kumar N. Patel (1964) obtained a 10.6-μm laser emission from carbon dioxide, today a high-power workhorse in industry. Sorokin and J. R. Lankard (1966) at IBM's Watson Research Center demonstrated the first organic dye laser, today a standard tool of laser spectroscopy. The first chemical laser, hydrogen chloride emitting at 3.7 μm, was demonstrated in 1965 by J. V. V. Kaspar and G. C. Pimentel.

The rate of discovery of new laser types slowed during the 1970s, with the most important discoveries being the excimer and free-electron lasers. The first members of the excimer laser family were rare gas dimers, pairs of rare gas atoms bonded together to form a "molecule" which existed only in the excited state. In the mid-1970s, interest shifted to rare gas halides, which are much more practical light sources and which have become a significant part of the laser business. In contrast, the free-electron laser, pioneered by John M. J. Madey at Stanford University (Deacon et al., 1977), remains in the laboratory, although there is widespread optimism that it will prove to be of practical importance.

The major research breakthroughs of the 1980s were dramatic extensions of the wavelength and power range of semiconductor lasers, development of new families of tunable solid-state lasers, and demonstration of x-ray lasers. The developments in semiconductor and tunable solid-state lasers have been commercialized rapidly, and are described in later chapters. In contrast, x-ray lasers remain in the laboratory.

Evolution of Practical Lasers

It always takes time to translate laboratory developments into practical products, and lasers have been no exception. For a while after Maiman demonstrated the first ruby laser, other laboratories around the world were busy building their own versions. Once they found out that the lasers worked, the researchers set out to see what could be done with them. The answers were not always clear, and one wag labeled the laser "a solution looking for a problem," a description that stuck long after its time had passed.

Maiman was one of the first to try to create a commercial laser industry. Soon after he demonstrated the ruby laser, he resigned from Hughes, and after spending nearly a year at the now-defunct Quantatron, formed Korad, a company which began making ruby and other solid-state lasers. Those early lasers generally went directly to research and development laboratories to see what could be done with them. Testing was sometimes improvised because suitable measurement instruments were not always available. Legend has it that the power of early pulsed lasers was measured in "gillettes"—the number of razor blades that could be pierced by a single pulse. However, it was not long before ruby lasers began finding applications ranging from drilling holes in diamond dies used in drawing wires to finding the ranges to military targets.

It took a while longer to commercialize the now-ubiquitous helium-neon laser. The first operation of that laser was at its 1.15-μm near-infrared line; it was not until 1962 that the now-standard 632.8-nm red line was demonstrated. The helium-neon laser has never been able to produce high powers, but it has proved to be easy to manufacture, and its low-power output at a visible wavelength has found many uses. Prominent early uses included laboratory demonstrations, production of holograms, and construction alignment; it also has been used for many types of information handling.

The solid-state neodymium laser took more time to perfect because of problems in finding a suitable crystalline host. Although glass was identified early, heat-dissipation problems have always restricted its operation. It took a number of years to settle on YAG as the best crystalline host for neodymium that could be grown in reasonable quality at a reasonable cost. Because of its excellent thermal and optical qualities, Nd-YAG has become the standard solid-state laser material, displacing ruby from many applications. However, several new solid-state lasers are becoming important, especially types with tunable output wavelength.

The carbon dioxide laser also took time to perfect. Much early effort went into finding the gas mixture that gave the optimum output power. New methods of exciting the laser gas have also been developed to enhance the output power and efficiency. Military-sponsored research pushed output powers to many kilowatts, with some demonstration lasers having produced hundreds of kilowatts. Compact CO_2 lasers can produce a few watts when the gas passes through a compact waveguide structure—a design that is attractive for many military and civilian uses.

The dye laser has evolved into a powerful tool for spectroscopy and atomic physics. The field has yielded impressive results, aided by dye laser developments pushed by laser pioneer Arthur Schawlow and Theodor Hänsch, now at the Max Planck Institute in Germany. A number of other laser spectroscopists have played key roles in advancing the field (not always with dye lasers), notably Nicolas Bloembergen of Harvard University and Vladilen S. Letokhov of the Soviet Institute of Spectroscopy. Schawlow and Bloembergen shared the 1981 Nobel prize in physics for their contributions to laser spectroscopy; the younger Hänsch and Letokhov may be candidates for future awards.

Argon and krypton ion lasers have also become standard commercial products, along with helium-cadmium lasers, rare gas excimers, and a few other types. While the technology for argon, krypton, and helium-cadmium lasers is fairly mature, that for excimer lasers is still developing.

The biggest success story of the past decade has been the emergence of the semiconductor diode laser. Although among the first types developed, the diode laser initially had big problems. Early versions required very high drive currents, which quickly heated the devices past their damage thresholds, limiting operation to short pulses at cryogenic temperatures. It was not until the mid-1970s that the first diode lasers capable of continuous operation at room temperature came on the market. Great advances in laser diode structure since then have led to better lasers with longer lifetimes, lower threshold currents, higher powers, a broader range of wavelengths, better beam quality, and more linear light-current curves. These advances and inexpensive mass-production methods have led to widespread applications of diode lasers in fiber-optic communications and information handling.

Current Trends

Laser technology is continuing to advance, but the progress is not spread evenly among the various types. Some types, notably helium-neon, ion, ruby, and carbon dioxide, are generally mature in design; efforts continue to refine performance and improve manufacturing, but dramatic advances seem unlikely. However, major progress continues in other areas.

Excimer laser manufacturers are trying to push their technology from the research laboratory into the industrial world. They face formidable problems because the need to use halogens in rare-gas-halide lasers makes it hard to achieve the highly reliable operation desired by industry. However, they are making significant progress. Government researchers are also making progress in pushing excimer laser output powers to higher levels.

The trends that have pushed semiconductor laser technology are continuing. New structures are making possible higher powers and higher-speed modulation. New materials and new structures are opening up new wavelength regions, most recently the red end of the visible spectrum. Many observers believe that diode lasers will take over many applications of helium-neon lasers.

Solid-state laser technology has benefited from both advances in diode lasers and development of new materials. High-power diode lasers are excellent pumps for some solid-state materials, allowing design of much more compact and efficient solid-state lasers. New solid-state laser materials are making new wavelengths available, as are new nonlinear materials which can change laser wavelength.

The free-electron laser seems destined to find an important place in the family of high-power lasers. Theoreticians spent years puzzling over the nature of the beast, but they now seem to understand it well enough to predict the results of experiments. A planned series of experiments will examine the prospects for reaching high efficiencies and high powers, and there is also work under way on building kilowatt-output free-electron lasers for laboratory research. The prospects for practical uses of x-ray lasers are less clear.

The laser industry per se is remote from military laser weapon programs; only a handful of aerospace contractors both conduct laser weapon research and build commercial lasers. However, a research program spending hundreds of millions of dollars a year on laser-related technology cannot help but impact the laser world. Even if laser weapons never shoot down a single unfriendly missile, some high-power laser technology developed in the laser weapon program has found many peaceful uses.

Bibliography

N. G. Basov and A. M. Prokhorov: "3-level gas oscillator." *Zh. Eksp. Teor. Fiz. (JETP)* 27:431, 1954.

Mario Bertolotti: *Masters & Lasers. An Historical Approach.* Adam Hilger Ltd., Bristol, U.K., 1983.

William B. Bridges: "Laser oscillation in singly ionized argon in the visible spectrum." *Applied Physics Letters* 4:128–130, 1964; erratum, *Applied Physics Letters* 5:39, 1964.

Joan Bromberg: *The Laser in America: 1950–1970,* MIT Press, Cambridge, Mass. (1991).

D. A. G. Deacon, et al.: "First operation of a free-electron laser," *Physical Review Letters* 38:892, 1977.

Albert Einstein: *Mitt. Phys. Ges. Zurich* 16(18):47, 1916. English translations appear in B. L. van der Waerden (ed.), *Sources of Quantum Mechanics,* North-Holland, Amsterdam, 1967; and in D. ter Haar (ed.), *The Old Quantum Theory,* Pergamon, Oxford and New York, 1967.

J. E. Geusic, H. M. Marcos, and L. G. Van Uitert: "Laser oscillations in Nd-doped yttrium aluminum, yttrium gallium, and gadolinium garnets," *Applied Physics Letters* 4:182, 1964.

Gordon Gould: U.S. patent 4,053,845, issued October 11, 1977; U.S. patent 4,161,436, issued July 17, 1979; U.S. Patent 4,704, 583, issued November 4, 1987; U.S. patent 4,746,201, issued May 24, 1988.

Robert N. Hall: "Injection lasers," *IEEE Transactions on Electron Devices ED-23*(7): 700–704, July 1976 (historical recollections).

Jeff Hecht: *Beam Weapons: The Next Arms Race,* Plenum, New York, 1984.

Jeff Hecht: *Laser Pioneers,* Academic Press, Boston, 1991 (updated edition of *Lasers & Applications Laser Pioneer Interviews,* High Tech Publications, Torrance, Calif., 1985).

Jeff Hecht and Dick Teresi: *Laser: Supertool of the 1980s*. Ticknor & Fields, New York, 1982.

Ali Javan, W. R. Bennett Jr., and D. R. Herriott: "Population inversion and continuous optical maser oscillation in a gas discharge containing a He-Ne mixture," *Physical Review Letters 6*:106, 1961.

L. F. Johnson and K. Nassau: *Proceedings of the Institute of Radio Engineers 49*:1704, 1961.

J. V. V. Kaspar and G. C. Pimentel: "HCl chemical laser," *Physical Review Letters 14*: 352, 1965.

Rudolf Ladenburg: "Untersuchungen über die anomale Dispersion angeregter Gase" [Investigation into anomalous dispersion in certain gases], *Zeitschrift für Physik 48*: 17–25 (April–May 1928).

Bela Lengyel: "Evolution of masers and lasers," *American Journal of Physics 34*(10): 903–913, October, 1966 (historical overview).

Theodore H. Maiman: "Stimulated optical radiation in ruby," *Nature 187*:493, 1960.

C. Kumar N. Patel: "Continuous-wave laser action on vibrational-rotational transitions of CO_2," *Physical Review 136A*:1187, 1964.

Arthur L. Schawlow, "Masers and lasers," *IEEE Transactions on Electron Devices ED-23*:773–779, July 1976 (historical recollections).

Arthur L. Schawlow and Charles H. Townes: "Infrared and optical masers," *Physical Review 112*:1940, 1958; also U.S. patent 2,929,922.

Elias Snitzer, "Optical maser action of Nd^{+3} in a barium crown glass," *Physical Review Letters 7*:444, 1961.

Peter P. Sorokin and J. R. Lankard: "Stimulated emission observed from an organic dye, chloroaluminum phtatocyanine," *IBM Journal of Research & Development 10*: 162, 1966.

Peter P. Sorokin and M. J. Stevenson, "Stimulated infrared emission from trivalent uranium," *Physical Review Letters 5*:557, 1960.

Charles H. Townes: "The laser's roots: Townes recalls the early days," *Laser Focus 14*(8):52–58, August, 1978 (historical recollections).

A. D. White and J. O. Rigden: "Continuous gas maser operation in the visible," *Proceedings of the IRE 50*:1796, 1962.

Laser Theory
and Principles

The term *laser* covers a variety of devices, as will be clear from later chapters. While many lasers bear obvious family resemblances to each other, some types are quite distinct. Helium-neon, ion, and carbon dioxide lasers have some important similarities because they all are electrically excited gas lasers, but at first glance they seem to have little in common with semiconductor diode lasers. However, the operation of all lasers relies on certain common physical principles, and it is those principles that will be covered in this chapter.

Laser physics is a discipline that draws on many other fields, among them quantum mechanics, optics, gas dynamics, semiconductor physics, atomic and molecular physics, and resonator theory. One of the most important factors is the transfer of energy among atoms and molecules, which is closely related to quantum mechanics. Laser characteristics also depend strongly on the optical design of the cavity of the laser medium, and that design, in turn, depends on the physics of the laser medium. This chapter gives an introduction to the fundamentals of lasers; readers seeking a more extensive treatment can consult a number of texts (Milonni and Eberly, 1988; O'Shea et al., 1977; Shimoda, 1984; Siegman, 1986; Svelto, 1989; Yariv, 1985).

Terminology

The formal name for laser physics is *quantum electronics*. The term dates back to the early days of maser research, and today it is used mainly in connection with academic research. Books with "quantum electronics" in the title tend toward an emphasis on theory rather than practical uses of lasers. The term remains attached to two institutions in the laser field: the biennial International Quantum Electronics Conference, traditionally the most research-oriented of the major meetings in the laser field, and the laser journal published by the Institute of Electrical and Electronics Engineers, the *IEEE Journal of Quantum Electronics*. This book will stay with the more common usage of laser physics.

Much laser physics terminology comes from spectroscopy and the physics of atoms and molecules. One of the most important concepts is the *energy level*, a quantum state of an atom or molecule. The energy of a particular level is measured relative to the ground level, or lowest possible energy level, which is arbitrarily assigned zero energy. The term *species* is used to avoid repeating the phrase "atom or molecule" constantly when discussing energy levels in general terms. The shift of a species from one energy level to another is called a *transition*. A shift to higher energy levels is an excitation to an excited state and requires the absorption of an amount of energy equal to the difference in energy between the two levels. A drop from a higher energy level to a lower one is called a *decay* or *deexcitation* and is accompanied by the release of the transition energy.

Species can absorb or release energy in the form of photons, quanta of electromagnetic radiation. In the laser world, the word "light" is used generically for electromagnetic radiation throughout the infrared and ultraviolet parts of the spectrum, as well as for true visible light. Light itself has a dual nature, as both a particle (a photon) and as a wave. The nature of light can be measured in three ways: as the wavelength of a light wave, as the frequency of oscillation of a wave, or as the energy of a photon. Wavelength is invariably measured in metric units, micrometers (μm), nanometers (nm), or the not-quite-legitimately metric angstrom (Å) (0.1 nm). Frequency and photon energy are actually equivalent measurements, according to the Planck relationship $E = h\nu$, where E is energy, h is Planck's constant, and ν is the frequency. Energy is typically measured in electronvolts (eV; 1 eV is the energy needed to move an electron through a potential of 1 V). Frequency is normally measured in hertz (Hz), the number of oscillations per second. Spectroscopists, particularly those used to working in the infrared, sometimes use the peculiar unit of inverse centimeters (cm^{-1}), which as the name implies is the reciprocal of the wavelength in centimeters; although the units are based on wavelength, they are proportional to frequency or photon energy.

Many other laser terms come from optics or electronics.

Energy Levels and Transitions

There are three basic types of energy levels encountered in laser physics, illustrated in Fig. 3.1:

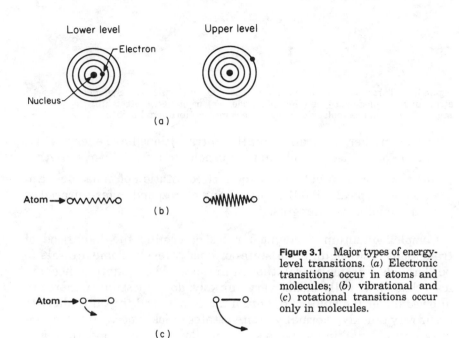

Figure 3.1 Major types of energy-level transitions. (*a*) Electronic transitions occur in atoms and molecules; (*b*) vibrational and (*c*) rotational transitions occur only in molecules.

- *Electronic levels* involving the electron configuration in a species. Electronic transitions are the ones portrayed in the Bohr picture of the hydrogen atom, in which the single electron gradually climbs the ladder of energy levels as it moves away from the nucleus, eventually attaining high enough energies to escape from the nucleus. As more electrons are added to the system in more complex atoms and molecules, the picture becomes increasingly complex, and various interactions create a multitude of electronic energy levels. The electronic transitions which normally occur in laser physics involve electrons in the outer shells and typically involve energies on the order of an electronvolt or more, corresponding to wavelengths in the near-infrared, the visible, and the ultraviolet. Inner-shell transitions, closer to the nucleus, involve higher energies corresponding to far-ultraviolet or x-ray photons and are less favorable for laser action.

- *Vibrational levels* involving vibration of the atoms in a molecule. Vibrations in various degrees of freedom are quantized, creating a series of levels for each possible vibration pattern. There are also multiple vibration patterns possible, as shown in Fig. 3.2, with the number increasing with the number of atoms in a molecule. Vibrational transitions occur when molecules shift from one vibrational state to another, which can involve a shift from one vibrational pat-

Figure 3.2 Three vibrational modes of the CO_2 molecule. (*a*) Symmetric stretching (ν_1 mode); (*b*) bending (ν_2 mode); (*c*) asymmetric stretching (ν_3 mode). More complex molecules have more complex vibrational modes.

tern to another. Typically vibrational transitions have energies of a fraction of an electronvolt and correspond to infrared wavelengths.

- *Rotational levels* involving the quantized rotation of molecules. Energies are typically 0.001 to 0.1 eV and correspond to far-infrared or submillimeter wavelengths.

Complex quantum-mechanical rules determine the likelihood of transitions between different states or energy levels. Some are said to be "forbidden" transitions, although in the world of quantum mechanics forbidden actually means very unlikely, *not* impossible. Others are allowed under the rules of quantum mechanics and thus are likely to occur very quickly. Normally excited states quickly decay to lower energy states by making one or a series of allowed transitions. However, there are also "metastable" excited levels, states in which a species becomes trapped because there is no readily allowed transition to a lower energy level. These metastable states are important in many lasers because they hold a species in an excited state for a long time on an atomic scale of things.

Although the electronic, vibrational, and rotational states of a species are independent, they can change simultaneously in a single transition. One such case is in the organic dye laser, described in Chap. 17. Electronic transitions of atoms or ions occur independently because those species do not have vibrational or rotational states. However, in molecules, changes in vibrational states are generally accompanied by changes in rotational states.

There is an important difference between transitions involving shifts in only one kind of energy level and those involving simultaneous shifts between two or more types. Transitions between two electronic energy levels of an atom are sharply defined at a specific wavelength. However, if a variety of shifts can occur simultaneously, the transition is spread out over a range of wavelengths, as in dye lasers or carbon dioxide lasers. This may produce a series of discrete lines, as in a CO_2 laser, or a continuum, as in a dye laser.

Excitation

Excitation is the process of raising a species from a lower energy level to a higher one and is a prerequirement for laser action. The transfer of energy into the laser medium can occur by several mechanisms, including:

- Absorption of photons
- Collisions between electrons (or sometimes ions) and species in the active medium
- Collisions among atoms and molecules in the active medium
- Recombination of free electrons with ionized atoms
- Recombination of current carriers in a semiconductor
- Chemical reactions producing excited species
- Acceleration of electrons

The probability of a species absorbing energy by some of these mechanisms is indicated by measurements of *absorption cross section*. Excitation probabilities or cross sections depend on factors such as the wavelength of the illuminating light, the speed of incident electrons, or coincidences in the energy-level structures of species. The excitation of a laser is often a multistep process. For example, in the helium-neon laser electrons passing through the active medium transfer energy to helium atoms, which in turn transfer the absorbed energy to the less-abundant neon atoms, which then emit laser light.

Stimulated and Spontaneous Emission

Excited species can release their excess energy by nonradiative processes, such as collisions with other atoms or molecules, or by emitting a photon. Emission of a photon can be spontaneous or stimulated. As the name implies, spontaneous emission occurs without outside intervention when a species drops to a lower energy level after a natural decay time (analogous to the half-life of radioactive isotopes), typically a tiny fraction of a second.

Emission of a photon and transition to a lower energy level can also be stimulated by the presence of a photon with the same energy as the transition. On first glance, it might seem highly unlikely that a photon of precisely the right energy would wander by in the normally short time before an excited species spontaneously drops to a lower energy level. However, in most cases there is a ready source of such photons: spontaneous emission by the same species. The first few spontaneously emitted photons can trigger stimulated emission of others, leading to a cascade of stimulated emission. Stimulated emission has a couple of special properties. It has the same wavelength as the original photon, and it is in phase (or coherent) with the original light.

Spontaneous emission generates the light we see from the sun, stars, light bulbs, flames, fluorescence, and other normal objects. Stimulated emission produces laser light. Strictly speaking, other light sources can contain small fractions of stimulated emission, but in practice the amount of stimulated emission is undetectable.

Population Inversions

Stimulated emission is rare because of the same thermodynamic fact of life that makes spontaneous emission occur: species tend to drop to the lowest available energy level. Under normal conditions, known as thermodynamic equilibrium, the population of a state tends to decrease as its energy increases, as shown in Fig. 3.3. This means that there should always be a larger population in the lower state of a transition than in the higher state, so a photon is far more likely to be absorbed by a lower-state species than to stimulate emission from one in the higher state. Under these conditions, spontaneous emission dominates.

For stimulated emission to occur, a *population inversion* would be needed—the population of the upper level of a transition would have to be higher than that of the lower level. Then a photon of the transition energy would be more likely to stimulate emission from the excited state than to be absorbed by the lower state. The result is laser gain or amplification, a net increase in the number of photons with the transition energy, corresponding to the difference between stimulated emission and absorption at that wavelength. Typically, gain is measured as percentage increase per pass through the laser medium,

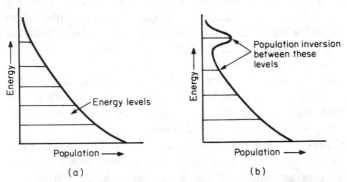

Figure 3.3 (*a*) Distribution of populations among energy levels at equilibrium; and (*b*) during a population inversion. Note that a population inversion does not have to exist for all possible transitions, only for the laser transition. Population inversions can only produce laser emission if they are above the equilibrium energy, the energy for which population is largest in equilibrium.

or per centimeter of distance through the laser medium. Sometimes gain is measured in the number of added photons generated for each centimeter of laser medium, for example, 4 cm^{-1}, which actually means four per centimeter, not the inverse centimeters of the spectroscopist. Note that a gain of 4 cm^{-1} actually implies that the input power is multiplied by a factor of 5.

Laser gain is proportional to the difference between the chance of stimulated emission and the chance of absorption. This means that the populations of both the upper and lower levels of the laser transition are important. Thus if laser action is to be sustained, the lower level must be depopulated as the upper level is populated, or the population inversion will end. That is indeed what happens in some pulsed lasers.

Laser amplification can occur over a range of wavelengths because no transition is infinitely narrow. The range of wavelengths at which absorption and emission can occur is broadened by molecular motion, vibrational and rotational energy levels, and other factors. Thus laser specialists speak of a material having a "gain bandwidth"—a range of wavelengths at which laser gain can occur. The value of the gain factor is highest in the center of the band and lowest at the edges.

Amplification and Oscillation

A population inversion alone is enough to produce "light amplification by the stimulated emission of radiation," but the result is only a coherent, monochromatic light bulb. Most of the useful characteristics of a laser come from oscillation of light within a laser cavity. To understand the difference between laser amplification and laser oscillation, it helps to consider examples of the two phenomena.

The amplification of stimulated emission can occur in nature, although the phenomenon was not discovered until after the invention of the laser. One of the more interesting examples occurs in the upper levels of the Martian atmosphere (Mumma et al., 1981). Solar radiation produces a population inversion in carbon dioxide in the tenuous upper layers of the planet's atmosphere. The gain is low because the gas pressure is low, but the volumes involved are so large that by human standards (although not by cosmic ones) the power emitted on the 10.4-μm CO_2 laser transition is high. However, the energy stored in those CO_2 molecules is extracted inefficiently, and the laser emission is dissipated randomly into space as shown in Fig. 3.4. As a result, the intensity reaching earth is so low that the existence of the laser emission was only discovered using sophisticated spectroscopic instruments.

An ordinary CO_2 laser operates at somewhat higher pressures and

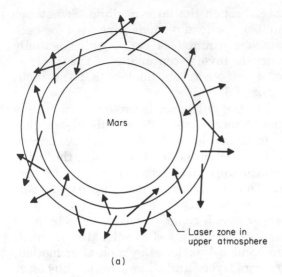

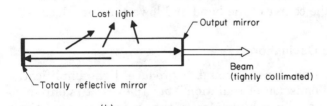

Figure 3.4   The key to production of a laser beam is directivity. (a) The upper part of the Martian atmosphere emits "laser light"—stimulated emission—randomly, making it hard to detect from earth. A man-made laser (b) includes a resonant cavity that generates a tightly collimated beam.

uses a much smaller volume of gas—but its real practical advantages are that it extracts energy more efficiently and concentrates it into a narrow beam. This is done by putting a pair of mirrors on either end of a cylindrical tube. If stimulated emission occurs on the axis between the two mirrors, it is reflected back and forth through the tube, stimulating emission again and again from CO_2 molecules. Stimulated emission in other directions is lost out of the laser medium. The result is that the stimulated emission is concentrated in a beam oscillating back and forth between the mirrors.

From an electronic standpoint, the mirrors provide positive feedback. A laser with a pair of totally reflective mirrors, like any positive-feedback system, in theory could amplify itself to infinity, but in practice cavity losses limit the positive feedback and the degree of amplification. In practical lasers one (or sometimes both) of the mirrors lets part of the light escape from the laser cavity, either around its edge, through a hole, or through a partially transparent section. The light that leaks out of the laser cavity forms the laser beam.

The fraction of the beam that is allowed out of the laser cavity depends on the gain of the laser medium. Once a continuous-output laser has started operating, the total gain must equal the sum of the cavity losses plus the fraction of the energy allowed out of the cavity. Thus the lower the gain, the smaller the fraction of light that can be transmitted out of the cavity. For example, the gain of a helium-neon laser is low, so the output mirror allows only 1 percent of the incident light to escape as the laser beam. If gain is high, transmission can be higher. In some cases, such as some excimer and nitrogen lasers, the gain is so high that a pair of cavity mirrors is not needed. (Such lasers are said to operate in superradiant mode.)

Another way to look at laser oscillation is as a threshold phenomenon that occurs when gain of the laser medium exceeds the sum of cavity losses and output-mirror transmission. In this case, gain is proportional to operating conditions, including energy input to the laser medium. As input power increases, laser output is zero until the threshold is passed, then increases steeply, as shown in Fig. 3.5. Thresholds differ for different cavity configurations and laser-medium conditions. In practice, different cavity designs and operating conditions are used to produce different output power levels.

The relationship between power inside the laser cavity (intracavity power) and in the laser beam depends on output mirror transparency. For low-gain media, where mirror transparency is low, the power levels inside the cavity are much higher than those outside. If high powers are critical, as in harmonic generation or some experiments, it may be desirable to put optical accessories *inside* the laser cavity to

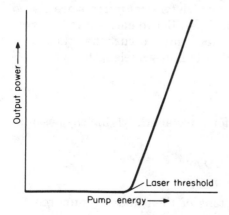

Figure 3.5 Laser threshold phenomenon—a laser does not generate significant optical output until the pump energy passes a threshold. At higher pump energies, the output power increases rapidly. In practice, each laser has limits on output, and eventually the output-input curve bends over.

take advantage of the higher powers available there. Some lasers are designed with intracavity space to allow placement of optics or samples inside the cavity.

With the exception of high-gain types that operate in superradiant mode, most lasers are oscillators, with cavities defined by a pair of reflectors. However, it is also possible to build laser amplifiers that boost the power produced by separate oscillators. Laser amplifiers have no cavity mirrors and do not generate a beam internally. Instead, they contain an active medium which amplifies the beam from a separate oscillator or other light source. Some optical amplifiers operate at lower power levels, to raise the power of an optical signal in a fiber-optic communication system by directly amplifying the light, as described in Chap. 26. Others are used to generate laser pulses with higher power than a simple laser oscillator could produce. This approach works best for laser media with reasonable internal energy storage and fairly high gain, such as neodymium-doped solid-state lasers described in Chap. 22. Massive amplifier chains can generate pulses with extremely high peak power for laser fusion experiments.

Resonators, Modes, and Beam Quality

Laser oscillation takes place in a resonant cavity defined by the mirrors at each end. Laser resonator theory draws on research on radio-frequency resonance, but the large difference in wavelength makes some important changes. Because cavity length is many times laser wavelength, it is possible to oscillate simultaneously on many modes. To keep oscillation modes limited, laser cavities are open with only a pair of small mirrors at opposite ends, unlike radio-wave cavities which can reflect from all sides.

Several configurations are possible for laser resonators, some of which are shown in Fig. 3.6. Conceptually, the simplest is the plane-parallel or Fabry-Perot resonator, in which two flat mirrors are placed at opposite ends of the cavity aligned parallel to each other and perpendicular to the axis of the cavity. Resonances occur when the cavity length D equals an integral number n of half-wavelengths ($\lambda/2$):

$$D = \frac{n\lambda}{2}$$

This expression can also be solved for wavelength, giving the possibility of oscillation at wavelengths

$$\lambda = \frac{2D}{n}$$

Because a typical laser cavity is tens of thousands or hundreds of

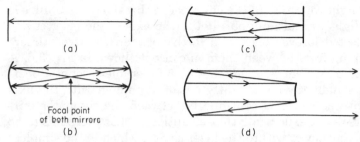

Figure 3.6 A sampling of laser cavity designs, showing how light is re-flected between the end mirrors. Note how design of (*d*) the positive-branch confocal unstable resonator leads to production of a beam with doughnutlike cross section. (*a*) Plane-parallel resonator (marginal sta-bility); (*b*) confocal resonator (stable); (*c*) hemiconfocal resonator (sta-ble); (*d*) positive-branch confocal unstable resonator.

thousands of wavelengths long, this might seem to permit oscillation at a very large number of wavelengths. However, oscillation is possi-ble only at wavelengths within the gain bandwidth of the laser me-dium, and special optics can further restrict the oscillation range.

The conceptual simplicity of the plane-parallel resonator is offset by the practical difficulty of aligning the mirrors precisely enough for stable operation of the laser. Similar principles apply to several de-signs that use one or two concave spherical mirrors to simplify align-ment, including

- *Concentric (or spherical) resonator:* Two identically curved spheri-cal mirrors separated by a distance equal to twice their radius of curvature, so the centers of curvature coincide.

- *Confocal resonator:* Two identically curved spherical mirrors sepa-rated by twice their focal length, so focal points coincide. (Because focal length of spherical mirrors is half the radius of curvature, this means that the center of curvature of one mirror is at the center of the other.)

- *Hemiconfocal resonator:* A spherical mirror separated from a flat mirror by its focal length.

- *Hemispherical resonator:* A spherical mirror separated from a flat mirror by its radius of curvature.

These resonators, and some similar designs with other mirror spac-ings, often are called *stable resonators*. The theory of laser resonators is complex, and strictly speaking, the "stable" label should not be ap-plied to the plane-parallel resonator or to the concentric resonator without further explanation.

The concept of a stable resonator can best be visualized by following the path of a light ray through the cavity. The threshold of stability is reached if a light ray initially parallel to the axis of the laser cavity could be reflected forever back and forth between the two mirrors without escaping from between them, a concept shown in Fig. 3.6. A functional and somewhat stronger condition for stable-resonator operation is that a light ray just slightly out of alignment with the axis should likewise be reflected continually between the resonator mirrors. Another way of expressing that condition is that there has to be some net focusing power within the laser cavity, which in the simplest case is supplied by the mirrors.

The plane-parallel resonator meets the threshold condition for stability, but by itself cannot provide the focusing power needed to meet the stronger condition. Thus, the plane-parallel resonator by itself is not truly a stable resonator, particularly because of the practical problem of aligning the mirrors precisely parallel to one another. However, it can be used if something else in the laser cavity provides the needed focusing power. This is the case in solid-state crystalline and glass lasers, where thermal focusing in the rod itself provides the focusing needed for stable operation.

The concentric resonator is another special case. It meets the theoretical stability criteria, but in practice does not give stable laser operation because it focuses the internal wavefront to a point at its midpoint. In practice, other designs with at least one concave mirror are generally used for low-power gas lasers operating continuously.

Resonators which do not meet the stability criteria are called *unstable resonators*, because the light rays diverge away from the axis. There are many variations on the unstable resonator. One simple example is a convex spherical mirror opposite a flat mirror. Others include concave mirrors of different diameters (so that the light reflected from the larger mirror escapes around the edges of the smaller one), and pairs of convex mirrors (Siegman, 1986; Svelto, 1989).

The two types of resonators have different advantages and different mode patterns. The stable resonator concentrates light along the laser axis, extracting energy efficiently from that region, but not from the outer regions far from the axis. The beam it produces has an intensity peak in the center, and a gaussian drop in intensity with increasing distance from the axis. It is the standard type used with low-gain and continuous-wave lasers, such as helium-neon types.

The unstable resonator tends to spread the light inside the laser cavity over a larger volume. In most cases, the output beam has an annular profile, with peak intensity in a ring around the axis, but a null on axis. This design collects laser energy from more of the volume in the laser cavity, typically leading to higher overall energy conversion efficiency than with a stable resonator, although the efficiency in extracting energy from the region of the laser axis is lower. This works best for high-gain pulsed lasers. One special advantage is the ability to use all-reflective optics, with the output beam escaping around a small mirror, important in working in the ultraviolet where there are few materials able to transmit high powers. Although the doughnut-shaped beam has an intensity null in the near field, it smooths out at greater distances from the laser, giving a more uniform energy distribution. In the far field, some unstable-resonator beams tend to have smaller divergence than beams from stable-resonator lasers.

Laser resonators have two distinct types of modes: transverse and longitudinal. Transverse modes manifest themselves in the cross-sectional profile of the beam, that is, in its intensity pattern. Longitudinal modes correspond to different resonances along the length of the laser cavity which occur at different frequencies or wavelengths within the gain bandwidth of the laser. A single transverse mode laser that oscillates in a single longitudinal mode is oscillating at only a single frequency; one oscillating in two longitudinal modes is simultaneously oscillating at two separate (but usually closely spaced) wavelengths. With the exception of some demanding applications in spectroscopy and communications, controlling the longitudinal modes so that oscillation is only at a single frequency is less important than controlling the transverse modes, which reflect beam quality.

A special terminology has evolved for transverse modes, based on theoretical work done in the early days of laser development (Fox and Li, 1961). Transverse modes are classified according to the number of nulls that appear across the beam cross section in two directions. The lowest-order, or fundamental, mode, where intensity peaks at the center, is known as TEM_{00}. A mode with single null along one axis and no null in the perpendicular direction is TEM_{01} or TEM_{10}, depending on orientation. A sampling of these modes, which are produced by stable resonators, is shown in Fig. 3.7. For most applications, the TEM_{00} mode is considered most desirable, but multimode beams can often deliver more power in a poorer-quality beam, and thus are acceptable for some uses. Somewhat different terminology has evolved for describing the modes of unstable-resonator lasers, but for most practical purposes unstable-resonator lasers can be considered to operate in their lowest-order mode, which produces a beam with annular cross section.

The different transverse-mode patterns described above are visible in the near field, close to the laser's output aperture. As the beam travels through space, the energy distribution becomes more uniform. Far from the laser, in what is called the far field, nonuniformities such as the intensity minimum at the center of an unstable-resonator beam tend to be smoothed out.

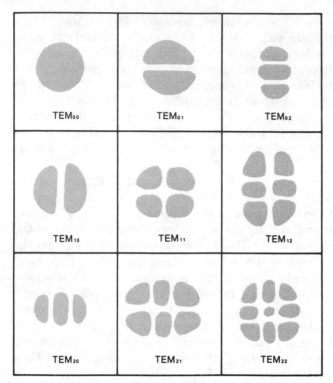

Figure 3.7 Lower-order modes that can be produced by a stable resonator. (*Courtesy of Melles Griot, from* Optics Guide 4.)

For most laser applications, the geometric properties of beam diameter and beam divergence are more important than mode structure. For a gaussian beam, beam diameter is generally measured to the points where intensity drops off to $1/e^2$ the peak intensity. The cavity optics determine beam diameter, which typically is measured at the output mirror. Use of the term *diameter* implies a circular cross section, but laser beams can also have oval or even rectangular profiles.

Beam divergence also depends primarily on cavity optics; it measures the angle at which the beam spreads out after it has left the laser. With the exception of semiconductor diode lasers, it is almost invariably less than 10 milliradians (mrad), and for many types it is typically on the order of 1 mrad. The lower limit on beam divergence is set by the diffraction limit, which says that the smallest angular spot which can be produced (measured in radians) is proportional to wavelength λ divided by output aperture D:

$$\text{Divergence} = 1.22\frac{\lambda}{D}$$

As this relationship indicates, beam divergence can be decreased by using larger output apertures or moving to shorter wavelengths. Lasers with divergence that approximates the diffraction limit are called *diffraction-limited*, a category which includes all lasers producing TEM_{00} beams.

Linewidth and Tuning

By ordinary standards, laser emission is monochromatic, but there are considerable variations in linewidth of the emitted light. The nature of the optical cavity and of the laser medium combine to determine the range of wavelengths emitted by a particular laser.

The simple model of a laser transition occurring between two precisely defined energy states is by its very nature an oversimplification. In some cases, the transition actually involves simultaneous changes in two or more quantum states, leading to a couple of distinct possibilities.

One, seen in carbon dioxide and chemical lasers, is emission on a series of closely spaced transitions. In these molecules, the primary vibrational transition is accompanied by a change in the rotational state. Because the spacing of rotational energy levels is a significant fraction of the energy of the vibrational transition, output is possible on many lines with wavelengths differing by 10 to 20 percent. Commercial carbon dioxide lasers, for example, can emit on over 100 lines at 9 to 11 μm.

The second possibility, seen in dye lasers, is a continuum of transitions, a band of wavelengths throughout which the laser can emit light. Pressure-broadening effects, which serve to blur out closely spaced energy levels, cause discrete emission lines to merge into a continuum when the carbon dioxide laser is operated at high pressures.

Lasers can also emit on a family of transitions, which often share some energy levels. Typically this occurs when the excitation process can populate two or more upper laser levels, or when a single excited level can decay in two or more ways that can produce laser emission. Such lasers may be designed to emit on a single line or to produce multiline emission.

Even if only a single laser transition is possible, some line-broadening effects come into play. In gas lasers, motion of the atoms and molecules in the gas causes doppler broadening of the emission lines. Other forms of energy transfer, such as atomic collisions and lattice vibrations in solids, can also cause some line broadening.

The overall result, looking at the laser medium, is a gain curve, a range of wavelengths at which laser gain can occur. If there are many separate lines, they may have separate gain curves, depending on the operating conditions.

The cavity optics set another set of constraints on the wavelengths at which a laser can oscillate. As mentioned earlier, laser resonators can support a number of different longitudinal modes at different wavelengths. Even in a naturally narrow-line laser such as the helium-neon type, several longitudinal cavity modes fall under the laser medium's gain bandwidth. Typically one longitudinal mode will dominate during oscillation, with others present at lower intensities, but the relative intensities at different lines can shift, a phenomenon called *mode hopping*, which can be hard to control in some lasers but is not a problem in others. The spacing of longitudinal modes depends on cavity length—the longer the cavity, the closer the mode spacing. However, except for the tunable semiconductor lasers described in Chap. 21, cavity lengths are generally so many wavelengths long that little can be gained by changing length slightly.

It is possible to insert special optical elements into the laser cavity to restrict oscillation to a narrower range of wavelengths. The usual choice is a Fabry-Perot etalon, a pair of glass plates aligned so they effectively form a resonator. Because the plates are closely spaced, the resonant modes are far apart, and oscillation can readily be restricted to a single longitudinal mode. The spacing can be adjusted by tilting the etalon, by moving one of the plates, or by changing gas pressure inside or outside of the etalon cell. Such adjustments make it possible to change the wavelength selected by the etalon. Oscillation can be limited to a less-narrow range of wavelengths by inserting a narrow-band optical filter that transmits only at those wavelengths into the cavity. Alternatively, the cavity mirrors can be coated to reflect light at only certain wavelengths, preventing oscillation from taking place at other wavelengths.

Where a broad range of emission wavelengths is possible, it is usually desirable to be able to tune or adjust the laser wavelength. Such tuning is usually done by inserting a prism or grating into the optical cavity to serve as a wavelength-selective element. The grating or prism is then tuned (typically by turning) so only light of the desired wavelength can pass back and forth along the optical axis of the laser cavity. Other wavelengths are directed away from the axis and hence do not oscillate. Tuning is commonest with dye lasers but is also used with CO_2 and chemical lasers when the ability to pick a specific wavelength is important.

Coherence and Speckle

The best-known special property of laser light is its coherence. Stimulated emission is in phase with the photon that does the stimulating, and as stimulated emission builds it remains in phase. The coherence is not perfect, but it is far better than that of other light sources.

One way of measuring coherence is as *coherence length*, the distance over which light remains coherent after it leaves the light source. This quantity varies for different types of lasers, depending on the frequency bandwidth of emission. As a rough guideline, coherence length L equals the speed of light c divided by the laser's frequency bandwidth $\Delta \nu$.

$$L = \frac{c}{\Delta \nu}$$

(Wilson and Hawkes, 1983). Most estimates of coherence length in this book have been derived using this simple relationship. It is actually based on the concept of temporal coherence, which measures coherence of a light wave emitted over time. The coherence length is equal to the coherence time multiplied by the speed of light.

There is also spatial coherence, which measures the area of a wavefront over which the light is coherent. This concept is independent of temporal coherence; the presence of one type of coherence does not necessarily imply the other.

Theorists have considered coherence in exhausting detail (Born and Wolf, 1975), but most of the theoretical details are of little import to laser users. Coherence is an ordering of light waves that lets them carry large quantities of information. Its major practical importance lies in holography, optical information processing, and development of coherent communication systems.

Coherence also causes a phenomenon called *speckle*, which derives its name from its appearance, as shown in Fig. 3.8. The granular

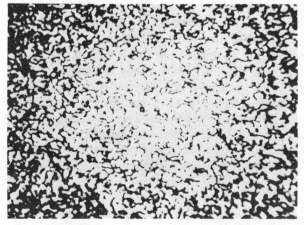

Figure 3.8 Photograph of laser speckle. (*Courtesy of F. P. Chiang.*)

speckle pattern is produced when laser light is scattered from a diffusing surface such as a sheet of paper. Speckle looks like noise but, in the words of Dennis Gabor, who received the 1971 Nobel prize in physics for his invention of holography, "It is not really noise; it is information which we do not want, information on the microscopic unevenness of the paper in which we are not interested" (Gabor, 1971). Speckle can be used in certain measurements, but it also can create problems in some measurement and communication applications, particularly those involving small-area receivers where the speckle pattern could cause fluctuations in received signal. Speckle gives a grainy texture to photographs and holograms recorded with laser light.

Efficiency Limitations

The overall efficiency of lasers is low by electrical standards, although comparable with that of other light sources. Typical commercial lasers convert 0.001 to over 30 percent of input electrical energy into energy in the laser beam. For comparison, only a few percent of the electrical energy used by an ordinary incandescent lamp is emitted as visible light, although much of the remainder is emitted in the infrared. The limitations on efficiency are inherent in the nature of a laser.

Energy-level structures of excited species create many losses. Typically the laser transition is only one in a series of transitions that a species undergoes in dropping from the upper excited level to the ground state, and each of those additional transitions leads to losses. For example, the 632.8-nm transition of the helium-neon laser, corresponding to 2 eV of energy, occurs about 20 eV above the ground state, so even in theory only about 10 percent of the excitation energy could emerge in the laser beam.

Excitation is also never 100 percent efficient. In the common cases of optical and electrical pumping, much of the energy put into the laser medium either leaks out without exciting anything or simply goes into heating the laser medium. In optical pumping with a flashlamp, for example, some of the wavelengths generated by the lamp may not be absorbed efficiently by the laser medium. Even if the energy is absorbed, it may not be extracted from the laser medium because of the cavity configuration and inherent losses in the optical cavity.

Losses are also inherent in converting the energy from ordinary electrical sources (i.e., wall current) to the form required by the laser, such as high-voltage discharges or flashlamp pulses.

Output Power

Output powers of commercial continuous-wave lasers run from under a milliwatt to tens of kilowatts, with comparable average powers available from pulsed lasers. Typically the range of powers available from any one type is much more limited because of differences in the internal physics. Gas kinetics, heat dissipation, and physical unwield-iness of tubes limit the helium-neon laser to power levels below about 75 mW. Heat dissipation and damage to optical surfaces limit semi-conductor laser powers to somewhat higher levels.

The usual problems are in energy-transfer kinetics. In gas lasers, the kinetic processes needed to sustain the population inversion and stimulated emission can break down for large volumes. In solid-state lasers, heating of the laser medium can cause serious problems. So far the highest continuous or average powers known to have been re-corded were produced by chemical lasers used in military laser weap-on research, reported to have produced a couple of megawatts (Hecht, 1984). However, few details other than the record power levels have reached the public eye.

Limits on individual types of lasers are described in more detail in the chapters on lasers that follow.

Time Scales of Operation

Lasers can operate continuously for many years, or generate ultra-short pulses lasting only a matter of femtoseconds [(fs) 1 fs = 10^{-15} s]. The differences reflect both the design of the lasers and the internal physics. Users should remember that the time scale of atomic physics is very different from that of the human race, being measured in nano-seconds or microseconds rather than seconds and minutes.

In the laser world, anything operating for a second or more at a time is called "continuous wave." Most commercial continuous-wave lasers can emit for far longer, sometimes many years. But whatever the hu-man perspective, from the standpoint of the physics it is legitimate to consider laser emission that is steady for a second to be continuous wave. Such short-term continuous-wave operation is most likely for experimental lasers, particularly high-power ones used in military la-ser weapon research.

Many types of lasers can operate only in pulsed mode, for a variety of reasons. In many solid-state lasers, the key problem is heat dissi-pation. It takes time for the excess pump energy delivered to the laser rod to make its way out as heat, and continuous-wave operation can cause heat to build up to laser-damaging levels. "Damage" in this case means not just physical damage to the laser rod but also disruption of the laser beam by thermally induced gradients.

Internal physics can also demand pulsed operation. In some cases, the only way to produce a population inversion is by using input-energy pulses with high power that cannot be sustained. In others, the population inversion is inherently short-lived, because of rapid decay of the excited state (as in excimer lasers), or rapid buildup of species in the ground state (as in ruby lasers). In high-pressure gas lasers, pulse length is limited by instabilities inevitable in electric discharges through high-pressure gas.

Pulse repetition rates also vary widely. Ruby and neodymium-glass lasers can generate only a few pulses per second, while copper vapor can produce several thousand and semiconductor lasers can pulse over a billion times a second [one gigahertz (GHz)]. Cavity dumping can produce tens of thousands of pulses a second, while modelocking can generate pulses at gigahertz repetition rates. Efforts are under way to increase repetition rates as a way to increase average power from certain lasers.

Pulse repetition rate, duration, and peak power can be adjusted by a variety of laser accessories, such as Q switches, modelockers, and cavity dumpers, described in Chap. 4. These give laser users added flexibility in matching laser characteristics to their requirements. Details on the pulse characteristics of individual lasers are described in later chapters.

Bibliography

Max Born and Emil Wolf: *Principles of Optics*, Pergamon Press, Oxford, England, 1975.

J. C. Dainty (ed.): *Laser Speckle and Related Phenomena*, 2d ed., Springer-Verlag, Berlin and New York, 1984.

A. G. Fox and Tingye Li: "Resonant modes in a maser interferometer." *Bell System Technical Journal 40*:453, 1961.

Dennis Gabor: "Holography 1948–1971," Nobel lecture, December 11, 1971, reprinted in K. Thyagarajan and A. K. Ghatak: *Lasers Theory & Application*, Plenum, New York, 1982, pp. 365–402.

Jeff Hecht: *Beam Weapons: The Next Arms Race*, Plenum, New York, 1984.

C. Breck Hitz: *Understanding Laser Technology*, PennWell Publishing, Tulsa, Okla., 1985.

Peter W. Milonni and Joseph H. Eberly: *Lasers*, Wiley-Interscience, New York, 1988.

Michael Mumma et al.: "Discovery of natural gain amplification in the 10-micrometer carbon-dioxide laser bands on Mars: a natural laser." *Science 212*:45–50, April 3, 1981.

Donald C. O'Shea, W. Russell Callen, and William T. Rhodes: *An Introduction to Lasers and Their Applications*, Addison-Wesley, Reading, Mass., 1977.

Koichi Shimoda: *Introduction to Laser Physics*, Springer-Verlag, Berlin and New York, 1984.

Anthony E. Siegman: *Lasers*, University Science Books, Mill Valley, Calif., 1986.

Orazio Svelto, *Principles of Lasers*, 3d ed., Plenum, New York, 1989.

Marvin J. Weber (ed.): *CRC Handbook of Laser Science & Technology*, 2 vols., CRC Press, Boca Raton, Fla. 1982.

J. Wilson and J. F. B. Hawkes: *Optoelectronics: An Introduction*, Prentice-Hall, Englewood Cliffs, N.J., 1983.

Amnon Yariv: *Quantum Electronics*, 3d ed., Holt, Rinehart & Winston, New York, 1985.

Matt Young: *Optics & Lasers*, 2d ed., Springer-Verlag, New York and Berlin, 1984.

Enhancements to Laser Operation

Laser operation can be enhanced either by modifying the laser device itself or by modifying the beam it generates. The distinction is significant in the world of commercial lasers. Modifications to the laser device typically are made by the manufacturer at the factory, or sometimes at customers' premises. However, it is the user who normally modifies the laser beam.

Factory modifications come in many forms, including substitution of different power supplies, choice of resonator optics, packaging of the device, and selection of control systems. Many data sheets include long lists of options offered by the manufacturers, which give users a considerable range of choices. However, these options are usually specific to a particular model or laser type, and the user has only limited control over them. The major options for each type of laser are described in Chaps. 7 to 29.

This chapter concentrates on modifications the user can make, although often they require hardware purchased from the laser manufacturer. All involve alterations to the laser beam, which can occur within as well as outside the laser cavity. These techniques offer users important flexibility in choice of laser wavelength, pulse length, and (in some cases) output power.

Wavelength Enhancements

Users can control laser wavelength either by altering the optical cavity or by altering the beam that emerges from the laser. Typically changes in the optical cavity are made using equipment supplied by the laser maker, but beam alterations can be made with other hardware. The boundaries are sometimes hazy; harmonic generators, for example, can be put inside or outside the laser cavity, and are offered both as options on certain laser models and as separate products.

Tuning and Wavelength Selection.   Fixed-wavelength laser emission is fine for many applications, but others would benefit from a choice of wavelengths. With a few lasers, there is no choice. With many, a choice is made by the manufacturer who assembles the laser cavity optics; one example is the helium-neon laser, which could emit several wavelengths, but is usually built with cavity optics that select the 632.8-nanometer (nm) red line. With other types, the user has some degree of selection, either by picking cavity optics or by adjusting a wavelength-selective element inside the laser cavity.

The choice of cavity optics works best for lasers which emit at widely spaced spectral lines. For example, one set of cavity optics can generate a blue beam from a helium-cadmium laser, while a separate set can produce ultraviolet output. In practice, the choice of cavity optics is generally an option.

Insertion of a wavelength-selective element inside the laser cavity allows much finer resolution, either for continuous tuning or for the selection of different laser lines that are not widely separated. The selection normally is made by a wavelength-dispersive element—a prism or diffraction grating—that is placed between the cavity mirrors or serves as a cavity mirror. The dispersive element spreads out a spectrum, and only a narrow range of wavelengths in that spectrum is reflected along the axis of the laser cavity so that it can oscillate. Light at other wavelengths is scattered out of the laser medium. The net effect is to give the optical cavity high losses at all but the desired wavelengths. The wavelength can be changed by turning the dispersive element so different wavelengths are aligned to resonate in the laser cavity. It is used in dye lasers (an example is shown in Fig. 17.3), vibronic solid-state lasers, and carbon dioxide lasers, and to produce single-line output from argon or krypton ion lasers. Such tuning is described in more detail in Chap. 17; it normally comes as an option with lasers that may require it.

Other tuning mechanisms may be used in certain cases. Application of a magnetic field can cause Zeeman shifts of spectral lines, an effect that was used in some early laser-tuning experiments. Internal energy-level structures of materials and refractive indexes of materials are both influenced by temperature and pressure. Shifts in energy levels directly affect transition wavelength. Changes in refractive index of materials in the laser cavity alter the effective length of the cavity and hence the resonant wavelength. However, the only one of these approaches used practically is temperature tuning of the lead salt semiconductor lasers described in Chap. 21.

Line Narrowing and Single-Frequency Operation. As indicated in Chap. 3, the doppler-broadened gain profile of even nominally monochromatic lasers normally spans several longitudinal modes, each with slightly different wavelength. This is adequate for most applications, but not for precision measurements. Line-narrowing accessories are also needed to restrict oscillation of dye lasers and other types with closely spaced lines to a limited range of wavelengths.

The basic idea of line narrowing is to insert into the laser cavity optical elements which restrict oscillation to a range of wavelengths so narrow that it includes only a single longitudinal mode. The commonest line-narrowing component is the Fabry-Perot etalon, typically a transparent plate with two reflective surfaces which together form a short resonator that can be inserted within the laser cavity. It can restrict the range of wavelengths, as shown in Fig. 4.1, by requiring that any wavelengths oscillating in the laser cavity meet resonance criteria both of the etalon and of the laser cavity as a whole. Effective spacing of the etalon plates can be changed by rotating the device (i.e., tilting it with respect to the axis of the laser), or by changing internal gas pressure inside a hollow version. In some cases a pair of etalons with different optical characteristics may be used in the same optical cavity to limit operating bandwidth still further.

Frequency bandwidths as low as about 500 kilohertz (kHz) are obtainable from commercial lasers (sometimes called *single-frequency* lasers). Even narrower linewidths [< 100 hertz (Hz)] have been demonstrated in the laboratory using special (and rather elaborate) techniques.

Harmonic Generation. Nonlinear interactions between light and matter can generate harmonics at multiples of the light-wave frequency,

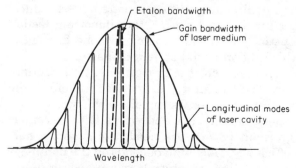

Figure 4.1 Transmission bandwidth of etalon compared with gain bandwidth of laser medium and the longitudinal modes of laser cavity. The etalon restricts oscillation to longitudinal modes of the laser cavity that occur within the etalon bandwidth.

or equivalently at integral fractions of the wavelength. High-order harmonics have been produced in the laboratory, such as the 28th harmonic of the 1064-nm output of a neodymium-YAG laser (Reintjes et al., 1977). However, for most applications only the second, third, or fourth harmonics are produced because the difficulty of harmonic-generation techniques, and their inherent energy losses, increases for higher harmonics.

Harmonic generation is a valuable technique because it greatly expands the range of wavelengths at which high laser powers are available, particularly in the ultraviolet and parts of the visible region. The commonest use of harmonic generation is with the 1064-nm neodymium-YAG laser, producing the 532-nm second harmonic, the 355-nm third harmonic, and the 266-nm fourth harmonic. Dye laser output is also often frequency-doubled to obtain tunable ultraviolet light; doubling the output frequency of GaAlAs semiconductor lasers gives blue light.

Commercial frequency doublers rely on nonlinear crystals. As a laser beam passes through the crystal, nonlinear interactions between the beam and the material generate an electromagnetic wave at twice the laser frequency. To obtain peak efficiency, the characteristics of the crystal must be matched precisely to the laser wavelength. This is relatively easy when working at a fixed wavelength, such as that of the Nd-YAG laser, but requires adjustment of the beam angle of incidence or the crystal temperature when wavelength is changed, as in a dye laser. Many nonlinear crystals are available, with the commonest including potassium dihydrogen phosphate (KDP, available in normal form or with deuterium substituted for hydrogen-1), lithium iodate ($LiIO_3$), lithium niobate ($LiNbO_3$), potassium pentaborate (KB5, KB_5O_8), β-barium borate, and ammonium dyhydrogen phosphate.

The choice of material depends on operating wavelength and power levels. The strength of the nonlinearity differs among materials, and for each material is a function of wavelength. Other important factors include transparency (especially at the harmonic wavelength) and the optical power level at which the material suffers damage.

The magnitude of the nonlinear effect is proportional to the square of the incident laser power. This has an important practical effect: the efficiency of harmonic generation increases rapidly with power levels. At low input powers, efficiency is very low and virtually no harmonic power is generated. High input powers are needed to generate harmonics with reasonable efficiency. How high depends on the nonlinear index of the material. Many nonlinear materials require power densities on the order of megawatts per square centimeter to generate significant harmonic power.

The critical parameter is the power per unit area (power density) inside the nonlinear material. Conventionally, the laser beam makes a single pass through the nonlinear material, which is placed outside the laser cavity. High power densities can be produced in different ways. One is to focus the laser beam in space, so its entire power passes through a small volume, where power density reaches the desired value. Another is to compress the laser pulse in time, so that the power level reaches a sufficiently high value for a short period. These approaches normally suffice if the laser source generates short, high-power pulses.

Alternatively, the frequency-doubling crystal can be placed either inside the laser cavity or inside another resonant cavity. This approach builds up the power level inside the crystal. (Recall that circulating power level inside a laser cavity is higher than the output beam power.) Intracavity crystals can be used to generate the second harmonic from a continuous-wave beam. Separate resonant cavities have been used to efficiently double the output of semiconductor lasers (Goldberg and Chun, 1989).

Generation of third and fourth harmonics using nonlinear crystals usually is a multistep process. The third harmonic is produced by first generating the second harmonic, then mixing it with the fundamental wavelength in another nonlinear crystal so the frequencies add together to produce the third harmonic. To produce the fourth harmonic, the output of a second-harmonic generator is passed through a second frequency doubler. Generation of the third and fourth harmonics requires even higher powers than frequency doubling and normally is done only with high-power pulses from neodymium lasers. A combination of poor efficiency and low ultraviolet transmission makes generation of higher-order harmonics in nonlinear crystals impractical.

Nonlinear interactions in gases or metal vapors, unlike those in crystals, can directly produce odd harmonics (Harris, 1975). As in crystals, the process is more efficient for lower harmonics, particularly the third. Because they produce higher harmonics and are transparent at much shorter wavelengths than crystals, gases or vapors can generate harmonics much further into the ultraviolet. For example, a group at AT&T Bell Laboratories has produced the 35.5-nm seventh harmonic of a krypton fluoride laser's 248-nm output by passing the beam through a pulsed supersonic jet of helium in a vacuum (Bokor et al., 1983); operation in a vacuum was needed to avoid absorption of the extreme-ultraviolet output. The demand for short-wavelength ultraviolet output is limited, however, and gas cells have found few harmonic-generation applications outside the research laboratory.

The chapters describing individual laser types discuss their suitability for use with harmonic generators. Although some nonlinear devices can be made fairly simple to use, the physics underlying their operation is so complex that few users will want to study it in any detail, although references are available that do so (Yariv, 1985; Yariv and Yeh, 1984).

Sum-and-Difference Frequency Generation. Harmonic generation is actually a simplified case of a more general nonlinear process, the generation of sum-and-difference frequencies as two (or perhaps more) light waves pass through a nonlinear material. In general, if the input frequencies are ω_1 and ω_2, the nonlinear process generates additional waves at frequencies $\omega_1 + \omega_2$ and $\omega_1 - \omega_2$. Typically one of these frequencies is selected for use, although the physical interaction generates both.

Sum-and-difference frequency generation can occur in solid or gas phases of suitable nonlinear materials. In practice the process is complex, and like harmonic generation requires high-power beams and careful alignment. Because of its complexity, its use remains much more limited than harmonic generation, although it can be a useful laboratory tool.

Parametric Oscillation. Parametric oscillation is another nonlinear phenomenon with complex theoretical underpinnings. It relies on an effect known as parametric amplification, which occurs when an intense "pump" beam at frequency ω_3 is passed through a suitable nonlinear crystal together with a weak "signal" beam at a lower frequency ω_1. Interactions in the crystal transfer energy from the pump frequency to the signal frequency, in effect amplifying the signal frequency. At the same time, an "idler" beam at a frequency $\omega_2 = \omega_3 - \omega_1$ is also generated (Yariv, 1985; Yariv and Yeh, 1984).

In practice, operation of a parametric oscillator differs somewhat from that theoretical picture. A principal difference is that the only optical input required is the pump beam. Parametric noise generated within the nonlinear crystal produces the weak signal beam.

The signal and idler frequencies are determined by the orientation of the crystal and the design of the oscillator cavity in complex ways that are of less relevance to users of parametric amplifiers than to designers of such systems (Svelto, 1989). What the user sees is output—the signal wave—that is continuously tunable in wavelength by adjusting the nonlinear crystal and/or the optical cavity.

The existence of a resonant optical cavity makes parametric oscillators superficially similar to lasers, and like lasers they generate coherent beams. However, there is no stimulated emission within the parametric oscillator cavity, so it is not really a laser.

All parametric oscillator cavities must have end mirrors transparent to the pump beam and reflective at the signal frequency, as shown in Fig. 4.2. There are two basic variations: designs which are doubly resonant and reflect both the signal and idler waves, and those which are singly resonant and reflect only the signal wave. Singly resonant cavities require the high pump power that is available only with pulsed excitation, but they provide more stable output than doubly resonant cavities, even though the latter can be pumped with continuous-wave lasers. Singly resonant cavities are the usual choice in commercial parametric oscillators.

Although parametric oscillation is in theory a general process, in practice it is more limited. Wavelengths of 0.5 to about 14 micrometers (μm) have been produced in the laboratory, but practical devices are limited to the near-infrared and use neodymium lasers as pumps. At one point parametric oscillators were considered promising tools for generating continuously tunable output in the near-infrared. However, other tunable infrared sources have proved more popular, and parametric oscillators are used mainly where other sources are not available.

Raman Shifting. Another way to change laser wavelength is *Raman shifting*. The phenomenon is named for Indian physicist Chandrasekhara V. Raman, who in 1930 received the Nobel prize in physics for discovering it. The Raman shift occurs when molecules scatter light "inelastically," in the process changing the photon energy. The scattering is inelastic because as it occurs, the molecule makes a vibrational and/or rotational transition. If the transition is to a higher energy level, the photon is scattered with a lower energy than it had when it arrived and thus has a longer wavelength than the incident light, a so-called Stokes shift. If the transition is to a lower energy level, the scattered photon carries away the excess energy, and thus has higher energy and longer wavelength than the incident light, an "anti-Stokes shift."

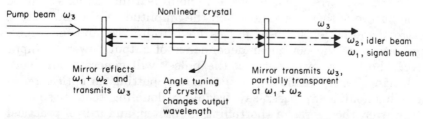

Figure 4.2 Parametric oscillator with a doubly resonant cavity. Mirrors transmit the pump wavelength, but reflect totally or partially the signal and idler beams. In a singly resonant cavity, the idler beam would not be confined to the cavity. The pump beam generates both signal and idler beams as it passes through the nonlinear crystal, which is normally rotated to change the angle of incidence and thereby the output wavelength.

Raman scattering of a single wavelength generates multiple lines corresponding to different rotational and vibrational transitions of the molecule. The frequency of each Raman line equals that of the input light plus or minus the frequency of the Raman transition. Each material thus produces its own characteristic Raman frequency shifts.

Like harmonic generation, Raman scattering is not likely to occur until high incident powers are reached. It found only very limited applications until the advent of the laser, which provided the needed source of high-power monochromatic light. Now laser Raman scattering serves as the basis of a type of spectroscopy, which has proved valuable because the scattered light shows the pattern of molecular vibrational transitions, with fine rotational shifts superimposed that are characteristic of particular molecules (Duley, 1983; Levenson and Kano, 1988).

Raman scattering can also be used to shift the wavelengths of high-power lasers to regions where they are more useful. Raman shifters have found particular use in downshifting the ultraviolet wavelengths produced by excimer lasers to longer wavelengths in the visible and near-ultraviolet. This greatly increases the number of wavelengths that can be generated from excimer lasers, although at a sacrifice in power.

Raman shifters have become available commercially. Although they can be used to move to either longer or shorter wavelengths, they are more efficient in generating longer wavelengths, and most current interest centers on "downshifting" frequency to longer wavelengths.

Changing Pulse Length

It is often desirable to change the "natural" duration of laser pulses. Sometimes the goal is to extend pulse length, but mostly it is to shorten the pulse, either to provide higher peak power or to provide better time resolution.

Some pulse shortening is inherent in nonlinear interactions because of the nature of their dependence on the amplitude of the interacting electromagnetic fields. Laser pulses have finite rise and fall times, and because of the nonlinear dependence of output power on input power, the lower-power parts of the pulses will generate only very weak and often insignificant output. Thus, putting an optical pulse through a nonlinear process, say, frequency doubling, tends to shorten it. However, the degree of shortening is limited, and it is of practical importance only in cases where a precise pulse duration is required after a nonlinear process.

Four techniques for producing short pulses are in widespread use. Three of them—Q switching, cavity dumping, and modelocking—operate by interacting with light inside the laser cavity, in essence clumping energy together to produce a pulse that is shorter in length and higher in peak power than unaltered laser emission. Such pulse shortening often costs energy, but for many applications that is an acceptable trade-off. The fourth technique is the simple but inelegant approach of switching the beam off and on outside the cavity using some type of shutter—either a mechanical device or an electro-optic or acousto-optic modulator. This does not serve to increase peak power but can produce short pulses; it falls under the heading of external optics and is described in Chap. 5.

Extension of pulse duration without fundamentally altering the structure of the laser is much harder than shortening pulses. Very few applications require such pulse stretching, and it will not be described here.

Q Switching. Like any oscillator, a laser cavity has a quality factor Q that measures the loss or gain of the cavity. This factor is defined as

$$Q = \frac{\text{energy stored per pass}}{\text{energy dissipated per pass}}$$

Normally, the Q factor of a laser cavity is constant, but modulating the Q factor raises interesting possibilities. If the Q factor is kept artificially low, say by putting a lossy optical element into the cavity, energy will gradually accumulate in the laser medium because the Q factor is too low for laser oscillation to occur and dissipate the energy. If the loss is removed suddenly, the result is a large population inversion in a high-Q cavity, producing a high-power burst of light, a few nanoseconds to several hundred nanoseconds long, in which the energy is emitted. This rapid change in cavity Q is called Q *switching*.

There are three basic variations on Q switching, as shown in Fig. 4.3. One is use of a rotating mirror or prism as the rear cavity mirror. Another is insertion of a modulator into the cavity. The third is insertion of a nonlinear lossy element into the cavity that becomes transparent once intracavity power exceeds a certain level. The first two techniques are called "active" Q switching because they require active user control; the third is considered "passive" because no special control is needed. Q switching of pulsed lasers cannot increase pulse energy, but it can squeeze much of that energy into a pulse a few nanoseconds to a few tens of nanoseconds long, and achieve peak powers in the megawatt range—higher than otherwise possible in most pulsed lasers. Although all types of Q switches can operate with pulsed lasers, only active Q switches can operate with lasers that are pumped by a continuous source of energy such as a continuous-arc lamp.

Operation of a rotating-mirror Q switch relies on the fact that most of the time the mirror is rotating it is not aligned properly with the

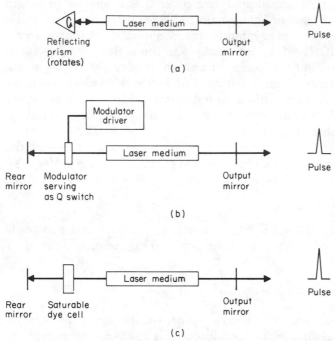

Figure 4.3 Types of Q switching used with lasers: (a) rotating mirror or prism; (b) active modulator; (c) passive. All cases show operation during a pulse. When the laser is not emitting, the rotating prism is turned so none of its faces parallels the output mirror. In other Q switches, the pulse is generated only during the brief interval that the modulator or saturable absorber cell is transparent.

output mirror, preventing oscillation. Periodically the rotating mirror passes through the point where it is properly aligned, causing cavity Q to increase abruptly and producing an intense Q-switched pulse.

When an optical modulator is used as a Q switch, it normally does not transmit light, blocking off one of the cavity mirrors. When the modulator is switched to transparency, the light in the cavity can reach the mirror, so cavity Q increases to a high level and the laser generates a Q-switched pulse. Normally the modulators used are electro-optic or acousto-optic devices, as described in Chap. 5, although they are packaged and sold as Q switches.

In passive Q switching, a material with nonlinear absorption characteristics (typically a dye in solution housed in a dye cell) is inserted into the laser cavity. At low power levels, the material has high absorption, blocking one of the mirrors from the laser medium. When the pump-energy source is pulsed, the amount of light in the laser cavity builds up, eventually reaching a level at which it bleaches out the nonlinear material's absorption. The dye cell thus becomes transparent, cavity Q abruptly increases, and a Q-switched pulse is produced. Typically the bleaching process takes on the order of a microsecond. Synchronization of passively Q-switched pulses is less precise than that of actively Q-switched pulses, but passive Q switches are simpler and cheaper. Their mode of operation limits passive Q switches to use with pulsed lasers, typically flashlamp-pumped solid-state types such as ruby and neodymium.

The saturation effect that permits passive Q switching is not particularly mysterious. The absorption is due entirely to one transition in the material. As intracavity intensity increases, more and more of the dye molecules absorb light, raising them to an excited state where they cannot absorb any more photons. Although the excited states eventually decay, at some point light intensity reaches a level at which as many species have been raised to the excited state as are in the lower state. At this point, emission is occurring as fast as absorption, so no net absorption can take place, and the dye is said to have been bleached or saturated.

One inherent limitation of Q switching is that it only works for a laser medium capable of storing energy for a time much longer than the Q-switched pulse duration. This requirement comes from the fact that the energy is stored by accumulating the excited species in its upper laser level, and this is done by suppressing the feedback needed for stimulated emission. This is possible only if the spontaneous emission lifetime is relatively long—otherwise the excited species will spontaneously drop down to a lower energy level. Thus Q switching does not work for all types of lasers.

Cavity Dumping. Cavity dumping operates on rather different principles than Q switching but produces somewhat similar results. The basic idea of cavity dumping is to couple laser energy directly out of the laser oscillator cavity without having it pass through an output mirror. In this case, both cavity mirrors are totally reflective and sustain a high circulating power within the laser cavity. This circulating power can be dumped out of the cavity by deflecting it. The basic concept can be seen in Fig. 4.4, where a mirror pops up into the laser cavity to deflect the circulating light out. The result is a pulse with length close to the cavity round-trip time.

In practice, cavity dumping is done in different ways. One approach is to put an acousto-optic deflector in the cavity. Normally the deflector transmits light to the cavity mirrors, but in the cavity-dumping mode the deflector switches light out of the laser cavity. Another approach involves an electro-optic modulator and polarizing beamsplitter. When the modulator is off, linearly polarized light is reflected

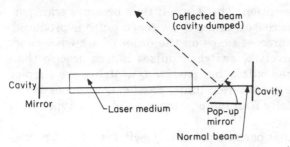

Figure 4.4 The basic concept of cavity dumping is dumping the energy out of the laser cavity by a means such as popping an intracavity deflecting mirror into the beam. Actual cavity dumpers are more complex systems that rely on acousto-optic deflectors or electro-optic modulators.

back and forth through the cavity and is not affected by the beamsplitter. However, when the modulator is turned on, light which passes through it is rotated in polarization, and that light is then reflected out of the cavity by the beamsplitter.

Cavity dumping can be used with continuous-wave lasers which do not store energy in their upper excited levels and hence cannot be Q-switched. It also can be combined with Q switching and modelocking to produce special pulse characteristics.

Cavity dumping of a continuous-wave laser can generate pulses of 10 to 50 nanoseconds (ns) at repetition rates of 0.5 to 5 megahertz (MHz). Typically pulses are shorter than Q-switched pulses from such lasers, are produced at higher repetition rates, and do not contain as much energy.

Flashlamp-pumped, Q-switched solid-state lasers can be cavity-dumped to shorten the normal 3- to 30-ns Q-switched pulses to 1 to 3 ns. One Q-switched, cavity-dumped pulse is produced per flashlamp pulse.

When used together with modelocking, cavity dumping can reduce the naturally high repetition rates of modelocked lasers to more manageable levels and increase the peak power in the pulses. In this case, the cavity-dumping frequency is synchronized with the generation of modelocked pulses, so periodically one modelocked pulse is switched out of the cavity. Typically, operating frequencies are chosen so the ratio of cavity-dumping frequency to modelocked pulse generation equals an integer plus one-half, say, 9½. This can build the peak power of modelocked pulses to as much as a factor of 30 higher than would otherwise be possible. The length of the modelocked pulses is not affected by cavity dumping, which only serves as a gating switch.

Modelocking. Pulses in the picosecond regime can be generated by modelocking. The simplest way to visualize modelocked pulses is as a group of photons clumped together and aligned in phase as they oscillate through the laser cavity. Each time they hit the partially transparent output mirror, part of the light escapes as an ultrashort pulse. The clump of photons then makes another round trip through the laser cavity before another pulse is emitted. Thus the pulses are separated by the cavity round-trip time $2L/c$, where L is cavity length and c is the speed of light.

The physics are actually much more complex (see, e.g., Svelto, 1989), and the details are well beyond the scope of this book. The basic idea is that modulation of the loss or phase of an optical element in the laser cavity causes many different longitudinal modes to oscillate together in phase. Modelocking occurs when the modulation frequency corresponds to the cavity round-trip time. The modulation can be active, by changing the transmission of a modulator, or passive, by saturation effects similar to those in a passive Q switch. In either case, interference among the modes produces a series of ultrashort pulses.

Modelocking requires a laser that oscillates in many longitudinal modes. Thus it does not work for many gas lasers with narrow emission lines. However, it can be used with argon or krypton ion lasers, solid-state crystalline lasers, semiconductor lasers, and dye lasers (which have exceptionally wide gain bandwidth). The pulse length is inversely proportional to the laser's oscillating bandwidth, so dye lasers can generate the shortest pulses because of their exceptionally broad gain bandwidths.

Both pulsed and continuous lasers can be modelocked. In either case, modelocking produces a train of pulses separated by the cavity round-trip time. For pulsed lasers, the result is a series of pulses that fall within the envelope defined by the normal pulse length. Single modelocked pulses can be selected by passing the pulse train through gating modulators that allow only a single pulse to pass.

Other Pulse Compression Techniques. Modelocking can generate subpicosecond pulses, but the limited gain bandwidth of laser media sets a minimum length on modelocked pulses because of the Fourier transform relationship between pulse duration and spectral bandwidth. The only way to generate shorter pulses is to spread out spectral bandwidth outside the laser. Laboratory researchers have found an ingenious technique for doing so in a way that lets them compress pulse length. An ultrashort modelocked pulse is passed through an optical fiber, a process that spreads out its wavelength spectrum. Spectral dispersion in the glass of the fiber also causes temporal dispersion, but it is an ordered dispersion. Because refractive index is a function of wavelength, the wavelengths for which the index is the smallest travel through the fiber fastest, creating a pulse in which wavelength varies with time. This pulse can then be compressed in time by passing it through a delay-line structure with differential wavelength delays that precisely compensate for the delays in the fiber. That technique (Fujimoto et al., 1984) has been used to generate pulses just 6 femtoseconds (fs) [0.006 picosecond (ps)] long (Fork et al, 1987).

Amplification

One emerging use of laser amplifiers is to increase the strength of weak optical signals, usually in a fiber-optic communication system. Such optical amplifiers can extend transmission distance or increase input to a receiver. Two types are under development, semiconductor laser amplifiers and optical-fiber amplifiers. Semiconductor amplifiers are similar to conventional semiconductor lasers, but their end facets are finished to suppress reflection, so they cannot generate a laser beam by internal oscillation. Instead, they must amplify the beam from an external laser. Fiber amplifiers are segments of optical fiber, with their cores doped with ions which can produce stimulated emission, and with ends coated to minimize reflection. You can consider them as miniature solid-state lasers. They are described in Chap. 26.

Larger-scale laser amplifiers are used to build up pulses containing high energy with high peak power. There are intrinsic limits on the maximum pulse energy that can be extracted from a particular laser oscillator. This pulse energy can be boosted by passing it through one or more laser amplifiers, because it is possible to extract more energy from a laser medium in amplifier configuration than when it functions as an oscillator.

Oscillator-amplifier configurations are used to extract high-energy pulses from optically pumped solid-state, gas, and dye lasers. Often, the oscillator and amplifier are packaged together in commercial laser systems. Normally commercial laser systems include no more than one or two amplifier stages. The most complex amplifier chains built are the ones assembled from neodymium-glass for laser fusion experiments at the Lawrence Livermore National Laboratory, where the pulse from the master oscillator can pass through up to 19 amplifiers.

Bibliography

J. Bokor, P. H. Bucksbaum, and R. R. Freeman: "Generation of 35.5-nm coherent radiation," *Optics Letters* 8(4):217–219, April 1983.

W. W. Duley: *Laser Processing and Analysis of Materials*, Plenum, New York, 1983.

R. L. Fork et al.: "Compression of optical pulses to six femtoseconds using cubic phase compensation," *Optics Letters* 12:483–487, 1987.

J. G. Fujimoto, A. M. Weiner, and E. P. Ippen: "Generation and measurement of optical pulses as short as 16 fs," *Applied Physics Letters* 44(9):832–834, 1984.

Lew Goldberg and Myung K. Chun, "Efficient generation at 421 nm by resonantly enhanced doubling of GaAlAs laser diode array emission," *Applied Physics Letters* 55(3):218–220, 1989.

S. E. Harris et al.: "Generation of ultraviolet and vacuum ultraviolet radiation," in S. F. Jacobs et al. (eds.), *Laser Applications to Optics & Spectroscopy*, Addison-Wesley, Reading, Mass., 1975, pp. 181–197.

C. Breck Hitz: *Understanding Laser Technology*, PennWell Publishing, Tulsa, Okla., 1985.

Marc D. Levenson and Satoru S. Kano, *Introduction to Nonlinear Laser Spectroscopy Revised Edition*, Academic Press, Boston, 1988; see chap. 4.

J. Reintjes et al.: "Seventh harmonic conversion of modelocked laser pulses to 38.0 nm," *Applied Physics Letters* 30(9):480–482, 1977.

Peter W. Smith, "Single frequency lasers," in Albert K. Levine and Anthony J. DeMaria (eds.), *Lasers*, vol. 4, Marcel Dekker, New York, 1976, pp. 74–118.

R. G. Smith: "Optical parametric oscillators," in Albert K. Levine and A. J. DeMaria (eds.), *Lasers*, vol. 4, Marcel Dekker, New York, 1976, pp. 190–307.

Orazio Svelto: *Principles of Lasers*, 3d ed., Plenum, New York, 1989.

A. P. Thorne: *Spectrophysics*, Halsted/Wiley, New York, 1974.

K. Thyagarajan and A. K. Ghatak: *Lasers Theory & Applications*, Plenum, New York, 1981.

Amnon Yariv, *Optical Electronics* 3d ed., Holt Reinhart, and Winston, New York, 1985; see chap. 8.

Amnon Yariv and Poichi Yeh: *Optical Waves in Crystals*, Wiley, New York, 1984.

External Optics and Their Functions

External optics are needed to focus and direct laser beams and to manipulate light in other ways that let it be put to practical use. The field of optics is a science in itself and can legitimately be considered to include laser technology as well as other diverse areas. This chapter cannot hope to tell everything about optics, but it can give an overview of the types of optics often used with laser systems. Readers seeking more detail can select from a number of handbooks (e.g., Department of Defense, 1962; Driscoll, 1978; Malacara, 1988; Melles Griot, 1988; Wolfe and Zissis, 1985) and textbooks (e.g., Hecht, 1987; Möller, 1988; Smith, 1990; Smith and Thomson, 1988).

Optical elements perform many tasks. Their simplest functions are the refraction and reflection of light, actions we normally associate with lenses and mirrors. In optical systems, other functions also can be important, such as selectively transmitting certain wavelengths of light, altering the polarization components of light, or breaking light up into its spectral components. These are generally considered "passive" optical functions because the components require no special control or input power to perform them. There are also "active" components which operate under active control or require input energy (other than light) to perform their function; examples include modulators which alter the intensity of a transmitted laser beam and scanners which move a beam.

This chapter concentrates more on passive optics because they are the most widely used. The principles of active optical devices can be extremely complex, so the description here will be limited to an overview of their functions. Although many lasers are used with optical instruments—assemblages of various optical components designed to perform particular functions—that field is far too large to be described properly in this sort of book.

Transmissive Optics

The most familiar types of optical components are transmissive optics, whose action depends on the refraction (bending) of light passing through them. The way in which transmissive optics refract light depends both on the shape of the component and on the refractive index—the crucial measurement of the transmissive characteristics of the material that makes up an optical component.

Refraction and Refractive Index. The refractive index was devised as a measurement of how much a material can bend light. Physically, its real meaning is as a measurement of how much a material slows down light traveling through it. The standard is the speed of light c in a vacuum, for which the index of refraction n is defined as equaling 1. For all other materials, the index of refraction n_{mat} is defined as the ratio

$$n_{mat} = \frac{c_{vacuum}}{c_{material}}$$

which normally is greater than 1.

Refractive index depends on the nature of the material and the wavelength of light. As shown in Table 5.1, there are considerable differences among materials. In general, refractive index tends to increase with material density, and for gases there is a definite relationship between gas density and refractive index. The refractive index of a gas at normal density is close to 1, so in practice the indexes of gases are measured relative to that of a vacuum. The index of air is small enough so that in practice the refractive indexes of solids are measured relative to air rather than to a vacuum. This not only simplifies the task of measurement but also gives a more useful value, because virtually all optical systems are used in air, and in such circumstances it is the ratio of refractive indexes of the material and air—rather than the absolute refractive index relative to a vacuum—that is relevant.

From the standpoint of optical design, the importance of the refractive index is its measurement of how strongly light is bent as it passes between two materials, as shown in Fig. 5.1. Note that the frequency of light-wave oscillation remains constant in all materials; the decrease in speed in a denser material translates into a decrease in wavelength. As light enters a material at an acute angle, as in Fig. 5.1, the wavefront is bent by an amount proportional to the differences

TABLE 5.1 Refractive Indexes of Selected Materials at Selected Wavelengths*

Material	Wavelength, μm	Refractive index, n
Air (1 atm pressure)	0.59	1.0002765
Pure water	0.59	1.33287
Zinc crown glass	0.59	1.517
Heavy flint glass	0.59	1.650
Vycor optical glass	0.58	1.458
Fused silica	0.59	1.4584
	0.214	1.5337
	0.89	1.4518
	2.325	1.4329
Crystal quartz	0.59	1.54424
As_2S_3	0.6	2.63640
	1.0	2.47773
	2.0	2.42615
	4.0	2.41116
	10.0	2.38155
CaF_2	0.59	1.43384
Germanium	2.06	4.1016
LiF	0.5	1.39430
NaCl	0.59	1.54427
	2.0	1.52670
	10.0	1.49482

* Note that in birefringement materials such as crystal quartz the refractive index depends on polarization of light rays. The values shown are for "ordinary" rays.
SOURCE: Data from Driscoll, 1978; Wolfe and Zissis, 1985; Weast, 1981.

in the wavelength of light in the two materials. For light going in either direction between materials a and b, the result is Snell's law of refraction:

$$\frac{n_b}{n_a} = \frac{\sin A}{\sin B}$$

where the n's are the refractive indexes in the two materials, A is the angle from the normal in material a, and B is the angle from the normal in material b. This can also be expressed in the form

$$n_b \sin B = n_a \sin A$$

When light travels from a high-index medium to one of lower refractive index, there is no solution to the Snell equation for large angles to the normal because the sine of an angle can be no larger than 1. This is the case of total internal reflection, in which all light incident on the surface between the two materials is totally reflected back into the denser material (for example, back into the glass at the glass-air in-

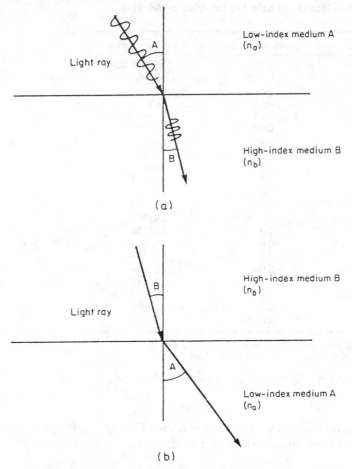

Figure 5.1 Refraction of a light ray passing (*a*) from a low-index medium into a high-index medium and (*b*) of one passing from a high-index medium to a low-index medium. Note that angles are measured from the normal to the surface.

terface of a prism). The smallest angle from the normal at which this occurs is called the *critical angle*; the concept is shown in Fig. 5.2.

The variation of refractive index with wavelength is sometimes neglected in optical calculations because it is normally small over a small wavelength range. However, it can have practical implications, particularly when dealing with a broad range of wavelengths. The variation in refractive index causes dispersion, which bends light of different wavelengths at different angles. The most vivid example is in a prism which intentionally spreads out the visible spectrum, but the same sort of spectral dispersion occurs whenever light passes

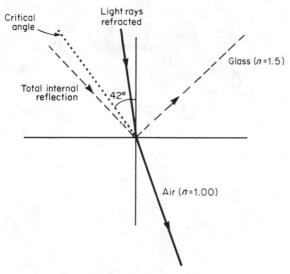

Figure 5.2 Total internal reflection occurs when light in glass strikes the interface with air at an angle from the normal greater than the critical angle, the angle of incidence for which Snell's law gives an angle of refraction of 90°. In the case shown, with glass having $n = 1.5$ and air having $n = 1$, total internal reflection occurs for angles of incidence greater than 42° to the normal.

through a medium with refractive index that varies with wavelength. For example, simple lenses show chromatic aberration, because they bring light of different wavelengths to slightly different focal points. And when light pulses covering a range of wavelengths travel long distances through optical fibers, spectral dispersion can spread the pulses out because some wavelengths travel faster than others.

Lenses. Lenses cause a beam of parallel light rays to converge or diverge. There are many different types; a sampling of positive and negative lenses is shown in Fig. 5.3. The most common types are radially symmetrical, so they focus light in two dimensions, bringing a circular beam down to a point. Less common are cylindrical lenses, with one or both surfaces curved in only one direction, and shaped like the sides of a cylinder. Cylindrical lenses focus light in only one dimension, bringing a circular beam down to a line. Radially symmetrical lenses often are called "spherical" because at least one of the two surfaces usually is a section of a sphere. However, some radially symmetrical lenses are made with other curvatures to avoid focusing limitations inherent with spherical surfaces.

Lenses are classified as either positive or negative. Positive lenses

Positive lenses

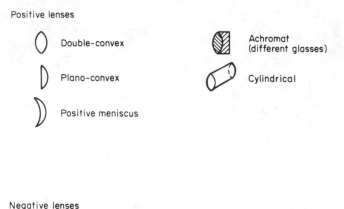

Negative lenses

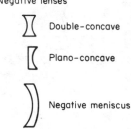

Figure 5.3 A sampling of positive and negative lenses. Achromats, shown only as positive lenses, can also take on other configurations. Cylindrical lenses, which can be positive or negative, are curved only in one dimension; other types shown are curved in two dimensions, but are shown in cross section only for simplicity.

bend parallel light rays so that they converge; negative lenses bend them so that they diverge. (The difference is easy to discern because positive lenses are thicker at the center than the sides, while negative lenses are thinnest at their centers.) The focusing power of a lens is normally measured as the focal length. The precise focal length is complex to define (see, e.g., Smith, 1978), but a reasonable approximation is to label it the distance from the lens to the focal point where rays initially parallel to the axis are brought to a point. That works for a positive lens because the rays converge. For a negative lens, the "focal point" is taken as the point behind the lens from which the rays appear to diverge, and the focal length is a negative number. Both cases are shown in Fig. 5.4 for ordinary two-dimensional spherical lenses. Note that strictly speaking the focal length is the distance from the center of the lens to the focal point, *not* from the surface of the lens.

Another difference between positive and negative lenses is in the formation of images. A positive lens can produce a real image—one that can be projected onto a screen—of an object at a suitable position.

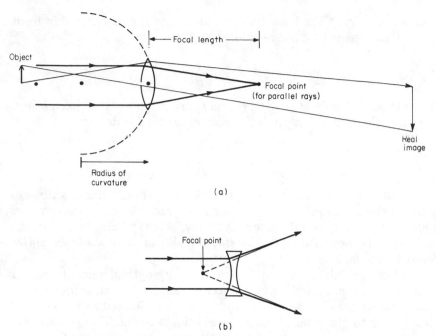

Figure 5.4 (*a*) Refraction by a positive lens bends parallel rays so they meet at a focal point. The real image produced by a positive lens is beyond the focal point. (*b*) For a negative lens, the focal point is the point from which initially parallel rays, diverged by passing through the negative lens, appear to originate. All lens surfaces have a radius of curvature, as shown only for one surface of the positive lens.

A negative lens cannot produce a real image, but only a virtual image—one that can be seen by the human eye but not projected onto a screen. (Positive lenses can also produce virtual images in addition to real images.) The difference between real and virtual images is important because only real images can be recorded photographically or by other means.

Focal length of a lens depends on its curvature and on the index of refraction of the lens material. By making some simplifying assumptions that the lens is thin compared to its diameter and the size of the optical system, the focal length f can be calculated for a spherical lens using the formula

$$\frac{1}{f} = (n - 1) \left(\frac{1}{r_1} + \frac{1}{r_2} \right)$$

where n is the index of refraction, and r_1 and r_2 are the radii of curvature of the lens surfaces.

This focal length, in turn, can be used in other simple calculations of optical characteristics. If an object is at a distance A behind a lens with focal length f, the lens will form an image at a distance B in front of the lens, defined by the relationship

$$\frac{1}{f} = \frac{1}{A} + \frac{1}{B}$$

If two lenses with focal lengths f_1 and f_2 are placed next to each other, the equivalent focal length of the pair of lenses f is defined by

$$\frac{1}{f} = \frac{1}{f_1} + \frac{1}{f_2}$$

As with the calculation of lens focal length, these relationships depend on the assumption that the lens is "thin." Much more complex equations are required if the lens is "thick." More precise formulations of these and other optical relationships can be found in standard optics references (e.g., Smith, 1978).

The focal length enters into calculation of another quantity useful in describing lenses and optical systems, the *focal ratio*, usually written as f/x, where x is the focal ratio value. It is defined as the ratio of focal length to diameter of the lens or optical system. This is the same focal ratio used in camera lenses and carries the same implications. Low focal ratios mean a short focal length and a working point close to the lens—and require a sharply curved lens. Curvature can be more gradual for a larger focal ratio, with longer focal length and greater working distance. As with cameras, the focal tolerances are greater for larger focal ratios.

Lenses have focusing imperfections known as aberrations, which are present even in optically perfect components. The two most important are spherical aberration and chromatic aberration.

A simple lens with a spherical surface does not bring parallel light rays entering it at different distances from its center (or axis) to precisely the same focal point. This spreading out of the focus is called *spherical aberration* and is most serious for lenses with low focal ratios (e.g., $f/1$). Spherical aberration can be reduced by using two-component doublet lenses, because the positive and negative components have opposite spherical aberrations that nearly cancel out. However, to overcome spherical aberration completely, optics must have nonspherical (aspheric) surfaces designed especially to avoid spherical aberration. These lenses are much harder to make than lenses with spherical surfaces and hence are much more expensive in most cases. Note, however, that spherical aberration becomes less important with increasing focal ratio and can be ignored in most systems with high f ratios.

Chromatic aberration is due to the variation of refractive index with wavelength. Because of this variation, a lens brings light of different wavelengths to different focal points. To compensate for chromatic aberration, multielement lenses are assembled from components of different glasses in which refractive index varies in different ways with wavelength. In such achromatic lenses, the spectral dispersion of one element balances out that of the other element or elements. The use of achromats is important in optical systems that cover the entire visible range, such as cameras or binoculars. However, simple lenses usually suffice for laser systems operating at a single wavelength, because the narrow spectral width of the laser makes chromatic aberration insignificant.

Normal lenses are made from material with uniform refractive index. Researchers are investigating the use of materials with non-uniform refractive index, called *graded-index* or *gradient-index lenses*. One approach already in use is the rod lens, a segment of thick optical fiber with refractive index that varies radially, letting it focus light (Iga et al., 1984). Another approach in development is to make larger lenses from materials with refractive index that varies through the lens thickness. Polishing a spherical surface onto the graded-index region would create a lens with the focusing power of an aspheric lens, but with an easier-to-make spherical surface.

Windows. A window is a flat piece of transparent material that should have virtually no effect on the light passing through it. Windows exist in optical systems, as they do in buildings, to separate one physical environment from another. This is typically done when different ambient conditions are required in one part of an optical system than in another; one example is when a laser beam is to be passed through a sample of a gas other than air. The window must be transparent at the wavelength used by the optical system, but, as described below, no optical material is transparent at all wavelengths.

Gas-laser tubes often are sealed with windows mounted at *Brewster's angle*, at which light linearly polarized in one direction is transmitted without surface reflective losses, although reflective losses do occur for orthogonally polarized light. Brewster's angle θ_B is defined relative to the normal to the window surface by

$$\theta_B = \arctan \frac{n_1}{n_0}$$

where n_1 is the window's refractive index, and n_0 is the index of the material (e.g., the laser gas) through which the light travels to reach the window. For glass with index of 1.54, the Brewster angle is 57°. A Brewster angle window does not eliminate light polarized in the orthogonal direction, but it introduces enough losses to make the laser output strongly linearly polarized.

Because all solids absorb some of the light passing through them, solid windows are vulnerable to optical damage when transmitting high-power beams. To avoid this problem, "aerodynamic windows" have been developed, in which gas flows by an opening fast enough to prevent mixing of gases on opposite sides of the opening. Such windows are required only for lasers with very high power output.

Prisms. Prisms serve two main functions in optical systems: dispersing light according to wavelength, or redirecting light without changing the size of the beam or the relative alignment of the light rays.

Dispersive prisms rely on the variation in refractive index with wavelength to refract different wavelengths at different angles, spreading out the spectrum. In some optical systems, a slit is used to pick a narrow range of wavelengths out of this spectrum. Alternatively, the prism can be included in a laser resonator aligned so only a certain wavelength will oscillate, as described in Chap. 4.

Beam-redirecting prisms generally rely on total internal reflection to bend light around corners, invert images, reverse images right to left, or perform combinations of these operations. A variety of prisms have been designed for such tasks, such as those shown in Fig. 5.5. Most of these types are used more often in imaging instruments such as binoculars than in laser systems. One exception is the retroreflective prism or "corner cube," designed to return an input beam exactly in the direction from which it came, which is often used with lasers; however, not all retroreflectors are prisms.

The term *prism* can also be used for polygonal blocks which are coated with reflective materials and function as mirrors. These reflect light from their outer faces, so the light never enters the "prism," which in fact need not be made of transparent material. Typically such prisms rotate so the different faces sequentially scan a laser beam, by reflection, across another surface.

Transmissive Materials

The characteristics of lenses, prisms, and windows depend on the transmissive materials from which they are made. The wavelengths at which these materials are transparent depends on the internal energy structure. Strong absorptions occur at wavelengths corresponding to energies at which the material can make a transition from its normal energy state to a higher one. The material is transparent at wavelengths corresponding to energies where there are no possible transitions. Energy-level structures differ, and so do transparencies,

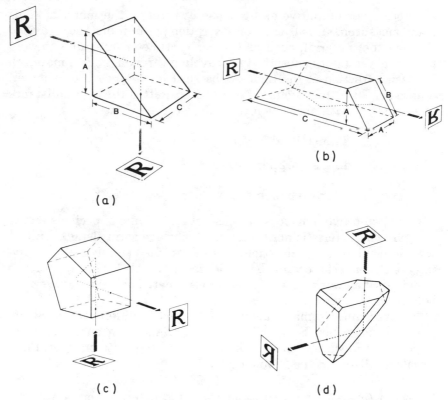

Figure 5.5 A sampling of prism types used for redirecting beams, showing their effect on images passing through them. (*a*) Right-angle prism; (*b*) dove prism; (*c*) penta prism; (*d*) roof ("amici") prism. *(Courtesy of Melles Griot.)*

but there is no such thing as a material transparent throughout the electromagnetic spectrum. Even air is functionally opaque in much of the ultraviolet and infrared spectral regions.

The amount of light transmitted equals the amount of incident light, minus losses from reflection at the surface and from absorption at the surface and within the material. Reflectivity can vary over a wide range, from nearly 100 percent for many metals in the visible and infrared to near zero for materials such as graphite that look black to the eye. Reflectivity tends to increase with the difference in refractive index between two adjacent materials. Coating the high-index material with a thin layer of material with intermediate refractive index can decrease reflectivity. It is also possible to increase or decrease reflectivity at particular wavelengths by building up multi-layer thin-film interference coatings in ways described below.

Absorption occurs both at the surface and within the material. Bulk absorption is cumulative as light passes through the material. This usual measurement is fractional absorption per unit distance traveled through the material, such as 0.1 per centimeter, meaning that 10 percent of the light is absorbed in traveling through 1 cm of material. The same fraction of the light is absorbed in the second centimeter, and so on. Thus the amount of power transmitted through a distance D of material with absorption constant k can be written

$$\text{Transmitted power} = \text{input power} \times e^{-kD}$$

The corresponding absorption law is

$$\text{Absorbed power} = \text{input power} \times (1 - e^{-kD})$$

It is often convenient to write absorption in units of decibels and to measure characteristic absorption in decibels per unit length. In this way absorption (or attenuation) can be obtained simply by multiplying the characteristic value by the distance traveled. This approach is standard when working with fiber optics, where normal units are decibels per kilometer.

The transmissive materials normally used in different spectral regions are described below with two materials often encountered in optics: air and water. Some books which tabulate data on optical materials are listed in the Bibliography.

Ultraviolet Optics.   The ultraviolet spectrum extends from 400 nanometers (nm) to a short-wavelength limit of 1 to 10 nm (definitions vary). The shorter-wavelength end of the ultraviolet spectrum is poorly explored because of experimental problems that increase in severity with shorter wavelength. Laser action also becomes increasingly difficult to produce at shorter wavelengths. A variety of commercially available lasers emit in the 200- to 400-nm range, but only two emit at shorter wavelengths: the argon fluoride excimer at 193 nm, and molecular fluorine at 157 nm. Shorter-wavelength lasers have been demonstrated experimentally but have yet to become practical.

Optical materials used in the near-ultraviolet, wavelengths of 300 to 400 nm, are familiar, because most optical glasses transmit in that region. Special types of quartz can be used at wavelengths shorter than 200 nm, beyond the range of other silicate materials (Driscoll, 1978). For shorter-wavelength operation, the usual choices are magnesium fluoride (transmissive to 110 nm), calcium fluoride (transmissive to 125 nm), or lithium fluoride (transmissive to 104 nm). The first two materials are more durable than LiF (Harshaw Chemical, 1982).

Optical damage thresholds tend to be lower in the ultraviolet than in the visible or infrared, so damage is a serious concern. It can occur in the bulk material or at the surface, particularly to coatings applied to enhance or reduce reflections. Nonsilicate ultraviolet windows are subject to some environmental degradation, although the problem is not as severe as for the more delicate materials used in parts of the infrared.

Material problems become increasingly severe at shorter ultraviolet wavelengths. As photon energy increases with decreasing wavelength, it eventually reaches a point sufficient to remove one of the electrons from a completed shell of an atom or molecule. At this threshold, ultraviolet absorption increases abruptly, and the material effectively becomes opaque. The atmosphere becomes opaque at about 200 nm, but rare gases are transparent at shorter wavelengths. Helium, with the highest ionization threshold, is transparent to about 58 nm (Zombeck, 1982). All known solid window materials become opaque by 100 nm. These transmission problems make experimentation in the extreme ultraviolet extremely difficult.

Visible Materials. Silicate glasses are the obvious choice for transparent visible-wavelength optics because of their low cost, durability, and high transmission. Standard optical glasses are widely used with lasers. On the other hand, plastic optics, although widely used in consumer optical systems, are rarely used with lasers. Major reasons include concerns about their vulnerability to scratches, and potential scattering from inhomogeneities within the material or from surface scratches. The economies of scale possible with high-volume consumer production are also harder to achieve with the smaller volumes of laser system production.

There is some haziness in the definition of the "visible" region in the laser world. Strictly speaking, the visible region is usually defined as 400 to 700 nm, the wavelengths where the human eye is most sensitive, although the eye can actually detect a somewhat broader range, particularly at longer wavelengths, with very low sensitivity. However, optical engineers may functionally define the visible as the region where common optical glasses and plastics can be used—from about 300 nm to 1.5 or 2 micrometers (μm). This reflects the fact that near-infrared optical systems, such as those designed for use with a 1.06-μm neodymium laser, normally will also function in the visible spectrum as well.

Infrared Materials. Many materials have been developed for infrared optics, thanks largely to military development of night-vision systems. Development has concentrated in three wavelength bands in which air is transparent: 1 to 2 μm, 3 to 5 μm, and 8 to 12 μm.

The 1- to 2-μm region is usually called the near-infrared, although sometimes that term is used to cover somewhat longer wavelengths, at times stretching to 15 μm. Most silicate glasses are transparent to at least 2 μm, so conventional optical glasses are normally used in this region.

The 3- to 5-μm band, sometimes called the *mid-infrared* (although that term has also been used for longer wavelengths), is interrupted by some strong atmospheric absorption lines. Chemical and carbon monoxide lasers emit in this region, although some of their lines coincide with strong atmospheric absorption bands. Some silicate glasses and quartz are transmissive at the short-wavelength end of this region, but their absorption increases rapidly with longer wavelengths, and they are not usable at the long-wavelength end. Thus the usual choices for such optics are more exotic materials such as those whose transmission ranges are shown in Fig. 5.6.

The 8- to 12-μm region is often called the *thermal* infrared, because those wavelengths correspond to the peak of blackbody emission at room temperature. The region is also known as the far-infrared, a term also widely used for infrared wavelengths of 10 μm and longer. It is important for laser applications because it covers the 10.6-μm wavelength of the carbon dioxide laser. Some important materials used in this wavelength range are shown in Fig. 5.6.

Newcomers to the infrared should be aware of two common features of infrared optical materials. One is that typically they are *not* transparent at visible wavelengths, so an optical system that focuses a CO_2 laser beam may be as clear as mud to the human eye. The other is that some infrared materials are hygroscopic and have such a strong tendency to absorb water from the air that if left unprotected they will slowly dissolve. Users should exercise considerable care in selecting infrared materials, being particularly careful of physical and mechanical characteristics. Important data on infrared materials are available in several references (e.g., Driscoll, 1978; Harshaw Chemical, 1982; Spiro and Schlessinger, 1989; Wolfe and Zissis, 1985).

Air. Air is almost inevitably a part of any optical system. It can present special problems at any wavelengths where extremely high powers are being transmitted, or where beams must be focused precisely over long distances. Clear air is transparent in the visible region, although dust, haze, smoke, fog, clouds, and precipitation can block visible light. Such problems are rarely significant on a laboratory or industrial scale, but other problems can arise at other wavelengths.

The atmosphere transmits light well at wavelengths as short as about 290 nm, where ozone (O_3) begins absorbing strongly. Fortunately, the main concentration of ozone is in the upper atmosphere,

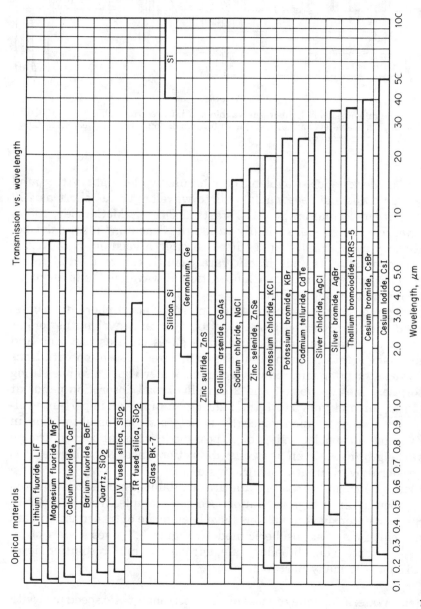

Figure 5.6 Spectral regions at which selected optical materials transmit adequately for optical applications. Note that different forms of silica, including quartz and optical glass, have different transmissions—a consequence of different levels of impurities. (*Courtesy of Janos Technology Inc.*)

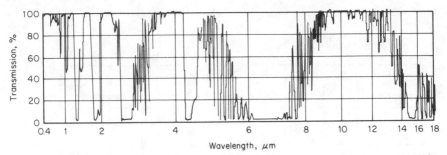

Figure 5.7 Transmission of light in the visible and infrared through a 304-m (1000-ft) horizontal path of sea-level atmosphere, with 5.7 mm of 26°C precipitable water. *(Courtesy of Sanders Associates Inc.)*

where it blocks short-wavelength solar ultraviolet light. At ground level, air transmits reasonably well to wavelengths as short as 200 nm, where molecular oxygen (O_2) starts absorbing (McCartney, 1983). Shorter wavelengths are usually called the "vacuum" ultraviolet because they are normally transmitted through a vacuum. However, if vacuum operation is undesirable, rare gas fills can be used at somewhat shorter wavelengths, with operation at wavelengths to about 58 nm possible by using helium.

Strong molecular absorption lines interrupt atmospheric transmission in the infrared and block it altogether at wavelengths longer than 13 μm. Carbon dioxide and trace gases such as methane and ozone make significant contributions to infrared absorption with lines scattered at shorter infrared wavelengths. Water vapor also has strong absorption lines between 1 and 2 μm plus broad absorption bands at 2.5 to 3.5 μm and from 5 to 7 μm, making humidity an important variable. Infrared absorption generally is not as strong as absorption in the vacuum ultraviolet, so it is often possible to transmit beams through short lengths of air even at strong absorption lines. Nonetheless, it is far easier to work in atmospheric windows.

Extensive compilations of data are available which show infrared absorption in air at high resolution (see, for example, Wolfe and Zissis, 1985), and extensive tabulations have been made of transmission data and predictions of atmospheric models. Figure 5.7 gives a view of infrared absorption on a broader scale, without the high resolution needed to show the many narrow lines.

Water. Water is opaque to most of the electromagnetic spectrum, with the notable exceptions being in the visible region and extremely low frequency radio waves. The absorption coefficient has a minimum of about 0.0001 per centimeter (cm^{-1}) near 500 nm in the blue-green, rising rapidly at shorter and longer wavelengths. For seawater, absorption reaches 0.01 cm^{-1} near 750 and 300 nm, and has values of 100 to 10,000 cm^{-1} at infrared wavelengths beyond 2 μm (Wolfe and Zissis, 1985). The detailed profile and precise absorption minimum vary depending on the type and purity of water, but in all cases maximum transmission is in the blue or blue-green.

Filters and Coatings

Filters and coatings can alter the transmission characteristics of optics and optical systems. Some filters are discrete optical components which derive their filtering characteristics from the nature of the bulk optical material, but many are actually filtering coatings applied to plain transmissive optical substrates. Modern coatings can be applied in many ways, giving designers great flexibility in selecting properties of an optical system. Coating and filter design are worthy of an in-depth treatment (see, for example, Department of Defense, 1961; Dobrowolski, 1978; Goldstein, 1988; Melles Griot, 1988); this section can provide only a brief overview of the field.

Types of Coatings and Their Functions. Coatings can be viewed in two ways, by the physical mechanisms of their operation or by their function. The same physical principles can be used for different types of coatings, and a single coating can perform multiple functions. Conversely, the same filtering function (particularly spectral filtering) often can be performed in two or more ways.

The major physical types of filters are

- *Interference coatings:* Types in which alternating layers of two (or sometimes more) different materials, each a fraction of a wavelength of light thick, are deposited on an optical surface. The thicknesses of the layers are carefully controlled. Internal reflections within the multilayer stack cause constructive and destructive interference for certain wavelengths and angles of incidence. Complex but well-quantified design rules let the designer pick the desired coating properties. Because interference effects depend on the distances light has to travel through each layer, the properties of interference coatings change with angle of incidence, and that angle must be known for proper system design. This effect also makes it possible to identify an interference filter, by tilting it and watching for a change in transmission. Interference coatings are made of nonconductive or "dielectric" materials, so they also are called *dielectric* coatings. Also known as *dichroic*, or *multilayer* coatings, they are widely used to perform a variety of functions.

- *Simple antireflection coatings:* Types in which a material with refractive index between that of air and an optical element is deposited on the optic to reduce surface-reflection losses.

- *Color-filter coatings:* Dyes applied to the surface of components which transmit light at some wavelengths and absorb it at others.

- *Neutral-density coatings:* Absorptive materials which absorb light uniformly across the range of wavelengths covered by the device or optical system.

- *Metal-film coatings:* Reflective metal layers applied to form mirrors.

- *Metal-dielectric coatings:* Metal-film coatings in which the reflectivity at a particular wavelength is enhanced by applying an overcoat of a transparent dielectric (i.e., nonconductive) material.

- *Protective coatings:* Thin films of materials applied to protect optical components from physical damage or corrosion, such as water-resistant coatings applied to hygroscopic materials, or transparent overcoats deposited on metal to prevent oxidation.

In some cases, notably protective coatings, the physical type also identifies the function, but this is not true for others, particularly interference coatings. From an optical standpoint, there are several different functions:

- *Spectral filters:* Types which selectively transmit and reflect certain wavelengths to separate them, described in more detail below.

- *Antireflection coatings:* Types designed to minimize surface reflection losses. In addition to the simple antireflection coatings described above, multilayer interference coatings can serve to reduce reflection losses at selected wavelengths.

- *Reflecting coatings:* Types designed to provide maximum reflectivity either at a single wavelength (interference coatings) or over a range of wavelengths (interference or metal-film coatings).

- *Polarizing coatings:* Types which selectively transmit light of one polarization. Typically light with one linear polarization is transmitted and that with orthogonal polarization is reflected.

- *Beamsplitting coatings:* Types which transmit and reflect light in a given ratio (say 1:1), thus splitting a single beam into two beams with a desired intensity ratio. Many types are polarizing, splitting light of one polarization in one direction and light of the orthogonal direction in the other direction.

- *Neutral-density coatings:*   Types which simply reduce the intensity of light transmitted throughout their working spectrum.

- *Graded coatings:*   Attenuating or neutral-density coatings in which the degree of attenuation varies across the aperture of the coating. In certain cases, this can serve an important function in improving beam quality.

Optical components are available either with coatings already applied or without coatings so that user-specified coatings can be custom-applied. Large optics suppliers normally stock components coated for important laser lines, while custom coating may be required for other wavelengths. The application of coatings is something of an art, and design and application must be performed by specialists.

Spectral Filters.   Diverse applications require selective transmission of certain wavelengths. The spectral filters that do this job, like optical coatings, fall into two sets of categories based on physical mechanisms and optical functions.

For laser applications, the commonest types are interference filters, in which a multilayer interference coating selectively transmits some wavelengths and reflects others. The reflection, rather than absorption, of rejected light is important because laser power density is high enough that absorption could heat absorptive filters enough to cause damage. The design flexibility possible with interference filters also makes it possible to specify a wider range of optical characteristics than for other types. Examples include narrower transmission and rejection bands and sharper changes in transmission with wavelength.

In other types of filters, rejected light is absorbed rather than transmitted, which can cause problems at high power densities. In colored-glass and crystal filters, absorption takes place in the bulk of the material. In gelatin filters, the spectral selection is a thin layer of gelatin containing an organic dye, which is sandwiched between a pair of glass plates. Colored-glass and gelatin filters are in widespread use because of their low cost, but their power-handling capability is limited.

Spectral filters fall into several functional categories. Their characteristics, described briefly below, are illustrated with typical spectral curves in Fig. 5.8.

- High-pass filters transmit wavelengths shorter than a given value and reject longer wavelengths. Examples include the "hot mirror" which transmits visible light but reflects in the infrared, and the "heat-absorbing filter," which transmits visible light and absorbs infrared.

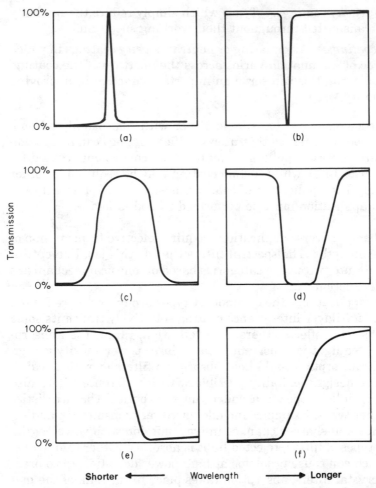

Figure 5.8 Transmission of representative types of filters, shown schematically (these are not actual plots). Wavelength increases to the right on all plots, a common convention in optics. (a) Narrow-line filter; (b) line-rejection filter; (c) bandpass filter; (d) band-rejection filter; (e) high-pass filter; (f) low-pass filter.

- Low-pass filters transmit wavelengths longer than a certain value and reject shorter wavelengths.

- Bandpass filters transmit light within a defined range (e.g., 500 to 600 nm) while rejecting longer and shorter wavelengths. A color filter that transmits only green light would fall under this heading.

- Band-rejection filters specifically reject light within a defined region but transmit other light in that spectral region.

- Edge filters are high- or low-pass filters with a sharp cutoff wavelength.

- Line filters transmit only a very narrow range of wavelengths around a certain value, for example, only light very close to the 632.8-nm helium-neon laser wavelength. The bandpass can be less than 1 nm wide, although peak transmission in such cases is limited.

- Line-rejection filters are the inverse of line filters, selectively rejecting light within a narrow range of wavelengths, such as blocking the helium neon laser wavelength.

Attenuators or Neutral-Density Filters.   Neutral-density filters or attenuators are designed to produce attenuation that is uniform regardless of wavelength. Of course, in practice the range over which attenuation is uniform is limited, typically to the visible spectrum. Transmission of a neutral-density filter is measured in units of optical density, defined as

$$\text{Optical density} = -\log_{10}(\text{transmission})$$

so a filter that transmits 0.001 (0.1 percent) of the incident light has optical density of 3.

Variable attenuators can be useful in many applications. One way to achieve variable attenuation is with a single optical element graded in optical density across its surface. Use of a single element, usually called a *wedge* although not necessarily wedge-shaped, is simple because linear motion or rotation can change optical density. However, the precision of calibration is limited. An alternative is the use of a set of multiple calibrated filters, providing greater accuracy at a cost in convenience.

Spatial Filters.   In laser applications, it is sometimes necessary to block the outer parts of the laser beam to remove stray light scattered by dust and lens imperfections, to reduce the effects of lens aberrations, and to improve wavefront quality. This function, sometimes called *aperturing*, is performed by spatial filters, which are essentially holes shaped to block unwanted portions of the beam. Some spatial filters are quite small in diameter and are called *pinholes*—an apt description of their appearance.

An alternative used in some applications is the "apodizing" or "graded" filter, in which optical density varies with the density from the optical axis—increasing as one moves outward. Instead of sharply cutting the beam off at a particular diameter, such a filter gradually reduces power density, an approach desirable in some applications.

Reflective Optics

As mentioned above, many mirrors are thin reflective coatings applied to substrates which have the desired flat or curved shape. Metal coatings are reflective over a broad range of wavelengths, while interference coatings are spectrally selective and reflect a limited range of wavelengths. Both types of coatings can be applied to many substrates, with glass a common choice because of its good optical qualities. Because no light need enter the substrate, glass can be used in any spectral range. Low-expansion materials can be used to avoid thermal distortion.

Mirrors also can be machined from solid metal, with the metal itself, or a reflective coating applied to it, providing the reflective surface. Metal mirrors are highly conductive and can be used with high-power lasers with suitable heat sinks or active cooling (such as water flowing through the substrate).

Metal mirrors are most often used in the infrared, where metal reflectivity is highest. Mirror surfaces can be polished, or machined with a diamond-tipped cutting tool. Diamond turning is particularly valuable for producing nonspherical surfaces, which are difficult or impossible to generate by conventional polishing techniques. Diamond turning does leave small ridges on the surface, but these are not optically significant if they are much smaller than the wavelength. For example, surfaces in which fine machining lines can just be detected by the human eye using visible light work fine for the 10-μm light from a CO_2 laser.

The commonest mirrors are flat or spherical; as for lenses, a spherical mirror is one with a surface that is a section of a sphere. Spherical mirrors can be concave or convex and have a focal length equal to half the radius of curvature. Normally these mirrors are totally reflective (with only a small fraction of the light absorbed and virtually none transmitted), but some flat mirrors are coated with partially reflective coatings that serve as beamsplitters.

Mirrors also can be manufactured with more complex surfaces. Aspherical mirrors are focusing mirrors with surfaces that are neither flat nor spherically curved. One common type is the parabolic mirror, a design used for reflecting telescopes to avoid spherical aberration. Some special applications require off-axis mirrors, in which the curved surface is a section of a three-dimensional surface but does not include the axis of rotation that defined the surface. Parabolic mirrors are fairly easy to produce, but other aspherical surfaces can be expensive to manufacture.

One special type of mirror used in a number of laser applications is the retroreflector or corner cube. This is a prism or a set of three mirrors with faces aligned at right angles with respect to each other, as shown in Fig. 5.9, to reflect any laser beam incident on the device back in the direction from which it came. This function is useful in alignment and measurement systems.

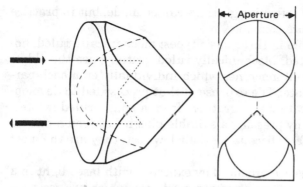

Figure 5.9 A solid-glass retroreflector, in which total internal reflection returns the incident beam to its origin. Corner-cube mirrors are similar, but mirrors aligned at right angles to one another form the retroreflector, rather than a glass block. *(Courtesy of Melles Griot.)*

Polygonal mirrors are used in some applications requiring scanning of laser beams across a surface. A multifaceted polygon, typically with 10 or more faces, is rotated about its axis, with each facet successively scanning the beam across the surface. The mirror may be a coated prism or a machined metal block, each with polished reflective surfaces around its perimeter.

Complex reflective surfaces, such as polygons and aspherical mirrors, often are produced by optical replication. This is essentially a mastering process in which a mold is generated from a master and used to form optical-grade surfaces, as described below. The replicated component is then coated with a reflective layer and can serve as a mirror.

Diffractive Optics

If light waves pass by a sharp edge, they are scattered or *diffracted* so that the rays no longer travel in a perfectly straight line. Diffraction is a phenomenon common to all waves, not just to light, and its precise effects depend on interference (as explained in most optics texts). Diffractive optics can manipulate light in ways similar to refractive and reflective optics. There are three major types: diffraction gratings, holographic optics, and diffractive or "binary" optics.

Diffraction Gratings. Diffraction gratings are surfaces covered by closely spaced parallel lines which diffract incident light. Interference among the light waves diffracted by the parallel lines makes the angle at which light leaves the grating a function of wavelength. Thus, a diffraction grating spreads out a spectrum like a refractive prism.

Both transmissive and reflective gratings can be made, but in practice reflective types are more common.

Gratings are available in three basic types: mechanically ruled, holographic, and replicated. Mechanically ruled gratings are the oldest type, produced by special machines which individually etch each parallel line into a flat plate. The expense of that process led to development of ways to replicate many gratings from a single ruled master. Replicas are produced by pressing a suitable plastic between a master grating and a flat surface; because of their low cost, they are the most common type.

Holographic gratings are produced by exposure with laser light in a way that produces a series of interference fringes, which are recorded to form the grating. Lines can be spaced more closely than in ruled gratings, and the process is less costly than ruling original gratings. Replicas also can be made from holographic originals.

Holographic Optical Elements. Holograms are best known for their ability to reproduce three-dimensional images. They do so by diffracting light from a two-dimensional interference pattern recorded on a light-sensitive plate or film.

Recently, interest has grown in another application of holograms—serving as optical elements comparable to conventional refractive or reflective optical elements (see, e.g., Hariharan, 1984). The concept is to create a hologram which scatters a beam of light in the same way that a conventional optical element would refract or reflect the beam. Computers can calculate the patterns required to scatter light in a particular way.

Holographic optical elements are attractive for applications that otherwise would require complex or bulky optical systems. A single holographic optical element can replace several conventional components. Their light weight and small size have led to their use in beam scanning and heads for optical disk storage systems. Holographic optics also can offer otherwise difficult-to-obtain capabilities, such as low-focal-ratio optical systems, at a lower cost than can conventional optics.

The diffraction angle of light scattered by a holographic optical element is a function of wavelength, so the element must be designed to operate at a particular wavelength. For the same reason, holographic lenses can focus only monochromatic light precisely—a requirement met by single-line lasers.

One important issue in holographic optics is diffraction efficiency, because only light diffracted by the hologram is focused to the desired point. Diffraction efficiencies of more than 80 percent have been claimed for commercial products, but such high efficiencies are not always attainable. Low diffraction efficiencies can mean both high losses and high levels of scattered light.

Binary Optics. Binary optics are diffractive optical elements whose diffractive power derives from microscopic patterns on a surface rather than from a hologram. The term *binary* originated from early versions, which had only two layers of surface relief, but current types are more complex. The surface relief is on the order of the wavelength of light, causing a phase delay of light which causes the diffraction. It is formed on the surface by photolithographic techniques similar to those used in semiconductor fabrication and can be replicated to make multiple copies.

Binary optics share many features with holographic optical elements, but advocates claim that they are more efficient. Thin, light-weight diffractive elements can have short focal lengths. They also can split or combine beams, sending different parts in different directions. Another possibility is dual-focus lenses, containing patterns which focus part of the incident light at one focal point and part at another focal point. Binary surface structures can be added to flat or curved surfaces of conventional optical elements, to correct for aberrations or perform functions that would require complex surfaces. On the other hand, because diffraction angle depends on wavelength, the spectral range and angular field of binary optics are limited.

Binary optics is an emerging technology at this writing, with few practical applications. However, it bears watching because of its potential to solve problems in some applications.

Special Components and Techniques

Some specialized components which do not readily fall into the reflective, transmissive, or diffractive categories are used in optical systems. A few types have been mentioned in passing above, but the major ones deserve brief explanations by themselves.

Beamsplitters. The function of a beamsplitter is to divide an incident beam into two parts, with one transmitted and the other reflected. Typically it is placed at a 45° angle to the incident beam, so reflected light is deflected at a right angle, while the transmitted light passes straight through. The beamsplitting ratio can be 1:1, or some other value.

Beamsplitters can take several forms, including partly transparent metal films, specially designed prisms, types which reflect one linear polarization and transmit the orthogonal polarization, and multilayer interference coatings which selectively transmit and reflect certain wavelengths. Although polarizing beamsplitters are the only types explicitly intended to affect beam polarization, other types may be sensitive to polarization. For example, certain types may reflect light of one linear polarization more strongly than that of the orthogonal polarization. That can lead to different beamsplitting ratios for differently polarized inputs, a potential source of problems in some optical systems. Users should take care to understand these effects in designing polarization-sensitive systems.

Fiber Optics. Optical fibers are best known for their use in signal transmission for communications. However, they were originally developed to serve as optical elements. Single fibers and bundles of randomly arranged fibers can serve as light pipes, conveying light from one point to another along sometimes convoluted paths.

Bundles of fibers in which the fibers maintain their relative alignments at the input and output faces can serve other functions. For example, a simple image inverter can be made by twisting the bundle 180°. A lens can do the same sort of image inversion, but not in such a short linear distance. Aligned or "coherent" fiber bundles can be made rigid or flexible, and both types have been used in many applications, such as in medical endoscopes to view inside regions of the human body not otherwise accessible without surgery. Although most recent writings on fiber optics concentrate on communications, there are some good references on the design and applications of fiber bundles as optical elements (e.g., Siegmund, 1978).

Guided-Wave and Integrated Optics. Researchers have worked for many years on optical waveguides and integrated optics, but the two closely related technologies have not moved much beyond the laboratory. Most research and development now is devoted to applications in fiber-optic communications.

Many developers have demonstrated planar optical waveguides, which confine light in a flat region, usually at the surface of a bulk optical material. Light is guided through the structure as it is through an optical fiber, by a difference in optical properties of the guide and the surrounding material.

Integrated optics seeks to combine several optical elements on a single monolithic substrate. The integrated components may be connected by optical waveguides, or they may themselves be optical waveguides with special properties. Developers also seek to include some of the active components described below. One conceptual goal is to replace bulk optical devices with integrated optical circuits, as integrated electronic circuits replaced discrete transistors, capacitors, and other electronic components.

Polarizing Optics. The electric and magnetic fields that make up electromagnetic radiation oscillate in directions transverse to each other as the wave travels in a direction orthogonal to both oscillations. Light in which the electric fields are aligned to oscillate in the same direction (and hence the magnetic fields are also aligned in the orthogonal direction) is said to be linearly polarized. Two linear polarizations are possible, orthogonal to each other and to the light ray.

Light can also be polarized circularly or elliptically. For the more general case of elliptical polarization, a stationary observer would see the polarization vector of the light rotating, with its endpoint describing an elliptical pattern. (In this case, direction of the polarization vector indicates polarization direction, while magnitude indicates amplitude of the polarized wave.) Physically, this means that the light ray consists of two linearly polarized components of unequal amplitude, one 90° in phase behind the other. In the special case of circular polarization, the two orthogonal polarization components are equal in amplitude, and the rotating polarization vector describes a circle.

Many polarizing components are available, which either alter the polarization of input light or transmit only light of a given polarization. The simplest are polarizers or polarizing filters, which transmit only light of one linear polarization. Depolarizers scramble the polarization of linearly polarized light, generating randomly polarized light. Polarization rotators rotate the plane of linearly polarized light. Waveplates delay the phase of the part of the input light with one linear polarization by a specified phase with respect to light of the orthogonal polarization. For quarter-wave plates, the delay is 90°, while for half-wave plates there is a 180° delay. Note that passing light through a quarter-wave plate polarizes it elliptically. The physics and operation of these and other polarizing optics are covered in more detail in a number of references and standard texts [see, for example, Stauffer (1988) for a scholarly discussion, or Melles Griot (1988) for a practical overview].

Replicated Optics. Optical replication, mentioned briefly above, is a process usable for production of some diffractive and reflective optics and a few transmissive components. The attraction of replication is its ability to produce components with unusual surface shapes at lower cost than conventional techniques.

Production of reflective replicas starts with fabrication of a substrate that approximates the shape of the finished piece, but lacks the required precision surface. The substrate is then coated with a plastic-epoxy material and pressed against a master piece to form a replica of the surface, which comes close enough to the quality of the original for many applications. The replica is then reflectively coated for use as an optical element. Transmissive replicas can be made by using trans-missive substrates and plastics, but they are used only rarely.

The practical advantages of replication come for quantity produc-tion of reflective components with nonspherical surfaces. Such compo-nents are hard to produce by conventional techniques, making them expensive singly or in quantity. However, a single original can be used to generate many replicas, at lower cost per component. One no-table disadvantage of replicated optics is lower power-handling capa-bility than conventional optics.

Active Optical Components

The optical components described so far have been passive optics, so called because their effect on the light they transmit or reflect gener-ally does not change with time. Other types of optics used with laser systems are active, in the sense that their action on light does change with time, typically because of active operator control. A sometimes hazy middle ground is occupied by "nonlinear" optics, which act on light in a way not linearly dependent on light intensity to perform functions such as harmonic generation (frequency doubling) even without active control.

Some active optical devices properly fall under the heading of beam enhancements and were described in Chap. 4. Those include compo-nents designed to change the length of laser pulses, or to alter laser frequency by nonlinear effects. Two other types of active optics widely used with lasers deserve brief mention here: beam intensity modula-tors and beam deflectors.

Modulators. Modulators change the fraction of incident light they transmit in response to external control signals. The modulation may rely on acousto-optic or electro-optic interactions in a suitable crystal, or on mechanical operation of a shutter or aperture.

In acousto-optic modulators, an acoustic wave sets up a pattern of density variations in a suitable material that functions as a diffrac-tion grating. This "effective diffraction grating" diffracts part of the laser light passing through the material at an angle away from the normal beam direction. The fraction of the light energy in the dif-fracted beam depends on the acoustic-wave strength. Modulation can be observed in both the diffracted and undiffracted beam; commercial devices are designed to use one of those beams as the output.

Commercial models are made mostly for the visible and near-infrared, but some are made for the 10.6-μm CO_2 laser wavelength. These devices have limited beam apertures and require radio-frequency (acoustic-wave) drive powers on the order of a watt. Modu-lation bandwidths normally are a few megahertz to a few tens of megahertz, but some models have bandwidths as large as 500 MHz. Made of materials including quartz, lead molybdate, and tellurium dioxide, commercial acousto-optic modulators typically sell for a few hundred dollars or up in small quantities.

Electro-optic modulators rely on the effects of electric fields on the refractive index of certain nonlinear materials. Inherent asymmetries within these crystals cause them to be birefringent, meaning that linearly polarized light with a vertical electric field experiences a different refractive index than linearly polarized light with orthogonal polarization. The application of an electric field changes this birefringence, in effect rotating the polarization of light. When the input light is linearly polarized, and a linear polarizing filter is used at the output, such a device can function as a modulator; the degree of polarization rotation determines how much light will be transmitted by the output polarizer. The entire assembly (with or without an electronic driver) is sold as an electro-optic modulator.

Prices of electro-optic modulators range from a few hundred dollars to several thousand, with the more costly versions typically including a driver. Typical bandwidths are on the order of 100 MHz, but peak modulation frequencies can exceed 1 GHz. Operating voltages are a few tens of volts to several kilovolts, depending on nonlinear material used. For visible wavelengths, the commonest materials are KDP (potassium dihydrogen phosphate, KH_2PO_4), and related compounds including ADP (ammonium dihydrogen phosphate, $NH_4H_2PO_4$), and some versions with deuterium substituted for hydrogen. A few binary semiconductors are used at longer infrared wavelengths. Certain liquids, such as nitrobenzene, can also be used.

A less elegant, but conceptually simpler, approach is mechanical modulation—interrupting the beam with a shutter. In most cases this is limited to off-on modulation, but it is possible to partly block an illuminated aperture. Several types of shutters are used, including a mechanical plate that slips into a slot, a rotating wheel, and a plate attached to a vibrating tuning fork or galvanometer. Safety regulations require the use of shutters to block output beams on many lasers during certain operating conditions.

The three standard types of modulators all operate on a single beam or aperture at once. Devices called spatial light modulators have been built to simultaneously adjust intensity of many points in an aperture. For example, some spatial light modulators can serve as variable transparencies. Such devices, now in developmental stages, are sought for optical computing and signal-processing systems, but remain high in cost and limited in performance.

Beam Deflectors. Beam deflectors or scanners (Marshall, 1985) are needed for applications in which a laser beam must be moved or scanned across a surface. Examples include printing and reading of bar codes. Both solid-state and mechanical types are available. Solid-state scanners include acousto-optic and electro-optic types, although the latter have found few applications. Mechanical types include rotating mirrors and holographic elements and resonant mirror scanners.

Acousto-optic deflectors, like acousto-optic modulators, rely on the interaction between a laser beam and an effective diffraction grating formed by an acoustic wave in a suitable material. As in a modulator, the grating diffracts a portion of the beam at an angle, while the rest of the beam passes straight through. The deflection angle depends on the grating spacing, which in turn is a function of acoustic-wave frequency, so scanning is accomplished by varying the acoustic drive frequency. Most solid-state scanners on the market are acousto-optic types. They are attractive because of the absence of mechanical parts, but their resolution typically is limited to on the order of a thousand spots per scan, keeping them from high-resolution applications.

Electro-optic deflectors have also been developed which rely on nonlinear electro-optic interactions. Although they offer better resolution than acousto-optic deflectors, they are more expensive and require high drive voltages, and hence have found few applications.

Rotating-mirror scanners rely on rapid rotation of a polygonal mirror around its axis. The beam arrives from a constant direction, and the reflected light is scanned across a line as rotation changes the angle of incidence on the mirror face. If the mirror faces are uniformly spaced around the polygon, each one scans the laser beam across the same line; thus a single rotation of a 12-facet mirror will scan the beam 12 times. This approach offers high resolution, but it imposes demanding mechanical tolerances on the optics.

Holographic scanners are similar to rotating mirror scanners in that they rely on rapid rotation of scanning optics about their axis. In this case, however, the scanning optics are a set of holographic optical elements mounted around a flat disk. Each holographic optic diffracts most of the incident light at an angle dependent on the angle of incidence, scanning the beam in a line. As with a rotating-mirror scanner, uniformity and symmetry of individual elements are critical for uniform scanning. This is a relatively new approach, and developers claim it offers advantages in light weight, potential low cost, and lower sensitivity to scanning deviations caused by wobble in the rotation.

In resonant-mirror scanners, a mirror is attached to a torsion rod that twists at its characteristic resonance frequency. With the beam incident from a constant direction, twisting of the mirror changes the angle of incidence, scanning the beam. Operating range is limited by the motor, scanning speed, and twisting angle, but such scanners are less expensive than rotating mirrors and do not require as rapid motion.

Bibliography

Department of Defense: *Military Standardization Handbook—Optical Design*, Mil-Hdbk 141, Defense Supply Agency, Washington, D.C., 1962. This handbook of classical optical design is still available from the Department of Defense and is still valuable as a practical introduction to conventional optics. It is available through the Naval Publications and Forms Center, 5801 Tabor Ave., Philadelphia, PA 19120, (215) 697-3321.

J. A. Dobrowolski: "Coatings and filters," in Walter J. Driscoll (ed.); *OSA Handbook of Optics*, McGraw-Hill, New York, 1978, sec. 8.

Walter J. Driscoll (ed.): *OSA Handbook of Optics*, McGraw-Hill, New York, 1978.

Fred Goldstein: "Optical filters," in Daniel Malacara (ed.), *Geometrical and Instrumental Optics*. Academic Press, Boston and San Diego, 1988, pp. 273–301 (review article).

P. Hariharan: *Optical Holography: Principles, Techniques, and Applications*. Cambridge University Press, Cambridge, 1984 (survey of holography).

Harshaw Chemical: *Crystal Optics*: Harshaw Chemical, Solon, Ohio, 1982 (compilation of data on optical materials from a major manufacturer, a standard data source in the industry).

Eugene Hecht: *Optics, 2nd ed.* Addison Wesley, Reading, Mass., 1987 (a widely used introductory optics textbook; note that the author is *Eugene* Hecht—the first edition was coauthored with Alfred Zajac).

K. Iga, Y. Kokubun, and M. Oikawa: *Fundamentals of Microoptics: Distributed-Index, Microlens, and Stacked Planar Optics*. Academic Press, Orlando, Fla., 1984.

Daniel Malacara, ed., *Geometrical and Instrumental Optics*. Academic Press, Boston and San Diego, 1988 (collection of review articles).

Gerald F. Marshall, ed., *Laser Beam Scanning*. Marcel Dekker, New York, 1985 (survey).

Earl J. McCartney: *Absorption & Emission by Atmospheric Gases*. Wiley-Interscience, New York, 1983.

Melles Griot: *Optics Guide 4*, Melles Griot, Irvine, Calif., 1988 (a combined handbook on practical optics and catalog from a major supplier of laser optics).

K. D. Möller: *Optics*. University Science Books, Mill Valley, Calif., 1988 (introductory textbook).

Edward D. Palik, ed., *Handbook of Optical Constants of Solids*. Academic Press, Orlando, Fla., and San Diego, 1985 (a detailed discussion of measurement techniques with extensive data tabulations, but lacking an index).

Walter P. Siegmund: "Fiber optics," in Walter J. Driscoll (ed.): *OSA Handbook of Optics*. McGraw-Hill, New York, 1978, sec. 13.

F. G. Smith and J. H. Thomson: *Optics*, 2d ed., Wiley, New York, 1988 (introductory textbook).

Warren J. Smith: "Image formation—geometrical and physical optics," in W. J. Driscoll (ed.), *OSA Handbook of Optics*, McGraw-Hill, New York, 1978.

Warren J. Smith: *Modern Optical Engineering: The Design of Optical Systems*, 2d ed., McGraw-Hill, New York, 1990 (introductory textbook, with strong practical emphasis).

F. G. Smith and J. H. Thomson: *Optics*, 2d ed., Wiley, New York, 1988 (introductory textbook).

Irving J. Spiro and Monroe Schlessinger: *Infrared Technology Fundamentals*. Marcel Dekker, New York, 1989 (overview of infrared technology for system designers).

Frederic R. Stauffer: "Optical polarization," in Daniel Malacara (ed.), *Physical Optics and Light Measurements*, Academic Press, Boston, 1988, pp. 107–166 (review article)

Robert C. Weast (ed.): *CRC Handbook of Chemistry and Physics*, 62d ed., CRC Press, Boca Raton, Fla., 1981.

William L. Wolfe and George J. Zissis (eds.): *The Infrared Handbook*, rev. ed., Office of Naval Research, Arlington, Va., 1985. An extensive and valuable compilation of information and data on infrared optics and design criteria. Available from the Environmental Research Institute of Michigan, P.O. Box 8618, Ann Arbor, MI 48107.

Martin V. Zombeck: *Handbook of Space Astronomy & Astrophysics*. Cambridge University Press, Cambridge, England, 1982. Includes some useful optical data.

6

Variations on the Laser Theme: A Classification of Major Types

Classifying lasers into major types is a useful step between the previous general discussion of lasers and the specifics of the rest of this book. There are many potential criteria for classifying lasers, but the two most useful ones are the type of active medium and the way in which it is excited.

This approach follows the usual practice in the laser world, where most devices are groups as gas, solid-state, or semiconductor lasers. A few important lasers are exceptions. From a practical standpoint, the most important is the tunable dye laser described in Chap. 17, in which the active medium is an organic dye dissolved in a liquid solvent. To match the categories with the usual states of matter, dye lasers can be called "liquid" lasers. This is good enough for most practical purposes, although a few other liquid lasers have been demonstrated, and a few dyes can lase in vapor form or solid hosts. Two other exceptions are free-electron and x-ray lasers, for reasons described below.

Gas lasers can be excited in many different ways, producing devices that differ greatly. For that reason, it is useful to break them down by excitation mechanism. However, for other lasers the nature of the medium limits excitation mechanisms so much that such a breakdown is not useful.

Gas Lasers

Gas lasers are many and varied. They dominate any list of commercially available types and have produced laser emission on more lines than have been observed in other media (see, e.g., Weber, 1982). Researchers have been aided in developing new gas lasers by the ease with which they can be built and tested. The basic elements are easy to assemble: a tube which can be filled with the desired gas mixture, a pair of resonator mirrors for the proper wavelength, and a suitable excitation source. This makes it possible to test new gas mixtures much more readily than new solid-state or semiconductor lasers, which must be custom-grown to the desired specifications.

Developers have also been aided by the flexibility that working with gases allows in finding the ideal mixture for use as a laser medium. The pressure of the laser gas can be changed readily, and extra gases can easily be added to study their effects on energy-transfer kinetics. This is helpful because in practice the "optimum" gas mixture for a particular type of gas laser varies considerably among models with different designs. Users studying specification sheets for lasers which need input gas, such as excimer types, will find that different manufacturers have different specifications—because of differences in operating conditions. Similarly, gas mixtures may differ among sealed-tube lasers of the same type, because of differences in operating conditions. Changes in pressure can lead to significant variations in laser operation. For example, carbon dioxide lasers can produce continuous beams at a small fraction of atmospheric pressure, but discharge instabilities occurring at atmospheric pressure and above limit operation in that regime to short pulses, although high peak power is possible.

The free movement of species in a gas mixture lets energy be transferred in many ways in a gas. Energy can be transferred by electrons, ions, or photons passing through the medium, and by collisions among species in the gas. Chemical reactions and gas-dynamic processes can produce excited species and population inversions. These mechanisms serve as the basis for the excitation mechanisms described briefly below, and in more detail in the chapters that follow.

Discharge Excitation.   In discharge excitation, an electric current flows through the laser medium. There are two fundamental variations, shown in Fig. 6.1. In longitudinal excitation, a bias is applied along the laser axis and the current flows the length of the tube. In transverse excitation, the discharge is applied perpendicular to the laser axis. As the electrons pass through the laser medium, they excite species in the gas. In some cases the laser species is excited directly, but more often the electrons transfer energy to another species, which then loses its excitation energy to the laser species in collisions. In the latter case, the gas that absorbs the energy directly from the electrons is normally at a higher pressure than the species which emits laser light. One example is the helium-neon laser described in Chap. 7.

Discharges can be continuous or pulsed, both of which require an

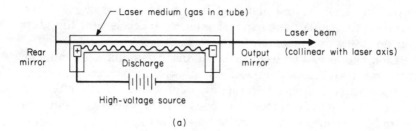

(a)

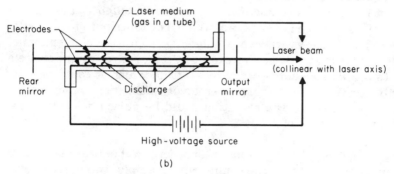

(b)

Figure 6.1 An electrical discharge exciting a gas laser can be applied (*a*) longitudinally, along the laser axis, or (*b*) transversely, perpendicular to the laser axis.

initial electrical breakdown of the gas. In a continuous laser, breakdown is initiated by a voltage spike several times the normal continuous level. After the current starts flowing, the voltage is quickly reduced to a lower level, typically ranging from a kilovolt (kV) to well over 10 kV. Except for lasers operating at high powers, the discharge currents are normally small. Low gas pressures are normally required for a continuous discharge to remain stable.

In pulsed gas lasers, the initial voltage spike—typically peaking in the tens of kilovolts—delivers energy to the laser gas. Pressures well over an atmosphere are possible, with discharge instabilities limiting pulse duration to tens or perhaps hundreds of nanoseconds at such pressures. Fast high-voltage switches are needed for many pulsed gas lasers; such components can cause problems because of high costs and limited lifetimes.

Electron-Beam and Ion-Beam Excitation. Excitation with an electron beam is essentially a variation on the pulsed-discharge approach. Electrons are deposited in the laser gas not by a discharge between a pair of electrodes, but by a pulse of electrons from an accelerator (an electron-beam generator). Electron-beam excitation has the advantage of quickly transferring plenty of energy to the laser gas. However, the need for a bulky and expensive electron accelerator limits its use to large lasers and research laboratories.

Pulsed ion beams can be used to excite laser gas mixtures in much the same way as electron beams. Excitation with proton beams has been demonstrated in the laboratory (Golden et al., 1978), but such ion-beam excitation is even more cumbersome than electron-beam excitation and is not in common use.

Radio-Frequency and Microwave Excitation. A laser gas can absorb energy from radio waves or microwaves passing through it, and in certain cases this absorbed energy can produce or contribute to a population inversion. The microwave or radio-wave excitation alone may not be sufficient for laser action. For example, microwaves might excite a species to an intermediate energy level, from which it could be electronically excited to the upper laser level much more readily than from the ground state. The radiation could be pulsed or continuous, depending on the type of laser.

Chemical Excitation. Energy from an exothermic chemical reaction in a gas can excite a laser. In the commonest example, two species react to produce a third species, which carries at least part of the exothermic energy of reaction as vibrational energy. If the reaction is rapid, the result can be a population inversion of the excited species produced by the reaction. The most important such chemical lasers are the family of hydrogen halides, particularly hydrogen fluoride, described in Chap. 11.

A chemical laser is an unusual device by laser standards. Gas flow is crucial; the fuels must be fed into the laser continually, mix smoothly, and flow rapidly out an exhaust so reaction products do not accumulate in the ground state and end the population inversion. Both pulsed and continuous operation are possible. Gas pressure in the laser chamber is normally quite low, which along with the toxicity of the hydrogen halides imposes some practical operational constraints on gas-flow techniques.

Although chemical lasers nominally draw their operating energy from a chemical reaction, an electrical "spark" may be needed to start the chemical reaction or to prepare one or both gas components for the reaction. The need for this modest electrical input is not a serious operational problem, but it can lead to quotes of electrical efficiency greater than 100 percent. In those cases, electrical efficiency is thermodynamically meaningless because the bulk of the energy in the laser was generated by the chemical reaction.

Gas-Dynamic Lasers. Another way to produce a population inversion is to expand a hot, high-pressure gas into a vacuum. The gas-dynamic expansion cools the gas, but enough of the molecules remain in the same energy level they occupied in the hotter, higher-pressure gas to produce a population inversion. The inversion lasts only until the gas can start approaching equilibrium, so, as in the chemical laser, rapid gas flow is vital. Within a gas-dynamic laser, the population inversion exists only for a short distance downstream of the expansion nozzles, and resonator mirrors must be placed there, as shown in Fig. 10.3.

The gas-dynamic laser principle is a general one, but because it relies on thermal energy distribution, its operation is limited to the infrared. In practice, gas-dynamic lasers are assumed to be carbon dioxide types, although other types have been demonstrated. The main use of gas-dynamic lasers is to produce powers of tens or hundreds of kilowatts for studies of high-energy laser effects; they are not available commercially.

Optical Excitation. A conceptually simple approach to laser excitation is optical pumping. In this approach, photons passing through the laser medium excite species to high energy levels, leading to a population inversion and thus to laser action.

Although some gas lasers are optically pumped, practical applications of the technique are limited by its low overall efficiency. The fundamental problem is that two energy-conversion steps are needed: one to produce the pump light and one to convert the energy in the pump light into laser output. As a result, the overall efficiency is the product of the efficiencies of the two steps, and thus is low, generally lower than possible when directly exciting the laser medium with an electrical discharge.

There are some exceptions. The major ones are the far-infrared gas lasers described in Chap. 15. Optical pumping is the best approach for these lasers because it offers the strongly state-selective excitation required to produce a population inversion on the closely spaced vibrational-rotational and rotational transitions (Coleman, 1982). Discharge excitation produces a more general excitation of the laser gas, populating many higher-energy states. Similarly, optical pumping is a valuable tool in laboratory demonstrations of laser action because it can excite specific states, which may be needed to produce laser action at specific wavelengths. However, even when pumping with such efficient light sources as the carbon dioxide laser, the general result is a laser with overall efficiency so low it is used only when its specific output wavelengths are required.

Nuclear Pumping. Laboratory demonstrations have shown that gas lasers can be excited by transferring energy from the products of a nuclear reaction to the excited species. In experimental devices, intense pulses of neutrons cause the fission of isotopes, including helium 3 (^{3}He), boron 10 (^{10}B), and uranium 235 (^{235}U). Ions produced by the fission reactions then transfer their energy to the laser gas, producing laser action. Research on the concept dates back to at least the early 1970s and was revived during the mid-1980s by the Strategic Defense Initiative (Morrison, 1988), but little progress has been made toward practical devices.

Solid-State Lasers

In the laser world, *solid-state* has a different meaning than in electronics. A solid-state laser is one in which the active medium is a nonconductive solid, a crystalline material, or glass doped with a species that can emit laser light. Semiconductor lasers are considered functionally different types, even though they are made of solid-state materials, because of fundamental differences in their operation.

In a crystalline or glass solid-state laser, the active species is an ion embedded in a matrix of another material, generally called the "host." It is excited by light from an external source. A variety of crystals and glasses can serve as hosts, with the main requirements being transparency, ease of growth, and good heat-transfer characteristics. Useful crystals include synthetic garnets and ruby. Silicate, phosphate, fluorophosphate, and other types of glasses can also serve as hosts. In commercial lasers, the active species functionally is an impurity introduced into the host during fabrication so that it makes up on the order of 1 percent of the finished material. Lasers have been demonstrated in which the active species is a stoichiometric component of the host crystal—for example neodymium in NdP_5O_{14}—but the active species is still at low concentration, and such lasers remain in the laboratory.

The active species in solid-state lasers are locked within the matrix of an insulator, making it impossible to excite it by passing a discharge through the laser medium. Optical pumping is the only practical approach, dictating the use of crystals which are transparent at the absorption and emission wavelengths of the laser species. The efficiency of optical pumping is enhanced by broadening of absorption lines of the active species, caused by the host matrix, so excitation with the broadband emission of a flashlamp can be reasonably efficient. However, flashlamps themselves do not efficiently convert electrical energy into light, so even the most efficient flashlamp-pumped solid-state laser converts on the order of only 1 percent of electrical input into optical energy. The best semiconductor and gas lasers are more efficient.

Greater efficiency is possible if the pump light is concentrated in an absorption band of the laser species. Unfortunately, most narrowband light sources are quite inefficient. The most important exception to that rule is the semiconductor diode laser, and diode lasers are becoming important pump sources for neodymium solid-state lasers, as described in Chap. 22.

Some solid-state lasers have absorption bands that cannot be pumped efficiently with flashlamps or diode lasers. The requirement for an external gas or solid-state pump laser limits their overall efficiency, and adds to their cost. However, those are considered acceptable trade-offs for wavelength tunability in some vibronic solid-state lasers described in Chap. 24 and for the color-center lasers described in Chap. 25.

Semiconductor Lasers

Practical semiconductor lasers are excited when current carriers in a semiconductor recombine at the junction of regions doped with n- and p-type donor materials. This occurs when current is flowing through a forward-biased diode made from certain semiconductor materials. At low current densities, recombination at the diode junction generates excited states which spontaneously emit light, and the device operates as an incoherent light-emitting diode (LED). If the current density is high, and if the semiconductor device includes reflective facets to provide optical feedback, the diode can operate as a laser.

Two other types of semiconductor lasers have been demonstrated in the laboratory, although neither has proved practical. In one, an electron beam pumps an intrinsic semiconductor material that does not contain a junction layer. Such devices have been packaged in a manner similar to a cathode-ray tube, with the electron beam striking a sheet of semiconductor material from the rear, stimulating the emission of a laser beam from the surface of the tube.

Optically pumped semiconductor lasers have also been demonstrated in the laboratory. Their main use is in research on semiconductor laser physics. The principal attraction is the ability to excite materials in which electrical excitation is impractical—generally because the current levels required to produce laser action would destroy the device either before or shortly after laser threshold was reached.

Liquid Lasers

Discussions of liquid lasers almost invariably start and end with the tunable dye laser, described in Chap. 17. The active medium in dye lasers is a fluorescent organic dye, dissolved in a liquid solvent. As in solid-state lasers, the only reasonable excitation technique is optical pumping, with a flashlamp or (more often) with an external laser. The low efficiency of laser pumping is tolerable because it offers better-quality output. The main attractions of the dye laser, its tunable output wavelength and ability to produce ultrashort pulses or ultra-narrow linewidth, are so important for many applications that its low overall efficiency is entirely acceptable. Laser action has been demonstrated from dyes in the vapor phase, or embedded in a solid host, but such lasers have not proved practical.

It is possible to produce laser action by optically pumping rare earth ions in solution (Samelson, 1982). However, such lasers have not found practical applications.

Laboratory Developments

A few lasers do not fit easily into the categories described above. The two most important types are the free-electron laser and the x-ray laser, described in Chaps. 28 and 29, respectively.

The active medium in the free-electron laser is a beam of high-energy electrons, passing through a spatially periodic magnetic field. As the electrons move back and forth in the field, they release energy and produce stimulated emission. The details are complex.

In x-ray lasers, the active medium is a highly ionized plasma which has a brief existence. The plasma is produced by depositing enough energy into a solid to vaporize and ionize it. It emits x rays, amplifies them by stimulated emission, and quickly dissipates.

Researchers also are working on other lasers which do not fit neatly into the categories used for commercial lasers. Some hope to develop γ-ray lasers or "grasers." Others envision such exotic creations as two-photon lasers, in which stimulated emission generates two photons simultaneously, or lasers in which there is no true population inversion. However, we will bypass those excursions into esoteric physics to concentrate on practical lasers.

Bibliography

Paul D. Coleman: "Far-infrared lasers, introduction," in Marvin L. Weber (ed.), *CRC Handbook of Laser Science & Technology*, vol. 2, CRC Press, Boca Raton, Fla., 1982, pp. 411–419 (review article).

J. Golden et al.: "Intense proton-beam pumped Ar–N_2 laser," *Applied Physics Letters* *33*(2):143, 1978 (research paper).

Jeff Hecht: *Understanding Lasers*, Howard W. Sams, Indianapolis, 1988 (elementary-level introduction to lasers).

C. Breck Hitz: *Understanding Laser Technology*, PennWell Publishing, Tulsa, Okla., 1985 (introduction to lasers).

Peter W. Milonni and Joseph H. Eberly: *Lasers*, Wiley-Interscience, New York, 1988 (textbook).

David C. Morrison, "DoE revives reactor-driven lasers," *Lasers & Optronics* 7(7):17–19, July, 1988 (news analysis).

Harold Samelson: "Inorganic liquid lasers," in Marvin L. Weber (ed.), *CRC Handbook of Laser Science & Technology*, vol. 1, CRC Press, Boca Raton, Fla., 1982, pp. 397–422 (review article).

Koichi Shimoda: *Introduction to Laser Physics*, Springer-Verlag, Berlin and New York, 1984 (textbook).

Anthony E. Siegman: *Lasers*, University Science Books, Mill Valley, Calif., 1986 (textbook).

Orazio Svelto: *Principles of Lasers*, 3d ed., Plenum, New York, 1989 (textbook).

Marvin L. Weber (ed.): *CRC Handbook of Laser Science & Technology*, 2 vols, CRC Press, Boca Raton, Fla., 1982–1990 (multivolume reference set).

Amnon Yariv: *Quantum Electronics*, 3d ed., Holt Rinehart & Winston, New York, 1985 (textbook).

Helium-Neon Lasers

The helium-neon laser probably is the most familiar of all lasers. The least-expensive gas laser, it has long been the standard choice to demonstrate laser physics in schools, colleges, and museums. In its most familiar form, the "He–Ne" emits a fraction of a milliwatt to tens of milliwatts (mW) of red light at 632.8 nanometers (nm). As such, it has long been the most common and most economical visible laser.

The helium-neon laser was among the first lasers demonstrated, and was the first gas laser (Javan et al., 1961). Initial versions emitted at 1153 nm in the infrared, but other researchers soon found that the same gas mixture could lase in the red (White and Rigden, 1962). Other lines were produced in the laboratory, but the strong 632.8-nm red line has long been the most important because it made up to about 50 mW available at a visible wavelength. Sales have climbed over the years, and were nearing the half-million mark in 1990 (*Lasers & Optronics*, 1990).

However, important changes are in the offing. Red semiconductor lasers were developed in the late 1980s. Although they cost more than near-infrared semiconductor lasers, the prices of red semiconductor lasers are dropping, and they have become competitive with red helium-neon lasers. Semiconductor lasers have important advantages, including smaller size, higher efficiency, and no need for the high drive voltage used in gas lasers. They cannot replace all red helium-neon lasers because the gas lasers have better coherence and beam quality, and at present offer a slightly shorter wavelength. However, most observers believe that semiconductor lasers will capture much of the market that now belongs to red helium-neon lasers.

Meanwhile, new visible wavelengths have been produced from He–Ne lasers. Green, yellow, and orange helium-neon lasers, as well as multiline versions, are being offered commercially. Infrared versions also are produced. Output powers are lower than on the usual red line, but the other wavelengths offer advantages for certain applications. While combined sales of these types remain small (only about 1 to 2 percent of all He–Ne lasers), their share of the market is likely to grow, especially as semiconductor lasers capture more of the traditional market for red He–Ne lasers.

Internal Workings

Basic Physics.   The active medium in a helium-neon laser is a mixture of helium and neon and total pressures of a fraction of a torr to several torr, with best working pressure depending on discharge-tube diameter. Power is highest when gas pressure (in torr) times discharge diameter (in millimeters) is 3.5 to 4 (Lehecka et al., 1990). Typically the gas mixture contains 5 to 12 times more helium than neon.

The energy in a helium-neon laser comes from an electric discharge, which passes a few milliamperes (mA) through the laser tube at a couple of thousand volts when the laser is in steady operation. [An ignition voltage of about 10 kilovolts (kV) is needed to start laser operation.] Electrons passing through the active medium collide with both helium and neon atoms, raising them to excited levels. The more abundant helium atoms collect most of the energy, then transfer that energy readily to neon atoms, which have excited states at about the same energy above their ground states. The neon atoms then lose their excitation energy and drop to lower energy levels via several transitions, as shown in Fig. 7.1. Emission is possible on several transitions indicated in the figure. The wavelength an individual laser emits depends on the choice of optics and operating conditions of the tube.

Internal Structure.   A helium-neon laser is a gas-filled tube, with internal electrodes exciting the gas to emit light. Mirrors on each end of the tube define the laser cavity. Early helium-neon lasers were quite simple, but over the years manufacturers have incorporated a number of refinements to improve performance.

Figure 7.2 shows the internal structure of a typical modern helium-neon laser with output in the milliwatt range. The discharge passes from the cathode at one end of the tube to the anode at the other, going through a capillary bore one to a few millimeters in diameter. The capillary structure concentrates the discharge, thus improving overall efficiency. The small diameter of the discharge bore also helps control laser beam diameter, mode, and beam divergence. Much effort goes into selecting electrode shapes to make the discharge uniform. The

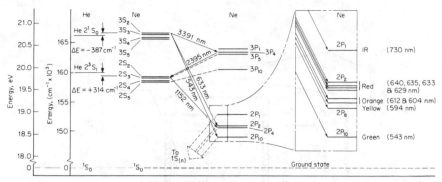

Figure 7.1 Energy levels in a helium-neon laser. The many possible laser transitions are shown by solid lines with arrows. An electric discharge initially excites both helium and neon atoms; collisions transfer energy from excited helium to neon atoms at essentially the same energy level. Laser emission is one of the steps as the excited atoms release their energy on their way to the ground state. The expanded view at right shows how transitions to different sublevels produces the family of visible He–Ne lines. The strongest transition is at 633 nm. (*Courtesy of PMS Electro-Optics; fig. 1 from Robert G. Knollenberg, "Prospects for the helium-neon laser through the end of the century," in* Proceedings of the SPIE, *Vol. 741*, Design of Optical Systems Incorporating Low-Power Lasers, *SPIE, Bellingham, Wash., 1987.)*

outer glass tube is much larger; a common diameter is 3 centimeters (cm) for a milliwatt laser. Higher-power lasers have larger tube diameters.

In early helium-neon lasers, windows were epoxied to the ends of the laser tube, and the cavity mirrors mounted externally. That approach presented problems because helium could diffuse out of the tube through the epoxy (the gas could also diffuse through thin tube walls of soft glass). Epoxy seals also let water vapor seep slowly into the tube, leading to degradation of performance after a few thousand hours of operation.

Now virtually all He–Ne lasers are made with hard seals, in which the glass is bonded directly to metal at high temperatures. Typically the laser cavity mirrors are bonded to metal end plates, which in turn are bonded to the glass laser tube, without the use of epoxies. Hard seals reduce the helium leakage rate to under 0.01 torr [1.33 pascal (Pa)] per year, and make contamination from water vapor and other materials insignificant (Palecki, 1982). This can extend operating life of He–Ne tubes to 20,000 hours or more.

In today's mass-produced He–Ne lasers, cavity mirrors typically are bonded directly to the tube, through the metal end plate, a simpler structure than possible if the mirrors are separate from the laser tube output windows. This approach exposes the mirror coating directly to the discharge inside the laser tube, but current hard coatings can withstand such conditions. Some He–Ne lasers are still produced with

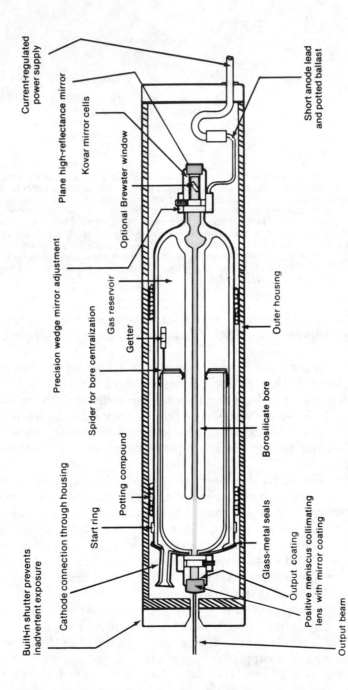

Current-regulated power supply

Plane high-reflectance mirror

Kovar mirror cells

Optional Brewster window

Short anode lead and potted ballast

Precision wedge mirror adjustment

Gas reservoir

Getter

Spider for bore centralization

Outer housing

Built-in shutter prevents inadvertent exposure

Cathode connection through housing

Borosilicate bore

Start ring

Potting compound

Glass-metal seals

Output coating

Positive meniscus collimating lens with mirror coating

Output beam

Figure 7.2 Internal design of a modern hard-sealed helium-neon laser. (*Courtesy of Melles Griot.*)

Brewster angle windows and external cavity mirrors, but these are typically special-purpose models produced in small quantities, often specifically for laboratory applications.

In standard helium-neon lasers, one mirror is totally reflective, while the one at the opposite end of the cavity transmits about 1 percent of the light which becomes the external laser beam. Cavity lengths range from around 10 cm for low-power models to a couple of meters for lasers with output in the 50-mW range.

Inherent Trade-Offs. The most obvious trade-off with helium-neon lasers is the increase in bulk and price of the laser when moving to higher powers. While the helium-neon laser can readily produce a few milliwatts at 632.8 nm, the internal physics make it much harder to produce higher powers. Output above 75 mW has been demonstrated in the laboratory but is not practical and is not available from commercial lasers.

Helium-neon lasers emitting at 632.8 nm are the least expensive and offer the highest power levels. Ironically, gain at 3.39 micrometers (μm) is 22 decibels per meter (dB/m), considerably higher than the 0.5 dB/m at 632.8 nm (Knollenberg, 1987). However, commercial interest has concentrated on the 632.8-nm red line, so manufacturers have invested much more effort in optimizing products for red emission.

The 632.8-nm line has gain 5 to 17 times higher than at other visible lines (Knollenberg, 1987), and provides more power than the other lines. The 543-nm green line at the weakest, with highest powers no more than about 1.5 mW. The need for special optics to overcome this low gain combines with limited production to make prices much higher for visible helium-neon lasers which operate at wavelengths other than 632.8 nm.

Variations and Types Covered. Most helium-neon lasers are linear tubes that generate 0.5 to 10 mW at 632.8 nm, although powers to 75 mW are available at that wavelength. There are three major variations on this basic theme: output on other visible lines, infrared emission, and ring lasers.

Although a rich spectrum of atomic neon lines has been seen in the laboratory (Davis, 1982), until 1985 there were only three lines available from commercial helium-neon lasers: 632.8 nm, 1.15 μm, and 3.39 μm. The two infrared lines were used mainly for research. Since then, a new family of He–Ne lasers emitting on other green, yellow, orange, red, and infrared lines have been introduced. Some of these lasers are built to operate at only one wavelength. Others can be tuned to emit on several visible and near-infrared lines from 543 to

TABLE 7.1   Helium-Neon Wavelengths and Power Levels, with Gains Relative to the 632.8-nm Line

Wavelength, nm	Maximum power, mW	Gain (relative to 632.8 nm)†
543.5	1.5	1/17
594.1	7.0	1/15
604	2.5	1/10
611.9	7.0	1/5
629	—*	1/5
632.8	75	1
635	—*	1/8
640.1	1.5*	1/5
730.5	0.3	1/8
1152.6	17.5	4–5
1523.5	1.5	—
2396	0.5	—
3392.0	24	44/1

*Asterisks indicate specifications were not available, or that lines were difficult to isolate from the stronger 632.8-nm transition.
†Knollenberg (1987).

730 nm, or to emit simultaneously on several visible lines. Specific wavelengths are listed in Table 7.1. Despite the proliferation of other lines, 98 to 99 percent of He–Ne lasers manufactured in 1990 emitted at 632.8 nm. Many of the lines are so weak that the lasers probably would never have been developed if helium-neon technology had not been already developed for red lasers.

Lasers designed to emit under 10 mW on a single line almost invariably use the same hard-seal mirror technology as mass-produced 632.8-nm He–Ne lasers. Higher-power He–Ne lasers are manufactured in smaller quantities, and often have interchangeable external optics which allow operation at 632.8 nm, 1.15 μm, or 3.39 μm.

The ring laser is a variation in which three (or sometimes four) mirrors define a "ring" path inside the laser resonator. This requires a triangular or square laser tube, such as is shown in Fig. 7.3. Ring lasers are used as gyroscopes to detect rotation about their central axis (Aronowitz, 1971; Chow et al., 1985). These lasers are specially built for use in military and commercial systems and are not sold separately as components on the open market. Because of their specialized nature, ring lasers are not covered in this chapter.

Beam Characteristics

Wavelength and Output Power. The standard wavelength of helium-neon lasers is 632.8 nm in the red. Commercial models emit continuous beams from a few tenths of a milliwatt to 75 mW, with most in the 0.5- to 7-mW range.

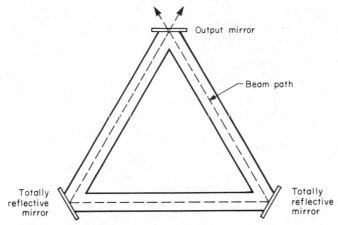

Figure 7.3 Outline of a triangular ring laser used as a rotation sensor. Light propagates in both directions around the ring, and subtle rotation-induced changes in wavelength are detected by observing interference effects at the output mirror.

All the other lines also are continuous-wave, but generate less power, as shown in Table 7.1. Some lasers have external optics which allow tuning between several visible lines or optics which can be interchanged for visible and infrared operation. (The 1.152- and 3.39-μm lines can lase at the same time in some tubes with suitable optics.)

Few characteristics of helium-neon lasers depend strongly on the operating wavelength. The descriptions that follow concentrate on lasers built to emit at 632.8 nm; important differences for versions operating at other wavelengths are noted where appropriate.

Efficiency. Overall efficiency of a helium-neon laser is low, typically in the range of 0.01 to 0.1 percent.

Temporal Characteristics of Output. Normally helium-neon lasers produce continuous beams. The continuous output can be modulated at rates to about 1 megahertz (MHz) by varying drive current in models designed for that purpose. He–Ne lasers are normally not operated in pulsed mode.

Spectral Bandwidth and Frequency Stability. The doppler-broadened gain curve of mass-produced helium-neon lasers spans around 1.4 gigahertz (GHz), the equivalent of 1.9×10^{-3} nm at the laser's red wavelength as shown in Fig. 7.4. The doppler gain curve encompasses several cavity modes, each much narrower than the broad doppler curve. The number of cavity modes depends on mode spacing—

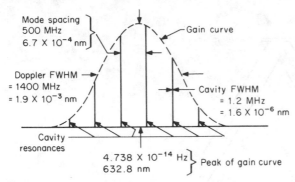

Figure 7.4 Doppler-broadened gain curve of a helium-neon laser, showing the much narrower cavity resonances that lie within that curve. FWHM = full width at half maximum. (*Courtesy of Melles Griot.*)

roughly 500 megahertz (MHz) on the average, but up to 1400 MHz for smaller cavities and as little as 90 MHz for larger cavities. The shorter the laser cavity, the fewer cavity modes falling under the gain curve.

For most practical purposes, the linewidth of 1 part in 300,000 is adequately narrow, and minor mode shifts do not cause noticeable effects. When narrower bandwidth is required, as in certain scientific applications, special optics can be added to restrict laser oscillation to one of the cavity modes within the gain curve. Frequency stability of such single-frequency lasers is on the order of ± 1-MHz drift [0.002 part per million (ppm)] during a 5-minute interval when operated at constant temperature. However, temperature-induced variations in the cavity length can cause the fundamental frequency to vary by around 5 MHz per degree Celsius change in ambient temperature, so temperature stability is a must for stable frequency output. Normally warm-up times of half an hour to an hour are needed to obtain stable single-frequency output, longer than needed to get stable output from a laser with a normally broad gain curve.

Amplitude Noise. Root-mean-square (rms) amplitude noise at important frequencies, between about 30 Hz and 10 MHz, typically is rated at 1 percent or less in mass-produced He–Ne lasers. Noise ratings of 2 percent are possible for lasers at the upper end of the commercial He–Ne power range. Long-term drift is typically rated at 2.5 or 5 percent over an 8-hour interval.

Beam Quality, Polarization, and Modes. Most helium-neon lasers produce a good-quality TEM_{00} beam with a classic gaussian intensity distribution, but some models emit multiple transverse modes. The beam may be linearly polarized or nominally unpolarized, but some unpolarized beams are *not* truly randomly polarized and may contain shifting proportions of light of different linear polarizations. Although this is acceptable for many applications, users who require an even distribution of polarizations or true random polarization may wish to add external depolarizers. Linear polarization is typically 500 to 1.

Coherence Length. Mass-produced He–Ne lasers have coherence lengths around 20 or 30 cm, adequate for holography of small objects if care is taken that the object and reference beams travel similar distances. Single-frequency He–Ne lasers have much narrower spectral bandwidths, and hence much longer coherence lengths, but are considerably more expensive.

Beam Diameter and Divergence. Beam diameters of mass-produced helium-neon lasers with TEM_{00} output in the milliwatt range are around a millimeter and tend to increase with output power. Divergence is on the order of 1 milliradian (mrad), dropping with increasing beam diameter because these lasers normally operate near the diffraction limit. Short, low-power models may have small-diameter beams and larger divergences—to about 0.34 mm and 3.0 milliradians (mrad). For longer, high-power models, beam diameters can reach a couple of millimeters, with divergence about 0.5 mrad.

Diameter and divergence of multimode beams from some lasers do not come close to the diffraction limit. For example, one 5-mW multimode laser has 2-mm beam diameter and 8-mrad divergence, far above the diffraction-limited beam quality of TEM_{00} lasers.

Stability of Beam Direction. Although many data sheets do not specify beam-direction stability, typical values for mass-produced He–Ne lasers are around ±0.2 mrad for operation from a cold start, and around ±0.02 to 0.03 mrad after a 15- to 20-minute warm-up period.

Suitability for Use with Laser Accessories. The low powers of helium-neon lasers effectively prevent their use with accessories requiring high laser powers, such as harmonic generators. Although modelocking and Q switching are possible, they are extremely rare in practice; other lasers are much better choices for modelocking or Q switching because they offer higher powers. The continuous beams from helium-neon lasers can be modulated by external modulators. Note, however, that these accessories can easily cost more than a typical He–Ne laser.

Operating Requirements

Input Power. Most packaged helium-neon lasers operate from ordinary 115-volt (V) wall current, with 230-V operation a standard option. Power consumption runs around 20 W for a 1-mW laser to over 400 W for a 50-mW version. Low-power versions are also available for battery operation.

Laser heads and tubes are available which require a few milliamperes (mA) of direct current at 1000 to 2000 V. The tubes themselves require a ballast resistance in the range of 45 to 100 kΩ for proper operation. Peak voltages of around 10,000 V are needed to initiate the discharge in the laser tube.

Cooling. Low-power helium-neon lasers rely on passive air cooling, while higher-power models have fans that provide forced-air cooling. Even the highest-power He–Ne lasers do not produce a tremendous amount of waste heat, no more than a few hundred watts.

Consumables. Practically speaking, there are no consumables required for He–Ne lasers.

Required Accessories. A complete packaged helium-neon laser requires no special accessories other than those needed for a particular application. Laser "heads" packaged without internal power supplies do require a separate module to convert wall current to the high-voltage direct current needed to power the laser tube. Bare laser tubes are also sold; these require both a high-voltage direct-current source and mounting accessories.

Operating Conditions and Temperatures. Although most helium-neon lasers are intended for a normal laboratory or industrial environment, they generally can withstand a broader range of conditions. Standard mass-produced He–Ne lasers are rated for operating temperatures of 0 to 40°C to −25 to +80°C. Storage-temperature extremes can be as broad as −40 to +80°C. The lasers are rated to operate at altitudes to about 3000 m (10,000 ft), and at humidities from 0 to 80 or 90 percent. Because helium-neon lasers are mass-produced and used in a wide range of conditions, specification sheets often provide much more detail on operating ranges than do data sheets for other types of lasers.

Some helium-neon lasers are ruggedized for use in hostile environments, such as mines, agriculture, and construction sites. These lasers generally are sealed to keep dust out of the laser head and to minimize high-voltage hazards.

Typically, a helium-neon laser begins producing light within a few seconds after it is turned on. However, it may take 15 to 60 minutes to warm up enough to give rated stability. Adequate warm-up times are particularly critical for single-frequency lasers, which require stable temperature for stable operation.

Mechanical Considerations. In recent years, some very compact helium-neon lasers have come on the market, with low-power tubes as short as about 10 cm (4 in) and as thin as 1.6 cm (5/8 in). Packaged laser heads are only slightly larger, although many users concerned with small size prefer to do their own packaging of the laser tube. Lasers producing a few milliwatts are typically a foot (0.3 m) long, while the most powerful models can run a couple of meters long. A bare tube can weigh as little as 0.1 kg, although weights of 0.25 to 0.5 kg are more common; packaged laser heads weigh about twice that much. Addition of a power supply can bring total weight to a kilogram or two for low-power lasers, while a complete laser capable of delivering a few milliwatts would weigh a few kilograms. The highest-power He–Ne lasers can weigh as much as 100 kg complete with laser head and power supply.

Safety. The basic rules for safe use of helium-neon lasers are to avoid staring into the laser beam and contacting the high voltages that power the discharge.

Laser powers of a few milliwatts or less are far from deadly and may not even be felt as a warm spot on your skin. However, the intensity of a helium-neon laser beam is comparable to that of sunlight (although it is spread over a much smaller area). Because both sunlight and a laser beam are made of parallel rays, the eye can focus both onto very small spots on the retina. If the eye stares at either the sun or a laser beam long enough, the intensity at the retina can be high enough to cause permanent damage. There are no hard-and-fast rules for exactly how much exposure will cause what type of damage, so caution should be the watchword. However, an inadvertent momentary glance into a milliwatt He–Ne beam is no more likely to cause instantaneous blindness than an accidental glance at the sun. The hazards increase with laser power, so particular care should be taken with higher-power He–Ne lasers.

Many applications require that helium-neon beams be scanned across surfaces to read information. Careful design can limit powers to levels low enough and maintain scanning speeds high enough that the beam can be scanned through regions accessible by people without presenting any danger. An excellent example is the use of laser scanning systems in automated supermarket checkout; although millions of people pass by such systems daily, stringent federal standards do not even require a "Caution" sign on the laser reader.

Reliability and Maintenance

Lifetime. Hard-sealed helium-neon laser tubes typically have rated operating lifetimes of 10,000 to 20,000 hours, and shelf lives of 10 years, although shelf lives can be lower for higher-power models. Tubes without hard seals typically have operating lifetimes of a few thousand hours.

Maintenance and Adjustments Needed. Helium-neon lasers are largely maintenance-free, and most should require no adjustments under normal operating conditions. However, laser heads which are dropped may require realignment, while those operated in particularly dirty environments may require periodic cleaning.

If extremely stable operation is required, it may be necessary to leave the laser running continually. Complete He–Ne lasers come with beam-blocking shutters (required by U.S. laser safety standards) which allow the laser to run continually without having the beam emerging from the head.

Mechanical Durability. He–Ne lasers generally are built to withstand moderately rough handling, and some are ruggedized specifically for use in mines, construction, agriculture, and surveying. Normal mass-produced He–Ne heads and tubes are reasonably rugged, although because they are made of glass, breakage is always possible if they are dropped.

Some manufacturers specify shock resistance for laser heads and tubes. One company specifies shock resistance of 15 times the force of gravity (15 g's) for 11 milliseconds (ms) for laser heads, and 50 g's for 1 ms for laser tubes. The different specifications reflect the different damage vulnerabilities. Laser heads can be knocked out of alignment, but the main damage to tubes is breakage.

Failure Modes and Causes. Loss of helium, poisoning of the helium-neon mixture by such contaminants as water vapor, or damage to the internal optics will cause gradual degradation in output power. Malfunction of the high-voltage system and shock damage to the laser tube or optics are other potential problems. Slow degradation of output power level is inevitable even with hard-sealed tubes, but users should realize that the rated lifetimes of helium-neon lasers are long compared with those of other light sources. Also note that reduced output level does not always prevent operation of a system or experiment.

Possible Repairs. Helium-neon laser tubes can be cleaned and refilled with a fresh gas mixture, but this costs more than buying a new mass-produced laser tube. (It may, however, be practical for high-power models.) Mass-produced tubes are cheap, and the simplest course is usually substituting a new one. Power supplies are also normally replaceable.

Commercial Devices

Standard Configurations. Helium-neon lasers are produced in large quantity and offered in several forms. The three commonest offerings, shown in Fig. 7.5, are

- *Laser tube:* A sealed glass tube containing the laser gas and electrodes, typically with integral cavity mirrors sealed to the ends of the tube.
- *Laser head:* A laser tube packaged into a housing, typically a cylindrical tube. The laser head includes optics and safety features as well as the protective housing, but the power supply which generates the high voltages needed to drive the tube is packaged separately.
- *Complete laser system:* The laser head plus power supply, packaged together with safety interlocks and all other equipment needed to permit operation. The laser is ready to run once it is plugged into a power outlet and switched on. The complete laser may be housed in one integral box or may consist of a head attached to a power supply and control box.

These different versions are intended for users with different needs. Complete laser systems are intended for those who want a ready-to-use laser, such as a demonstration laser for a physics laboratory or a light source for holography. Laser heads and laser tubes are intended for incorporation into other equipment, such as bar-code readers or laser printers. Tubes are often used as replacement components. Power supplies, which convert a low-voltage source of direct or alternating current to the high voltages needed for powering the laser tube are also sold as separate modules which users can connect to laser heads or tubes.

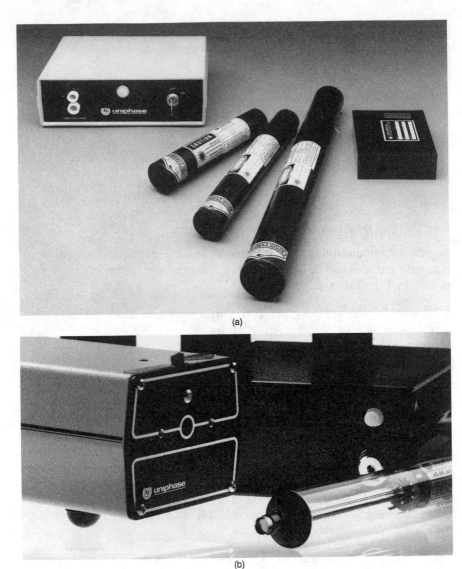

(a)

(b)

Figure 7.5 Packaged helium-neon lasers: (*a*) cylindrical laser heads with separate power supplies; (*b*) complete laser package (left) and unmounted He–Ne tube (right). (*Courtesy of Uniphase.*)

Options.   Many options for helium-neon lasers concern packaging. In addition to those listed above, they include ruggedization, hermetic seals, and battery operation to permit use in the field or under hostile conditions. There are also other options that deal more with the laser's operating mode, including

- Choice of polarized or unpolarized output
- Single-frequency output
- Type of power supply
- Operation at different wavelengths
- Wavelength tuning
- Intracavity space for insertion of laser accessories or experimental samples (This option is available only with laboratory lasers that lack integral optics and offers higher power than is available in the beam that emerges from the laser cavity.)

Special notes. Red He–Ne lasers with output of a few milliwatts or less are generally mass-produced, but those with powers of 10 mW or more are made in smaller quantities and often packaged primarily for laboratory applications. All He–Ne lasers emitting at wavelengths other than 632.8 nm are made in small quantities; they are generally similar to higher-power red lasers, but use different optics.

Pricing. At this writing, single-unit list prices for red helium-neon laser tubes run from $135 for 0.5-mW models to over $2000 for the most powerful tubes. Prices can drop to as low as about $60 per tube for quantity orders. Cylindrical laser heads, without power supplies, sell for two to three times the price of a bare laser tube. Separate power supplies run from under $200 to about $800 for all but the most powerful lasers, which can require power supplies costing well over $2000. Single-unit prices for complete red He–Ne lasers, ready to plug into the wall and operate, run from about $100 to $300 for low-power units to over $10,000 for the most powerful units. Surplus dealers sometimes offer tubes, heads, or lasers at much lower prices than laser makers.

Prices are significantly higher for He–Ne lasers emitting on other lines, which are produced in much smaller quantities. Heads emitting the green, orange, or yellow lines are $600 to $2000, depending on power level; a power supply typically is $100 to $300 more. Tunable multiline lasers are $2000 to $6000, depending on power and lines available. Infrared He–Ne lasers are priced at $1500 to $15,000, with the most expensive lasers those generating more than 10 mW.

Suppliers. Many companies in the United States, Japan, and Europe supply helium-neon lasers, although a few companies have left the business in recent years. A few companies offer broad product lines, but most specialize in certain market niches, such as laboratory lasers or mass-produced lasers for industrial users. Everyone offers low-power red He–Ne lasers, and many companies offer low-power green models. Fewer companies offer lasers emitting on the other visible and infrared lines, or red He–Ne lasers generating more than about 25 mW.

The market is competitive but not cutthroat. Unit sales rose during the late 1980s, and the industry shed the overcapacity which had caused large-quantity prices to drop earlier. There is a widespread expectation that recently developed visible semiconductor lasers will soon start to replace mass-produced red helium-neon lasers in many applications. If this proves true, it could lead to another round of price cuts, or to a contraction in supply if some companies decide to leave the business.

Red He–Ne tubes, heads, and lasers sometimes come on the surplus market, where they are offered at prices much lower than through laser manufacturers.

Applications

For many years, the helium-neon laser was the standard choice for applications which require an inexpensive laser producing power in the milliwatt range. During the 1980s, the advent of inexpensive infrared semiconductor lasers changed this market. Semiconductor lasers captured the big new markets of the 1980s, compact-disk audio players, low-cost laser printers for personal computers, and fiber-optic communication systems. However, helium-neon gas lasers remained dominant for applications which required visible light.

The emergence of semiconductor lasers operating at 635 to 670 nm (see Chap. 18) is changing the market for the traditional red helium-neon laser. Semiconductor lasers are much smaller and need much less power than helium-neon, and do not require the high voltages needed to drive the gas discharge—crucial advantages for the many applications where engineers want to limit size and power consumption. Helium-neon lasers, on the other hand, have much better coherence and beam quality, and the human eye is much more sensitive to their 632.8-nm output than to the longer red wavelengths. Helium-neon lasers also can offer output in the green, yellow, and orange regions, wavelengths not yet available from semiconductor lasers, as well as at certain infrared lines. Pricing at this writing is competitive, but if historical trends are any indication, visible semiconductor lasers will soon be significantly less expensive than red helium-neon lasers.

It is likely that red semiconductor lasers will replace 632.8-nm He–Ne lasers in the many applications where the semiconductor laser beam is acceptable. The pace of replacement will depend on the design of new systems, because major changes are needed to accommodate a semiconductor laser. This also means that a market will exist for replacement tubes as long as He–Ne systems remain in operation (although it should be noted that the 20,000-hour rated lifetime of a He–Ne tube translates into 2¼ years of continuous operation).

Today, helium-neon lasers have a broad range of applications. Some of these will remain strong, while others may not. They fall into several broad categories:

- *Alignment and positioning:* The use of a laser beam to draw a straight line for optical or electronic detection to aid in aligning or positioning. Construction workers use instruments with scanning He–Ne beams to define a plane or line while they are building walls and hanging ceilings. He–Ne lasers draw straight lines to keep sewer pipes and tunnels on straight courses. He–Ne lasers define straight paths for machine tools and help medical personnel position patients in x-ray imaging systems. He–Ne beams are used to align grading equipment used in heavy construction and agricultural irrigation. Infrared diode lasers have replaced He–Ne lasers in some systems which use electronic sensors. He–Ne lasers offer better beam quality and visibility than visible semiconductor lasers, but their larger power requirement is a disadvantage.

- *Reading and scanning:* The tightly focused beam from a helium-neon laser can be scanned across a surface to read bar codes or other symbols. The biggest single application for helium-neon lasers has been in bar-code readers for supermarket checkout, where a laser underneath the counter reads codes on packages scanned rapidly by a window about 0.5 m from the laser head. The beam repetitively scans a well-defined pattern, and variations in the scattered light are decoded to read the symbol. This application requires red light and good beam quality, so helium-neon lasers are likely to remain dominant in the near future. However, visible diode lasers and other red light sources (such as LEDs) can be used in hand-held readers that pass directly over bar codes. Bar codes are spreading to other applications, so the market should grow.

- *Writing and recording:* A modulated laser beam can be scanned across a light-sensitive surface to record information. In laser printers, the beam scans a photoconductive drum, discharging the electrostatic charge held by the surface at points where the beam is "on." The resulting pattern is then printed on paper by a copierlike process. Lasers also can encode data as a series of dots on light-sensitive disks for computer data storage. Helium-neon lasers initially were used for these applications, but today all but the most expensive high-speed systems use semiconductor lasers. Helium-neon lasers also are used in some printing and publishing systems for applications such as color separation.

- *Holography:* The good coherence, low cost, and visible output of red helium-neon lasers have made them the standard choice for recording holograms of stationary objects since the first laser holograms were made in the early 1960s.

- *Medicine:* Red helium-neon lasers have been used in several types of medical therapy, primarily unconventional treatments such as laser acupuncture, biostimulation, stimulation of wound healing, and alleviation of pain, but they are being replaced by visible diode lasers in many of these treatments. The more orthodox medical establishment uses helium-neon lasers as pointers for infrared surgical lasers, and in diagnostic instruments which sort cells and perform other biological measurements. The green and yellow lines can excite fluorescent dyes, and also can be used for cell sorting and counting.

- *Measurement:* Many measurement instruments use helium-neon lasers. Their good coherence and beam quality make red helium-neon lasers the most common choice for interferometric measurements of surface contours. They also are used to measure light scattering, as are some He–Ne lasers operating at other visible wavelengths. Ring lasers are used as rotation sensors, or "laser gyroscopes" for aircraft navigation; they are standard on Boeing 757 and 767 airliners.

- *Demonstrations and displays:* Red helium-neon lasers are the standard choice for school and museum demonstrations of how lasers work because of their good beam quality, coherence, and modest cost. Similarly, they are used in laser displays where their red color and milliwatt power levels will suffice. Other visible helium-neon lines also can be used in displays, taking advantage of the eye's greater sensitivity at shorter wavelengths.

- *Research and development:* Like all other kinds of lasers, helium-neon lasers are used in many laboratories. They provide inexpensive sources of tightly collimated, coherent beams for many types of research. Although red He–Ne lasers are common, those operating at other wavelengths also are widely used in research. For example, the 3.39-μm line is absorbed strongly by carbon-hydrogen bonds in hydrocarbons, important for spectroscopy. Single-frequency lasers may be used in laboratory measurements of time and frequency.

- *Sensing:* Helium-neon lasers can be used as light sources for various optical sensors. The 1.523-μm line is valuable for measurements of optical fibers.

- *Communications:* A few red helium-neon lasers are used for some through-the-air communication which requires narrow beams, but the numbers are small. Although the 1.523-μm line can be used to measure the properties of optical fibers, semiconductor lasers and light-emitting diodes are the standard light sources for fiber-optic communications.

Bibliography

Frederick Aronowitz: "The laser gyro," in Monte Ross (ed.), *Laser Applications*, vol. 1, Academic Press, New York, 1971, pp. 134–200 (review paper).

W. W. Chow et al.: "The ring laser gyro," *Reviews of Modern Physics* 57(1):61–104, 1985 (scholarly review paper).

Christopher C. Davis: "Neutral gas lasers," in Marvin J. Weber (ed.), *CRC Handbook of Laser Science and Technology*, vol. 2, CRC Press, Boca Raton, Fla., 1982, pp. 3–68 (review article; includes tabulations of He–Ne lines).

Jeff W. Eerkens and William M. Lee: "New He–Ne laser with green (543 nm) and fiber-optimum IR (1523 nm) outputs," paper WM16 presented at Conference on Lasers and Electro-Optics, Baltimore, May 21–24, 1985.

Ali Javan, W. R. Bennett Jr., and D. R. Herriott: "Population inversion and continuous optical maser oscillation in a gas discharge containing a He–Ne mixture," *Physical Review Letters* 6:106, 1961 (first report of He–Ne laser).

Robert G. Knollenberg, "Prospects for the helium-neon laser through the end of the century," in *Proceedings of the SPIE Vol. 741, Design of Optical Systems Incorporating Low-Power Lasers*, SPIE, Bellingham, Wash., 1987 (overview of He–Ne technology).

T. Lehecka et al.: "Gas Lasers," pp. 451–596 in Kai Chang (ed.), *Handbook of Microwave and Optical Components*, vol. 3, *Optical Components*, Wiley-Interscience, New York, 1990 (overview of all gas lasers).

Gerald S. Palecki: "Helium-neon lasers, reliability & lifetime," *Lasers & Applications* 1(1):73–75, Sept., 1982 (short review article).

D. L. Perry: "CW laser oscillation at 5433 Å in neon," *IEEE Journal of Quantum Electronics* 7:107, 1971 (first report of green line).

Anthony L. S. Smith: "He–Ne lasers are alive and well," *Laser Focus World* 25(7):75–86, July, 1989 (review of He–Ne technology).

A. D. White and J. D. Rigden: "Continuous gas maser operation in the visible," *Proceedings of the IRE* 50:1796, 1962 (first report of red line).

Noble Gas Ion Lasers

The label *ion laser* is a vague one that the laser world applies specifically to types in which the active medium is an ionized rare gas. Argon, with strong lines in the blue-green and weaker lines in the ultraviolet and near-infrared, is the most important type commercially. Krypton ion lasers are also marketed, despite their weaker output, because they offer a wider range of visible wavelengths, plus additional ultraviolet lines. A neon ion laser which emits in the ultraviolet was introduced recently. (The helium-neon laser emits on *neutral* neon lines.) A pulsed xenon ion laser, with somewhat different characteristics than other noble gas ion lasers, is described briefly in Chap. 16.

The prime attraction of argon and krypton lasers is their ability to produce continuous-wave output of a few milliwatts to (from argon) tens of watts in the visible and up to 10 watts (W) in the ultraviolet. The technology is not easy to master. Ion lasers carry price tags of a few thousand to as much as $100,000, have tubes with operating lifetimes limited to 1000 to 10,000 hours, and tend to be delicate. However, there are few alternative sources of those power levels and wavelengths. Frequency doubling of solid-state neodymium-YAG lasers pumped by semiconductor lasers (see Chap. 22) offers an attractive alternative, but so far output powers are limited and prices are high (because of the high cost of high-power semiconductor laser arrays).

Recent years have seen considerable improvements in rare gas ion lasers. The operation of high-end rare gas ion lasers has been simplified by microprocessor-controlled systems which automatically stabilize the laser. Tube lifetimes have been improved, and new ultraviolet lines have been added. Ion lasers may not meet the system designer's ideal specifications, but they can offer cost-effective solutions to problems in building systems for medical, printing, entertainment, and data-storage applications, as well as maintaining a place as a laboratory laser.

Introduction and Description

Active Medium. In an ion laser, the active medium is a rare gas from which one or more electrons have been stripped to form a positive ion. In argon and krypton lasers, pure gas is used, with normal operating pressure around 1 torr [133 pascals (Pa)]. The two gases can also be mixed in what is called a *mixed-gas* laser, which emits on spectral lines of both ions. The laser transitions are between upper energy levels of the ions. For argon and krypton, singly ionized species emit in the visible and near-infrared, and doubly ionized species emit in the near-ultraviolet. Singly ionized neon emits in the ultraviolet.

The most common member of the ion-laser family is argon, which emits on the lines shown in Table 8.1. In larger argon lasers, the strongest line is the 514.5-nanometer (nm) line of singly ionized argon (Ar II in spectroscopic notation), followed by 488.0 nm. In smaller, air-cooled argon lasers, the 488.0-nm line is stronger. Several other weaker lines appear in untuned blue-green emission from argon, or when the laser is tuned specifically to those wavelengths. (Ar II also has a near-infrared line.) Doubly ionized Ar III emits in the ultraviolet, with the most useful wavelengths at 275, 300–305, 334, 351, and 364 nm. For ultraviolet operation, powers tend to be higher at the longer wavelengths.

Krypton is used less often because it is a less-efficient laser gas, with lower gain than argon; typically, similar tubes will generate only 10 to 30 percent as much power with krypton as with argon. Krypton's advantage is that it has laser lines at wavelengths where argon does not, as shown in Table 8.1. The strongest line of singly ionized krypton (Kr^+) is at 647.1 nm in the red, but other lines in the yellow, green, and violet can produce nearly half as much power. Doubly ionized krypton (Kr^{2+}) has three lines in the near-ultraviolet. Continuous-wave krypton lasers generating a total of 2.5 W at 242 to 266 nm, and wavelengths as short as 219 nm, have been reported recently (Petersen, 1990), but have yet to reach the commercial market. They appear to be from Kr^{3+}.

Argon and krypton can be combined in a single tube to emit on visible lines of both ions. These mixed-gas lasers will be described below.

The neon ion laser is a commercial newcomer, generating a multiline output of about 1 W, primarily on three ultraviolet lines:

332.4 nm   0.5 W

337.8 nm   0.2 W

339.2 nm   0.3 W

The ultraviolet lines of neon are emitted by singly ionized atoms, Ne^+, while those of argon and krypton come from doubly ionized atoms.

TABLE 8.1 **Major Argon and Krypton Lines and Their Intensities from a Large Water-Cooled Laser***

Wavelength, nm	Relative power (fraction of argon multiline visible output)	Absolute (line) power, W
Argon		
275.4 ⎫ 300.3 ⎪ 302.4 ⎬ 305.5 ⎭	0.025–0.050 (combined)	0.5–1 (combined)
334.0 ⎫ 351.1 ⎪ 351.4 ⎬ 363.8 ⎭	0.15–0.25 (combined)	3.0–5.0 (combined)
454.6	0.03	0.6
457.9	0.07	1.4
465.8	0.03	0.6
472.7	0.05	1.0
476.5	0.12	2.4
488.0	0.375	6.5
496.5	0.12	2.4
501.7	0.07	1.4
514.5	0.45	9.0
528.7	0.071	1.4
1090.0	Not specified	Not specified
Krypton		
337.4 ⎫ 350.7 ⎬ 356.4 ⎭	0.075 (combined)	1.5 (combined)
406.7	0.03	0.6
413.1	0.06	1.2
415.4	0.0075	0.15
468.0	0.02	0.4
476.2	0.0125	0.25
482.5	0.0125	0.25
520.8	0.0225	0.45
530.9	0.05	1.0
568.2	0.04	0.8
647.1	0.1	2.0
676.4	0.03	0.6
752.5	0.04	0.8
799.3	0.01	0.2

*Generating 20-W multiline output from argon in the blue-green (emitting on all lines from 454.5 to 528.6 nm); relative intensities are compared to the 20-W multiline argon output for both argon and krypton. (Krypton emits over such a wide range of wavelengths that combined output on all visible lines normally is not specified.) Figures are approximations based on specifications for commercial lasers.

Higher currents are required to doubly ionize atoms, so neon lasers have a lower threshold for ultraviolet output than do argon or krypton lasers. However, total ultraviolet powers remain lower.

Energy Transfer. Ion lasers are excited by a high-current discharge that passes along the length of the laser tube and is concentrated in the small-diameter bore or center of the tube. An initial spike of a few thousand volts breaks down the gas, then voltage drops to between about 90 and 400 V and discharge current jumps to 10 to 70 amperes (A). The high current densities in the center of the tube both ionize the gas and provide the energy that excites the ions to the upper laser levels. Stimulated emission brings the excited species to the lower laser level, which quickly decays to the ground-state ion, emitting extreme-ultraviolet light.

A simplified view of Ar^+ kinetics is given in Fig. 8.1: the details are complex and far beyond the scope of this discussion (see, e.g., Bridges, 1982). Three major mechanisms raise Ar^+ to the upper laser level. One is energy transfer from an electron to a ground-state argon ion. Another is collisional transfer of energy from an electron to Ar^+ in an excited metastable state. The third is decay of higher levels produced by electron excitation. The contributions of the various processes ap-

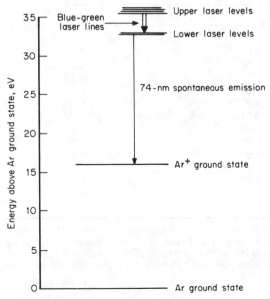

Figure 8.1 Energy levels of singly ionized argon, showing only the blue-green laser lines. Actual energy-level structure is considerably more complex, as indicated by the many possible emission lines.

pear to depend on operating conditions and are difficult to separate because all three lead to the observed variation of output with the square of current density in the tube.

Even this simplified view of the Ar^+ energy-level structure indicates some of the problems in building ion lasers. The upper laser levels lie nearly 20 electron volts (eV) above the ground state of Ar^+, and nearly 36 eV above the ground state of the argon atom. This means that powerful excitation is needed.

Another problem, not obvious from Fig. 8.1, is the very short lifetime of the lower laser level, which drops to the ground state of Ar^+ spontaneously with emission of a 74-nm extreme-ultraviolet photon. This short lifetime helps sustain a population inversion despite the relatively inefficient ways in which the upper laser levels are populated. However, it also means that production of a laser photon with roughly 2 eV of energy is accompanied by the loss of about 18 eV in spontaneous emission—limiting efficiency severely. The short-wavelength ultraviolet light also can damage many optical materials.

As might be expected, the ionized plasma is hot—about 3000 kelvins (K), according to spectral measurements of argon laser lines. Together with the high current, this creates extremely hostile conditions inside the laser tube, causing sputtering of material from tube walls and electrodes even when high-temperature materials are used. These and other factors combine to limit ion-laser efficiency and to make design of ion-laser tubes a very difficult problem.

The energy-level structure of argon is more complex than is shown in Fig. 8.1, because there are many possible upper laser levels, some of which share the same lower level as the 488.0- and 514.5-nm transitions. There is also a separate set of energy levels for Ar^{2+}, in which the upper laser levels are 25 to 30 eV above the Ar^{2+} ground state, which itself is 43 eV above the atomic ground state. Thus even higher current densities are needed for laser emission on the ultraviolet lines of Ar^{2+} than on the blue-green Ar^+ lines, and efficiencies and operating lifetimes are lower.

The picture is qualitatively similar for krypton. However, energy levels for the heavier atom are somewhat lower and more closely spaced. Most upper laser levels of Kr^+ are 16 to 19 eV above the ion's ground state, and the wavelengths of laser transitions are somewhat longer than those of the corresponding argon transitions. Krypton is a less-efficient laser medium than argon because its transitions have lower gain, but the spacing of energy levels for Kr^+ gives a broader range of laser wavelengths. As in argon, higher current densities are needed to produce the doubly charged ions which generate the ultraviolet lines. The discharge can drive krypton ions into the walls of the bore, reducing tube pressure. The pressure change is a concern because krypton output is quite sensitive to pressure, making a large gas reservoir important.

The upper levels of the major Ne^+ ultraviolet transitions are about 31 eV above the ground state. This is considerably higher than the major transitions of singly charged argon or neon, but those transitions are in the visible. The ultraviolet Ne^+ transitions come out better in comparison with the ultraviolet transitions of Ar^{2+}, which are at comparable wavelengths but considerably higher than the ground state of the neutral atom. In practice, this means that lower discharge currents are needed to generate ultraviolet emission from neon than from argon, although the currents are higher than needed to generate visible emission.

Internal Structure

The years since the first argon laser was demonstrated by William B. Bridges at Hughes Research Laboratories (Bridges, 1964) have seen a gradual evolution of ion-laser design. Designers have settled on ceramic and metal-ceramic tubes with rated lifetimes ranging up to 10,000 hours. Although different manufacturers have taken different approaches, common elements are present in many designs. This reflects the need to overcome the same technological obstacles.

To attain the high current densities needed for excitation, the discharge applied along the length of the laser tube must be confined to a narrow "bore" one to three millimeters in diameter in the tube's center. The high current densities create a hot plasma in the laser tube and lead to sputtering of the bore or tube materials by the plasma, which can erode elements in the tube, introduce contaminants into the laser gas, and deplete the gas supply by trapping some atoms. The strong electron flow in the discharge tends to push neutral atoms toward the positively charged anode, while ions migrate toward the negative cathode, creating a need for gas circulation in the tube. A supply of extra laser gas is needed to replenish gas depleted during operation. In addition, the low efficiency of laser operation requires some way of getting rid of waste heat. These factors have led to several common design elements: use of high-temperature ceramics and metals resistant to sputtering in the tube, a gas-flow path separate from the central laser bore, a gas reservoir attached to the tube, and active cooling with flowing water or forced air.

A common design for low- and moderate-power argon lasers is shown in Fig. 8.2. The discharge current passes from the cathode in the fat part of the tube along the narrow bore to be collected at the anode. The "necks" at each end of the tube are optical connections ending in Brewster angle windows, which allow linearly polarized light to

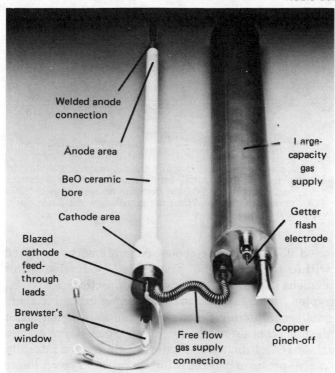

Figure 8.2 Structural components of a ceramic-bore ion laser, showing the characteristic long narrow discharge bore and fatter cathode region. The tube at right is a gas reservoir. (*Courtesy of Lexel Lasers.*)

escape without losses and serve as part of the optical path with an external resonator. The bore is typically made of beryllia (BeO), which is chosen for its ability to withstand high temperatures, low sputtering, good insulation, high thermal conductivity, good mechanical strength, and low porosity. Cooling is by water flowing along the ceramic outer surface of the bore, or by air forced to pass over it. A magnet which creates a magnetic field parallel to the bore axis may be used to help confine the discharge current to the center of the bore, increasing the ion density without seriously affecting the average electron energy.

Builders of high-power argon lasers have largely turned to an alternative (shown in Fig. 8.3). This approach relies on conductive metal disks brazed to a large-diameter ceramic tube. The bore is defined by central holes in tungsten disks, which are brazed to copper cups, which in turn are brazed directly to the ceramic tube envelope. The tungsten is chosen for its resistivity to sputtering and high temperature, and the copper is used because of its high thermal conductivity, essential for transferring heat to the envelope and thus to the cooling

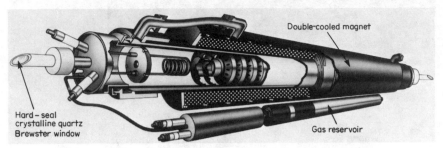

Figure 8.3 Hybrid metal-disk ceramic ion laser, in which the discharge passes through the bore defined by the central holes in the internal metal disks. Return flow is through holes outside of the central region. Heat is conducted from the central tungsten disks through copper cups to the ceramic material and thus to the cooling water flowing over the ceramic. (*Courtesy of Coherent Inc.*)

water flowing around it. Holes in the copper cups allow the required return gas flow within the laser tube. The use of conductive water cooling is a key element in making this approach practical.

The design of krypton tubes is similar to that of argon tubes, but there are a few differences worth noting. Because krypton is depleted faster than argon, the tube must contain much more reserve gas. Krypton lines other than the dominant 647.1-nm wavelength vary strongly in intensity with pressure, so active pressure control is important in krypton lasers designed to operate at those lines. The lower efficiency of krypton also requires a higher drive current density to achieve the same output power as argon. For similar reasons, most krypton lasers are water-cooled, although that is not an absolute must.

Ion lasers designed for ultraviolet operation are similar to krypton in having low efficiency and generally needing water cooling. Production of the ultraviolet lines of the doubly charged ions requires a discharge current density which can exceed 1000 A/cm^2, larger than needed to produce the longer-wavelength lines of singly charged ions, leading to a greater reliance on current-confining magnetic fields. In addition, optics resistant to ultraviolet damage are very necessary in lasers emitting on the ultraviolet lines. Ultraviolet output also requires higher internal magnetic fields than visible output, as shown in Fig. 8.4.

Optics. Most ion-laser tubes have "stems" at each end, terminating in Brewster angle windows which transmit linear polarized light to external cavity optics. The Brewster angle windows normally are fused to the tube with hard seals. Fused silica is used for Brewster angle windows in some low-power ion lasers operating at visible wavelengths. However, crystalline quartz windows are required for operation at high powers or in the ultraviolet because the short-wavelength

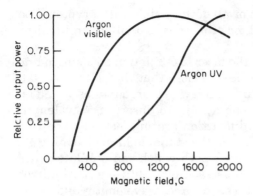

Figure 8.4 Effects of current-confining magnetic field on gain on singly ionized argon lines in the visible and on doubly ionized ultraviolet lines. (*Courtesy of Spectra-Physics Inc.*)

ultraviolet light produced in ion-laser plasmas causes formation of color-center defects in fused silica.

The intense short-wavelength ultraviolet from the discharge can erode crystalline quartz by decomposing SiO_2, causing surface absorption. Deposition of contaminants also can increase surface absorption. To limit this damage, laser manufacturers may apply protective coatings to the inside surface of the Brewster angle window, and use quartz crystals with certain crystalline orientations. In bonding the window to the tube, they must also ensure that thermal expansion coefficients of the window and tube segment match. This has led to elaborate designs for the ends of the tube, as shown in Fig. 8.5.

Cavity optics may be separate from ion-laser tubes or may be assembled in an integral unit with the tube. Integral optics simplify the laser, but they may restrict laser operation to certain lines. In practice, integral optics are more common on lower-power ion lasers, but they are available even on water-cooled models generating up to 5 W.

Standard optics for ion lasers are reflective over much of the visible spectrum, allowing multiline oscillation in the blue-green with argon,

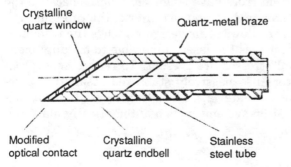

Figure 8.5 Cross section of "endbell" assembly which seals a Brewster angle window to the end of a high-power ion-laser tube. (*Courtesy of Spectra-Physics.*)

or multiline oscillation in much or all of the visible with krypton. Special reflective optics are needed for either gas to oscillate on its ultraviolet lines.

Ion laser optics also can limit the laser to oscillation on a single line. (Note, however, that some lines are so close that they often oscillate simultaneously; examples include ultraviolet argon lines at 333.6, 334.5, and 335.8 nm.) One approach is to insert a wavelength-dispersive element, a prism or diffraction grating, between the tube and one of the cavity mirrors. As in other tunable lasers, turning the dispersive element selects the wavelength which can oscillate in the laser cavity. An alternative is to use a wavelength-selective mirror, which has high reflectivity at the desired wavelength and lower reflectivity elsewhere. This makes net cavity gain high enough for oscillation at the desired wavelength, but not at other wavelengths which experience higher losses.

Ion-laser media have higher gain than do He–Ne and He–Cd lasers, but the round-trip gain still is low enough to require a highly reflective output mirror. Typically only 5 to 10 percent of the light is transmitted, so circulating power within the laser cavity is considerably higher than the power emerging in the beam.

Cavity Length and Configuration. Compact air-cooled ion lasers can have cavities as short as a quarter meter, with high-power models having cavities to about 2.4 m long. Models intended for industrial use generally have little extra room inside the cavity, but those intended for laboratory applications typically leave some space between the laser tube and the cavity optics for insertion of a mode-selecting etalon. Modelockers directly replace the tuning-prism assembly. For cavity dumping, an external module is added on the laser's output coupler end.

Stable resonators are the rule for continuous-wave, low-gain ion lasers. Normally the cavity mirrors have long focal lengths because the gain region is the long, narrow bore in the center of the tube through which the discharge current flows. This design matches the resonator and gain volumes, but it leaves the laser vulnerable to misalignment if mechanical or thermal stresses are present. To reduce such problems, some manufacturers use passively stabilized resonator structures made of low-expansion materials. The advent of microprocessors has led to active stabilization systems, which automatically maintain alignment (Peuse, 1988).

Inherent Trade-offs

Ion lasers confront both users and designers with several inherent trade-offs:

- Spectral purity of single-wavelength oscillation vs. higher power of multiline operation.

- Greater range of visible wavelengths from krypton vs. higher power and longer life of argon.

- An increase in output power is generally accompanied by increases in price and complexity of the laser and a decrease in tube lifetime.

- The simplicity of forced-air cooling vs. the greater effectiveness of heat removal with convective flowing water cooling.

- Ultraviolet output, more effective for some applications, comes only at higher cost, shorter tube life, and lower output power and tube efficiency than visible-wavelength output.

- The potential versatility of mixed-gas lasers emitting on both argon and krypton lines is offset by the faster depletion of krypton and accompanying loss of the krypton lines, making the mixed-gas tubes short-lived.

Variations and Types Covered

Argon is by far the most important ion laser commercially, accounting for most of the units sold, with krypton in second place, accounting for about 10 percent of the market. Sales of mixed-gas argon-krypton lasers are smaller, but that type is included here because of its close relationship to argon and krypton lasers. The first neon laser was introduced only recently. No other rare gas ion laser is commercially significant.

This chapter focuses on argon and krypton lasers in their various forms, including ultraviolet and visible versions, air- and water-cooled models, and types emitting on one or multiple lines.

Beam Characteristics

Wavelength and Output Power.   Multiline visible output of commercial argon lasers ranges from a few milliwatts to about 50 W. The laboratory record is 500 W, but such high powers have not proved practical. The less-efficient krypton laser is limited to lower powers; the most powerful commercial version can generate multiline output of 14 W, with highest power on the 647-nm red line. Mixed-gas lasers generate only a little more power than do krypton lasers. (They are designed so that output on the argon and krypton lines is roughly equal.)

Much work has gone into enhancing ultraviolet output in recent years (Petersen, 1987). Commercial argon lasers can produce up to 7 W on the ultraviolet lines combined, and up to 1.5 W on the deep-ultraviolet argon lines at 275.4 to 305.5 nm. At this writing, the only commercial neon ion laser emits 1 W combined on three ultraviolet lines.

Argon, krypton, and neon lasers all can emit on individual lines or on closely spaced groups of lines, as well as oscillating simultaneously on many lines. The major lines from commercial argon and krypton lasers are listed, with relative power levels, in Table 8.1. The wavelengths available from any particular laser depend on the cavity optics, tube design, and wavelength selection features.

Efficiency.   Ion lasers are inefficient. The best of them, argon, has typical overall (or "wall-plug") efficiency of 0.2 to 0.001 percent for multiline visible output. Values are lower for single-line output, for krypton or neon lasers, or for ultraviolet operation.

Temporal Characteristics.   Commercial ion lasers produce continuous-wave output normally. Modelocking or cavity dumping can produce short pulses of light from an ion laser in which the discharge is operating continuously. Because of the time needed to build up a stable high-current discharge, it is not wise to pulse the laser by turning its electrical power on and off. If modelocking and/or cavity dumping cannot provide the desired pulse modulation, the best course is use of an external modulator.

Spectral Bandwidth and Frequency Stability.   The spectral bandwidth of ion lasers is a function of their mode of operation. Multiline emission from argon lasers is concentrated at six wavelengths between 457.9 and 514.5 nm: multiline krypton emission can span an even broader wavelength range. Spectral bandwidth of a single line is about 4 to 12 gigahertz (GHz), or roughly 0.004 to 0.01 nm, on the order of 1 part in 10^5 at the 6×10^{14}-Hz frequency of the blue-green argon lines. The 4- to 12-GHz linewidth, in turn, is the doppler bandwidth and contains 40 to 120 longitudinal modes, each with much narrower linewidth. Linewidth of a single longitudinal mode of an ion laser is several megahertz, which can be obtained by using an etalon.

Frequency stability becomes an important concern with ion lasers only for operation in a single longitudinal mode. In a well-designed ion laser, temperature-induced variations in the optical path length length of the etalon or the laser cavity can cause frequency shifts, with changes in etalon cavity size normally dominant. Temperature stabilization of the etalon can keep frequency stable to within several megahertz over a 1-s interval, and to within several tens of megahertz over 10 hours. In most cases, such stabilization is good enough to prevent hopping between longitudinal modes, normally spaced 80 to 250 MHz apart, which could cause rapid frequency shifts.

Amplitude Noise. Control of the current flowing through the laser tube can limit root-mean-square (rms) amplitude noise level to from 0.4 percent to a few percent. Peak-to-peak variations, specified by some manufacturers, are inevitably larger.

Much better control of amplitude noise is possible by monitoring the optical output and using that information to control input current. Some ion-laser makers claim rms noise values of 0.1 percent using light control, although typical specifications are somewhat higher. Active stabilization is not recommended for single-frequency lasers.

Specifications of amplitude noise do not differ greatly for single- or multiline operation of ion lasers. All assume that the laser has been warmed up before measurement.

Beam Quality, Polarization, and Modes. Brewster angle windows on ion-laser tubes lead to linearly polarized output; the typical design calls for vertical polarization, specified at a ratio of 100 to 1 or more. TEM_{00} is the commonest transverse mode, especially on scientific lasers. Because somewhat higher powers are possible with multiple transverse modes—albeit at a sacrifice in beam quality—some models seeking to deliver maximum power for minimum cost produce multimode output. A few models emit in the doughnut-shaped TEM_{01}^{*} mode.

Coherence Length. Coherence length of an ion laser emitting on multiple lines is very small. For single-line oscillation with typical 6-GHz bandwidth, coherence length is about 50 mm. Use of an etalon to limit oscillation to a single longitudinal mode can extend coherence length to tens of meters or more.

Beam Diameter and Divergence. Typical values of beam diameter and divergence of TEM_{00} ion lasers are 0.6 to 2 mm and 0.4 to 1.2 milliradians (mrad), respectively, varying somewhat among models, with the product of the two quantities typically being about 0.9. Multiline, multimode lasers may have beam divergence to 3 mrad. Both beam diameter and divergence are slightly larger for krypton lasers emitting in the red than for blue-green argon lasers.

Stability of Beam Direction. The directional stability of any ion-laser beam is affected by random fluctuations within the laser tube and in the intracavity spaces, including absorption by water vapor and ozone. Wavelength-selecting prisms in single-line lasers are an additional source of beam-stability problems, which can arise because of variations in their refractive index. Variations caused by prisms are larger; one manufacturer has reported a change of 11 arcseconds (0.05 mrad) per degree Celsius, and warned that temperature changes as small as

10°C can completely detune the laser. These variations can be reduced by temperature compensation of prism wavelength selectors. An alternative that avoids this instability is the use of wavelength-selective output mirrors, which do not induce beam-direction errors because they are in line with the rest of the laser cavity. Many manufacturers do not specify beam-pointing stability, but one specified figure for high-power water-cooled ion lasers is pointing stability of about 0.01 mrad/°C.

Suitability for Use with Accessories. The most important accessories used with ion lasers are

- Modelockers, to produce trains of picosecond pulses from lasers emitting in multiple longitudinal modes. With the typical 5-GHz linewidth of ion-laser lines, the resulting pulses are about 90 to 200 picoseconds (ps) long, produced at repetition rates in the 75- to 150-MHz range.

- Cavity dumpers, which can produce high peak power pulses in the 5- to 20-ns range, or which can be used with a modelocker to slow down the repetition rate and increase the amplitude of modelocked pulses.

- Etalons, used to narrow spectral width to a single longitudinal mode.

- Intracavity second-harmonic generators, which can produce up to about 100 mW continuous-wave at 257.25 nm, the second harmonic of argon's 514.5-nm line. The harmonic generation module is aligned in front of the laser, and the output coupler moved to place the nonlinear crystal within the laser cavity. This allows harmonic generation with the high level of power within the laser cavity, rather than the lower power in the external beam.

Wavelength conversion also is possible by using an ion laser to pump another laser, usually one tunable in output wavelength. The most common types are dye lasers, described in Chap. 17, and titanium-doped sapphire, described in Chap. 24. The short upper-state lifetimes prevent energy storage in ion-laser media, rendering Q switching impractical.

Operating Requirements

Input Power. Input electrical requirements for ion lasers range from 10 or 15 A single-phase current at 120 V to 60 or 70 A of three-phase current at 460 V, with input requirements increasing with output power. An argon laser with multiline output of 3 to 5 W typically would require electrical input of 8 to 26 kW. Similar electrical input would generate about 0.75 W of multiline krypton output. Up to 26 kW of electrical input is needed to generate 0.4 W in the ultraviolet from argon or 0.15 W of ultraviolet from krypton.

The power supply converts the electrical power from the line supply into the high direct currents needed to drive the laser. The discharge is triggered by a peak voltage of a few thousand volts. The current then rapidly builds to 10 to 70 A, concentrated in the bore, while the voltage drops to a steady value of 90 to 560 V, depending on the length of the tube. Light output is roughly proportional to the square of current density, and hence increases faster than the input electrical power.

Cooling. Ion lasers require either forced air or water cooling, with water cooling needed for higher-power models. Specified tap-water flow rates for laser heads are 5.6 to 22.7 liters per minute (L/min), and higher-power models require water cooling for their power-supply modules. Water cooling is used for some argon lasers delivering as little as 100 mW of multiline output. Generally, water flows along the outside of the laser tube, as shown in Fig. 8.6, conducting away excess heat.

Forced-air cooling can be used for argon-laser tubes delivering multiline outputs as high as a few watts, although at that power level both gas flow rate and exhaust temperature must be high. One 4-W multiline argon laser requires 250 ft^3 (7 m^3) of air per minute and can heat the exhaust to 180°F (80°C); specified tube lifetime is 10,000 hours. Air cooling is the rule for very low power argon lasers.

Consumables. Outside of cooling water, there are no true consumables associated with ion lasers, with the possible exception of the tube itself. Tubes have lifetimes rated in the thousands of hours and account for about a third of the price of the complete laser. As described below, refurbishment is an alternative to replacement with a new tube.

Required Accessories. Pulsed operation of an ion laser requires a cavity dumper, modelocker, and/or external modulator. Single-frequency operation requires an etalon with temperature or mechanical stabilization. Temperature stabilization of the laser can improve stability in frequency, beam direction, and output power. Active stabilization systems can keep output power and beam direction constant by actively maintaining alignment of the optical resonator. Similar stabilization systems can keep an ion laser operating in a single longitudinal mode by controlling etalon temperature, automatically compensating for

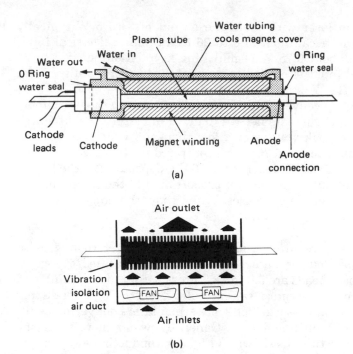

Figure 8.6   In water-cooled ion laser (top) the water flows around the outside of the laser tube to remove heat. In air-cooled types (bottom) fans force air past the tube, and the air removes excess heat from cooling fins attached to the tube. (*Courtesy of Lexel Lasers.*)

any deviations. Special optics generally are needed to obtain output on the weaker or less-common laser lines, and are required for ultraviolet operation of either argon or krypton lasers.

Operating Conditions and Temperature. Warm-up of ion lasers is needed to establish a stable current discharge within the tube and to reach proper temperatures for stable output power, wavelength, and beam pointing. Models intended for demanding scientific applications give specifications for stability of operation measured after as much as 2 hours of warm-up time, although 30 minutes is more typical. Active stabilization can reduce warm-up requirements to as little as 1 minute. Lasers intended for industrial applications have warm-up times measured in seconds or minutes.

Water-cooled ion lasers require tap-water input and drainage for their flow-through cooling systems. Air-cooled models require adequate ventilation for exhaust fans. Otherwise, ion lasers are designed to operate under normal laboratory conditions or under the temperature-controlled conditions typically found in an office or computer room.

Mechanical Considerations. Ion-laser sizes vary widely, increasing with output power. The heads of the largest types weigh about 100 kg, are about 2 m long, and are about 20 by 20 cm in cross section. They require separate power supplies, often including water connections and controls, that measure up to about 70 by 65 by 60 cm and weigh up to about 220 kilograms (kg).

On the other end of the scale, a compact air-cooled version delivering 10 milliwatts at 488 nm has a 5-kg head 28.5 cm long that measures 15.8 by 12 cm in cross section. It comes with an 8.5-kg power supply that measures 34.6 by 22.0 by 15.8 cm.

Safety. Argon lasers fall under Class IIIb or Class IV of the federal laser safety code, and users should take care to avoid direct exposure to the beam, particularly eye exposure. Standard safety goggles are available which strongly attenuate argon's blue-green lines, and the laser does not produce the single pulses of high peak power that have proven particularly dangerous with other types of laser. However, the unattenuated, narrow-divergence beam from an argon laser can penetrate the eyeball and permanently affect the retina.

Krypton lasers, although generally lower in power, fall into the same safety classes. Multiline visible output from krypton is a particular problem because the laser wavelengths are scattered across so much of the spectrum that goggles made to block the laser lines would also block most visible light. This makes it vital for users of multiline krypton lasers to exercise special care in avoiding eye exposure.

The ultraviolet lines of argon, krypton, and neon are strongly absorbed by the lens of the eye, and only a small fraction of such light reaches the retina. There are few continuous-wave lasers emitting at these wavelengths, and eye hazards are not well quantified, so the use of ultraviolet-absorbing goggles would be a prudent precaution. Care should be taken with the krypton lines at 752.5, 793.1, and 799.3 nm in the near-infrared; those lines can be weakly perceived by the human eye, but users should not make the mistake of judging their intensity by eye.

Like other discharge-driven lasers, ion lasers need high voltages, but only to initiate the discharge. After the discharge is established, voltage drops to the 75- to 400-V level, with currents of 10 A or more flowing through the tube. The hazards presented by such levels of current capacity are different than those at higher voltages, resembling the dangers of an open wall socket.

Special Considerations. Ion-laser alignment is delicate and can be affected by changes in temperature and mechanical stresses. Manufacturers design their resonators to stay rigid in the face of reasonable mechanical and thermal stresses, but they cannot be expected to take countermeasures to protect against abuses such as improper mounting, incorrect adjustment, or placement next to a hot-air vent.

Reliability and Maintenance

Operating lifetime of an ion laser is limited by the tube. Typical rated lifetimes range from 1000 to 7000 hours for water-cooled argon lasers and from 3000 to 10,000 hours for air-cooled argon, both for operation at visible lines, but not all these ratings are backed by manufacturer warranties. Lifetimes are shorter for krypton lasers, or for operation on ultraviolet lines, with krypton lifetime limited by gas depletion and ultraviolet lifetime limited by the high operating current used to produce ultraviolet emission. The operating life, defined as how long the tube meets specifications, is shortest for argon-krypton mixed-gas lasers, because the krypton is depleted rapidly, leaving a tube that emits only on the argon lines. Often the gas-fill system for mixed-gas lasers adds only krypton to reduce this problem.

Tubes degrade by gradually falling in output power, reaching the end of their useful life when output falls below application requirements or the tube fails altogether for other reasons. Average life falls within the rated ranges, although there is considerable variation between individual tubes. Laser operation can be restored by replacing the tube. Unless the tube is extremely worn from long use or has suffered severe physical damage, it can be refurbished, as described below, to once again produce rated output.

Maintenance and Adjustments Needed. Periodic realignment may be needed because of the sensitivity of ion lasers. Cooling and electrical systems should be checked regularly. As mentioned above, tube replacement or refurbishment is required after prolonged operation. Optics not sealed to the tube must be cleaned periodically, particularly for high-power lasers.

Mechanical Durability. Ceramic and glass ion lasers tend to be brittle, making them vulnerable to mechanical damage. New designs have offered some improvements, but the need for ceramics to withstand hostile conditions in the laser tube means that fragility remains a concern.

Failure Modes and Causes. Mechanical damage or failure of electrical components can cause failure of an ion laser. Under normal circumstances, however, the most likely problem is degradation of the laser tube caused by several factors, most related to the high discharge currents in the tube bore.

- Depletion and contamination of the laser gas, caused by trapping of gas atoms under material sputtered from the laser tube, and by sputtering of material into the gas. Getters are used to collect some contaminants. Many lasers have automatic gas replenishment.

- Erosion of material from the laser bore, which may be deposited elsewhere in the tube, perhaps blocking the bore.

- Contamination and erosion of the cathode by ion sputtering.

- Contamination of the optics, causing excessive optical losses.

- Degradation of optics exposed to hard ultraviolet radiation produced in the discharge.

- Misalignment of the resonator, caused by temperature or mechanical effects, increasing internal losses until the laser can no longer oscillate.

All six types of degradation can occur simultaneously. Mechanisms related to sputtering, such as bore erosion and contamination, tend to increase in severity with current density in the bore.

Possible Repairs.   The best course for ion-laser users with a dead tube is to replace it, but that does not automatically mean buying a new tube. Several companies now offer repair, refilling, and refurbishing of ion-laser tubes. Refilling with fresh laser gas is far from enough. A thorough refurbishing requires disassembly of the tube, replacement of some components, repair and cleaning of others, then careful reassembly and refilling with fresh gas (Wiedemann, 1984). Generally, the refurbished tubes are burned in for several hours. A well-refurbished tube can give 80 percent or more of the operating life of a new tube for about half the cost, and in many cases can be repeatedly refurbished if mechanically sound.

Commercial Devices

Standard Configurations.   Some elements of ion-laser design are common to many manufacturers, such as the separation of the tube-containing laser head from the power supply. Otherwise, however, configurations and design approaches can vary widely.

One variable is the type of cooling. Water cooling is essential for high-power operation and is common in moderate-power lasers because of its simplicity. It is also the usual choice for the comparatively inefficient krypton laser. Air cooling is the rule for argon lasers under 200 mW and can also be used at moderate powers, up to a few watts with sufficiently fast air flow. The simplicity of air cooling makes it attractive for industrial applications.

Lasers tend to be packaged either for laboratory applications or for incorporation into laser systems for applications such as exposure of printing plates, high-speed printing, or medical treatment. Laboratory models normally are designed for maximum flexibility and leave intracavity space for accessories such as etalons. Industrial versions tend to be packaged functionally, for incorporation inside a larger system package, and tend to make little or no provision for inclusion of optional equipment. Many laboratory models are intended to be tunable so they can emit at any of several possible single wavelengths and are packaged with the option to oscillate in a single longitudinal mode. Industrial versions may emit at one or several wavelengths, but they are rarely tunable and have little call to oscillate in a single longitudinal mode.

Some companies take a modular design approach that lets them offer many different models. The same power supply may be used to drive several different tubes, often including tubes with different active gases. The same laser head housing may hold different tubes or be used with different optics to obtain different performance from similar or identical tubes.

As the industrial market has grown, manufacturers have shown increasing interest in designing lasers that could directly replace products made by other companies. However, optical, mechanical, and electrical characteristics are still far from standardized.

Options.   Principal options offered on commercial ion lasers include

- Modelocking.
- Cavity dumping.
- Choice of TEM_{00} or multiple transverse mode operation.
- Selection of a single wavelength or line for laser oscillation.
- Etalons to limit oscillation to a single longitudinal mode.
- Temperature stabilization of components to enhance laser stability.
- External power monitors for light stabilization of output.
- Choices of power supplies and (in some cases) of cooling method.
- Polarization rotators.
- Operation on ultraviolet or deep-ultraviolet wavelengths. (Normally this choice must be made when the laser is purchased.)

- Cavity mirror stabilization.

- Mode stabilization for single longitudinal mode operation.

Krypton lasers cost much more per watt than argon: a 0.75-W multiline krypton laser has about the same price as a 5-W multiline argon laser. Mixed-gas lasers have prices comparable to a krypton laser of equivalent power. The option to produce ultraviolet output from argon or krypton can add considerably to the price.

Special Notes.   The ion-laser market has long been highly competitive and segmented. Some companies specialize in mass production for industrial markets, where pricing is particularly important. Others specialize in high-performance lasers, where they compete in offering sophisticated features. This has led to continued refinement of ion-laser design, particularly for ultraviolet operation.

Developments in solid-state laser technology have a major impact on the ion-laser market. The emergence of the tunable titanium-sapphire laser has been good news for makers of high-power ion lasers because the best pump source is multiline visible output from argon lasers. On the other hand, improvements in solid-state laser technology could give ion lasers serious competition for low-power continuous-wave output in the blue and green parts of the spectrum. Direct frequency doubling of new high-power semiconductor lasers can generate blue light, although beam quality remains a concern. Alternatively, high-power diode lasers can pump neodymium lasers, generating 1064-nm output that can easily be frequency doubled to 532 nm. Prices of high-power semiconductor lasers at this writing are too high for all solid-state visible sources to compete with low-power ion lasers on price alone, but that could change.

Pricing.   The prices of low-power ion lasers have dropped somewhat over the last few years, while those of high-power water-cooled models have increased with their sophistication. Representative prices as of mid-1990 for argon lasers are listed in Table 8.2. The variations re-

TABLE 8.2   Representative Prices for Argon Lasers

5 mW, multiline blue-green	$3000
5 mW, 488 nm	$4400
10 mW, 488 nm	$5000–$6000
40 mW, multiline blue-green	$7000
50 mW, multiline blue-green	$5500
5 W, multiline blue-green	$12,300–$30,000
25 W, multiline blue-green	$50,000–$80,000
7 W, ultraviolet	$100,000

flect differences among manufacturers, performance, lifetimes, and other factors.

Suppliers. Over a dozen companies make argon lasers. They range in size from small firms that specialize in ion lasers to overseas giants. Some of the companies which have been in the field longest are best known for their scientific lasers, but other veteran companies have concentrated on the industrial market—a market now being targeted by several new companies as well.

Fewer companies make krypton and mixed-gas lasers than argon types, reflecting the much smaller market. There is only a single maker of neon-ion lasers at this writing.

Applications

Ion lasers have become standard sources of several milliwatts to tens of watts of visible light for a variety of applications. The blue-green output of argon is particularly valuable because many materials respond more strongly to it than to the less-energetic red photons from krypton or helium-neon lasers. Argon competes with helium-cadmium lasers at powers of several milliwatts but is unrivaled at much higher powers. In the red, krypton can offer higher powers than helium-neon. Some of krypton's weaker lines are also valuable because they are otherwise unobtainable except from dye lasers. Noble gas ion lasers are also valuable sources of continuous-wave ultraviolet light. Major applications of ion lasers include

- Production of halftone patterns during color separation for printing of full-color material
- High-speed printing of computer output
- Exposure of printing plates, films, and other materials used in printing and publishing
- Mastering disks for videodisk and audiodisk players
- Medical treatment, particularly in ophthalmology, where argon laser light can penetrate readily to the retina to treat diabetic retinopathy
- Measurement in medical laboratories and research applications, including DNA research and ultraviolet fluorescence studies
- Pumping dye lasers and tunable solid-state lasers to obtain tunable continuous-wave output or picosecond pulses
- Entertainment and display applications requiring visible light

- Measurements of fluorescence produced by visible or ultraviolet light
- Detection of latent fingerprints by observing the fluorescence produced when the prints are illuminated by the blue-green light from argon lasers

Bibliography

William B. Bridges: "Laser oscillation in singly ionized argon in the visible spectrum," *Applied Physics Letters* 4:128–130, 1964, erratum; *Applied Physics Letters* 5:39, 1964 (first report of ion lasers).

William B. Bridges: "Ionized gas lasers," in Marvin J. Weber (ed.), *CRC Handbook of Laser Science & Technology*, vol. 2, *Gas Lasers*, CRC Press, Boca Raton, Fla., 1982, pp. 171–269 (review paper).

Alan B. Petersen: "Enhanced CW ion laser operation in the range 270 nm $\leq \lambda \leq$ 380 nm," *Proceedings of SPIE*, vol. 737, *New Developments and Applications in Gas Lasers*, SPIE, Bellingham, Wash., 1987 (research on new ultraviolet lines).

Alan B. Petersen: "Continuous wave krypton ion laser operation: 219–266 nm," Paper CTUH10 presented at *Conference on Lasers and Electro Optics, CLEO '90*, May 21–25, 1990, Anaheim, Calif.; technical digest available from Optical Society of America, Washington, D.C. (research on new ultraviolet lines).

Bruce Peuse: "Active stabilization of ion laser resonators," *Lasers & Optronics* 7(11):61–65, November 1988 (description of technique).

Orazio Svelto: *Principles of Lasers*, 3d ed., Plenum, New York, 1989 (textbook description of ion-laser operation).

Rudi Wiedemann: "Reprocessing argon-ion lasers," *Lasers & Applications* 3(2):81–84, February 1984 (description of reprocessing).

Helium-Cadmium Lasers

The helium-cadmium laser is the best known member of a family of lasers which emit on lines of ionized metal vapors. One of the first metal vapor lasers to be discovered, He–Cd can produce continuous powers up to about 150 milliwatts (mW) at 442 nanometers (nm) in the blue, or powers to about 50 mW on an ultraviolet line at 325 nm. These characteristics are attractive for many applications and have stimulated further development that has extended operating lifetimes of He–Cd tubes to six to ten thousand hours.

The metal vapor ion lasers were among the first types discovered in a systematic series of experiments intended to search for laser lines from a variety of materials. William Silfvast, then a graduate student at the University of Utah, and his faculty advisor Grant Fowles placed various metals in a discharge tube and observed what happened when electrical pulses were passed through the tube. Their first paper reported laser action on lines of six metals (Silfvast et al., 1966). Later research by Silfvast and others at Bell Laboratories showed that cadmium was a particularly efficient laser material and that addition of helium enhanced efficiency and permitted continuous-wave operation.

Metal vapor laser development has concentrated on helium-cadmium because of its useful wavelengths and high efficiency. Silfvast also developed the related helium-selenium laser, which emits on over 20 lines, mostly in the visible (Silfvast, 1973). He–Se was put on the market briefly in the 1970s, but it has never received as much development as He–Cd and is no longer marketed actively. Reference tabulations (e.g., Bridges, 1982) list many other metal ion lasers which have never received much development.

Internal Workings

Active Medium. Singly ionized cadmium vapor has over a dozen laser lines in and near the visible and can operate as a laser under a variety of conditions. Operation is most efficient when several millitorr of cadmium vapor is sealed in a tube with helium gas at a pressure of about 3 to 7 torr [400 to 930 pascals (Pa)] and an electrical discharge passed through the mixture, exciting the helium and ionizing the cadmium, which has a much lower ionization threshold. This is the usual configuration in commercial helium-cadmium lasers, which contain cadmium metal that must be heated to about 250°C to produce adequate vapor pressure.

Although laser operation is possible over a range of conditions, long-term continuous-wave operation at reasonable efficiency requires the proper balance of helium and cadmium pressures. Helium uniformly fills the tube because it is effectively an inert gas, but the positively charged Cd^+ ions migrate toward the negative electrode (cathode). He–Cd tubes are designed so that this migration spreads the Cd^+ ions more uniformly through the tube, without depositing metal in places where it could impair laser operation, such as on the cavity optics.

Energy Transfer. Excitation energy for the helium-cadmium laser comes from a direct-current discharge passing through the laser tube. Typical discharges are around 1500 volts (V), with current densities in the small-diameter bore on the order of 4 amperes (A) per square centimeter of cross section. Helium atoms in the laser gas absorb energy from the discharge and then transfer that energy to cadmium ions.

The energy levels of cadmium and helium involved in the principal He–Cd lines are shown in Fig. 9.1. The 441.6-nm blue transition and the 325-nm ultraviolet transition, which are closely related, are the easiest to produce and the most widely available. Considerably more energy is needed to raise Cd^+ ions to the upper laser levels of the red and green transitions (which themselves are closely related). In practice, different discharge conditions are necessary. The blue and ultraviolet lines can be generated in the positive column of the discharge, the region from the middle of a discharge to the anode where the electric field is not steeply graded and the electrons are not accelerated rapidly. The higher energy needed to produce the red and green lines is available only in the region near the cathode, where there is a larger change in electric field and a stronger electron acceleration. This difference means that special tube designs are needed to generate the red and green lines.

Three mechanisms can excite Cd^+ ions to the upper laser levels; their relative importance depends on tube design. The most important for the narrow-bore tubes used for blue and ultraviolet lasers is Penning ionization. Secondary processes are charge-transfer ionization and direct electron collisions.

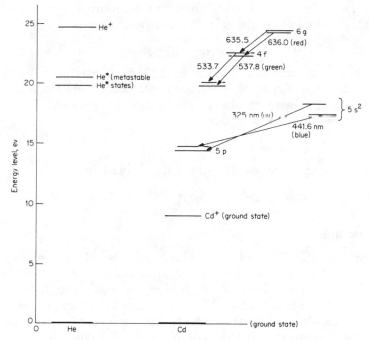

Figure 9.1 Energy levels in helium and cadmium, showing the laser transitions in the ultraviolet, blue, green, and red (wavelengths are in nanometers). The blue and ultraviolet lines are largely excited by energy transfer from the metastable helium excited states near 20 eV above ground level. Higher energy is needed to excite the red and green lines, which are emitted in a cascade process.

In Penning ionization, energy from an excited helium atom ionizes a cadmium atom:

$$He^* + Cd \rightarrow He + Cd^+ + e^-$$

Because helium's lowest excited energy level is well above the ground state of Cd^+, Penning ionization produces excited cadmium ions. The metastable helium excited states transfer energy to $2D$ levels of Cd^+ which have lifetimes on the order of 100 nanoseconds (ns). The lower $2P$ levels of Cd^+ also receive some energy but are quickly depopulated because of their 1-ns lifetime. The result is a population inversion between $2D$ and $2P$ states of cadmium, allowing laser action on either the 442- or 325-nm lines.

Charge-transfer ionization is of secondary importance in the traditional positive-column He–Cd laser, although it is dominant in some metal vapor ion lasers. In this case, a helium ion captures an electron from a cadmium atom to create an excited cadmium ion

$$He^+ + Cd \rightarrow He + (Cd^+)^*$$

This is the dominant excitation process in the helium-selenium laser, because a selenium energy level is resonant with the helium ion state.

A third alternative is direct electron excitation in two steps. First, an electron ionizes neutral cadmium

$$e^- + Cd \rightarrow Cd^+ + 2e^-$$

Then a second electron excites Cd^+ to the upper laser level

$$e^- + Cd^+ \rightarrow (Cd^+)^* + e^-$$

Internal Structure

The internal structure of common blue and ultraviolet helium-cadmium lasers bears a general resemblance to helium-neon and noble gas ion lasers. In all three types, discharges pass through long, thin bores between cathode and anode which confine the current to increase current density and hence increase laser efficiency. The interior of the bore houses the positive column of the He–Cd discharge. The gain and output power of blue and ultraviolet He–Cd lasers are higher than those of He–Ne lasers but lower than those of the major blue-green argon lines.

The need to maintain adequate concentrations of cadmium vapor gives some novel aspects to designs of positive-column He–Cd laser tubes, such as the one shown in Fig. 9.2. A reservoir of cadmium metal is required, which must contain about a gram of the metal for each thousand hours of tube operation. This reservoir is placed at the anode end of the discharge bore and is heated to provide the necessary metal vapor pressure. As the discharge ionizes the metal vapor in the tube, the positively charged ions migrate toward the negative electrode, the cathode, a phenomenon called *cataphoresis*. This process helps to dis-

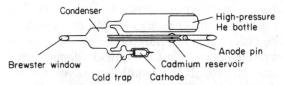

Figure 9.2 Simplified diagram of a He–Cd tube with high-pressure helium bottle to the side and a condenser and cold trap to catch cadmium metal. This tube has Brewster angle windows, but integral mirrors also can be attached. (*Courtesy of Liconix.*)

tribute cadmium vapor uniformly in positive-column He–Cd lasers, and causes deposition of metal toward the cathode as the vapor cools.

Metal deposition in the wrong places can cause serious problems, including obstruction of the cavity optics, blockage of the discharge bore, and poisoning of the cathode. To limit these problems, He–Cd tubes normally include a condenser outside the laser bore. In the tube design shown in Fig. 9.2, an additional cold trap for cadmium ions is placed in front of the cathode (Dowley, 1982).

As in the helium-neon laser, the optimum helium pressure (for maximum power) depends on bore diameter, with the product of pressure times bore diameter staying roughly constant. In a typical He–Cd laser with a 1.5-mm bore, output is highest when helium pressure is about 6 torr (800 Pa) and when cadmium pressure is around 0.1 percent of that level, i.e., several millitorr. Metal is depleted by the processes described earlier. Helium can also be depleted, by cathode sputtering, burial under condensing metal, and wall diffusion. Thus a reservoir is needed for helium as well as for cadmium. These reservoirs may be in separate branches to the tube, as in Fig. 9.2, or in structures coaxial with the discharge bore, as shown in Fig. 9.3.

The side-tube and coaxial tube designs each have advantages. Tubes with separate side branches do not require forced-air cooling. The newer coaxial design does require forced-air cooling and is more efficient in generating power per unit length because temperature along the bore is more uniform. The cylindrical structure also makes it easier to mount. However, the two types otherwise perform equivalently, and both are used in commercial lasers.

An entirely different tube structure is required to generate the green and red lines (which can lase simultaneously with the blue line), because they can be produced only in the negative glow region near the cathode. The preferred structure is a cylindrical hollow cathode that runs the length of the tube, with multiple anodes inserted

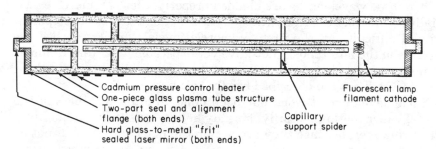

Cadmium pressure control heater
One-piece glass plasma tube structure
Two-part seal and alignment
flange (both ends)
Hard glass-to-metal "frit"
sealed laser mirror (both ends)

Fluorescent lamp
filament cathode

Capillary
support spider

Figure 9.3   Coaxial tube design, with the bore suspended in the center of a cylindrical tube; optics are bonded directly to the tube. Helium and cadmium reservoirs are within the outer cylinder. (*Courtesy of Omnichrome.*)

through holes in the side of the cathode. With this design, most of the volume within the laser cavity is in the negative glow region where Cd^+ can be excited to levels sufficiently high to generate the red and green lines.

The optics on He–Cd lasers are similar to those on He–Ne lasers. Hard-sealed optics have become standard on most models, with internal Brewster angle windows if polarized output is desired. A few models have interchangeable optics for blue or ultraviolet operation, but most come with integral optics for one line. Lasers which emit on the red and green lines have separate broadband optics which allow simultaneous oscillation on the major visible lines, with colors balanced to simulate white light. As in other low-gain lasers, the output mirror transmits only a small fraction of the incident light—typically 1 to 3 percent—and the other cavity mirror is a total reflector. Stable resonators are the rule, with typical cavity lengths 40 cm to a little over a meter.

Inherent Trade-offs

Although the ultraviolet line of He–Cd is more potent than the blue line in producing some effects, output is weaker in the ultraviolet. If the same tube design is used at the two wavelengths, 325-nm power will typically be about half that at 441.6 nm.

Variations and Types Covered

Virtually all commercial He–Cd lasers operate on the 325- or 441.6-nm transitions, but Cd^+ also can emit at several other visible wavelengths. So-called white-light He–Cd lasers have long intrigued developers (e.g., Fujii et al., 1975; Otaka et al., 1981; Schuebel, 1970; Wang and Reid, 1981) because they emit at a combination of red (635.6 and 636.0 nm), green (533.7 and 537.8 nm), and blue (441.6 nm) lines. When those wavelengths are blended properly, they can match essentially all the colors seen by the human eye. Commercial development proved difficult, but the first commercial models finally reached the market in the late 1980s. They have yet to become important commercially. As noted above, they require tube designs different from those for He–Cd. They are described briefly below.

Some development work has been done on other members of the metal vapor ion-laser family, notably helium-selenium and helium-zinc. However, because they are not on the market, they will not be described here. Commercial copper vapor and gold vapor lasers emit on lines of the neutral species and operate in ways rather different from He–Cd and related lasers; they are treated separately in Chap. 12.

Beam Characteristics

Wavelength and Output Power. Most He–Cd lasers emit at 441.6 nm in the blue or 325 nm in the ultraviolet. The cavity optics, which may be permanently bonded to the tube or externally mounted, select one wavelength. Continuous output on the blue line ranges from the milliwatt range to about 50 mW in TEM_{00} mode, and to above 150 mW for multimode output. Rated continuous output on the less-efficient ultraviolet line is up to about 50-mW multimode. Multiline output from white-light lasers is up to 50 mW.

Efficiency. Overall wall-plug conversion efficiency for blue emission from a He–Cd laser runs from 0.02 to about 0.002 percent, higher than argon ion but below helium-neon. Efficiency of He–Cd on the ultraviolet line is about one-half that in the blue. As in many other gas lasers, efficiency tends to increase with output power in commercial He–Cd lasers, until other factors begin to limit output power.

Temporal Characteristics. Helium-cadmium lasers normally emit continuously. The need to heat the tube to about 250°C to achieve the required cadmium pressure discourages turning the laser off for short intervals; He–Cd lasers may have a "standby" mode that leaves the tube nearly ready to operate but does not produce an output beam. External modulators are used if the beam must be turned off and on quickly and often.

Spectral Bandwidth. He–Cd lasers have oscillation bandwidths of about 5 gigahertz (GHz), corresponding to 0.003-nm spectral linewidth at 442 nm. Mode spacing depends on cavity length and is typically 130 to 370 MHz. Thus over a dozen longitudinal modes fall within the oscillation bandwidth. He–Cd lasers are very difficult to operate in a single longitudinal mode.

Amplitude Noise. Typical root-mean-square (rms) amplitude noise between 10 Hz and 10 MHz is specified under 2 to 2.5 percent for a single-line He–Cd laser, but in some cases that may exclude higher noise in frequency ranges where resonances occur. Specifications for peak-to-peak noise at dc to 100 kHz are somewhat higher, up to 10 percent. Output power stability, which measures long-term fluctuations, may be specified at 1 to 10 percent. Note that intervals, frequency ranges, and other definitions of conditions may differ among manufacturers.

Beam Quality, Polarization, and Modes. He–Cd lasers have good beam quality, with TEM_{00} output normal at 442 nm except for the most powerful lasers. Ultraviolet versions generate output in TEM_{00}, TEM_{01}^*, and multiple modes. Tubes with Brewster angle windows produce strongly linearly polarized beams; specifications are typically a 500 to 1 ratio, but polarizations of greater than 2000 to 1 are common.

Coherence Length. Coherence length for a He–Cd laser emitting with its natural 5-GHz linewidth is about 10 cm.

Beam Diameter and Divergence. TEM_{00} 442-nm beam diameters are 0.2 to 1.2 mm, with the smallest diameters measured at a beam waist at the output mirror. Divergences at 442 nm are 0.5 to 3 milliradians (mrad). Divergence at the 325-nm line is similar, 0.4 to 3 mrad, with diameters of 0.135 to 1.2 mm, again with the smallest values measured at a beam waist at the output mirror. As with other lasers, beam divergence is inversely proportional to beam diameter, so the smallest diameters come only at a cost of comparatively high divergence, and thus cannot be realized away from the laser except with focusing optics.

Stability of Beam Direction. Because He–Cd lasers are often used in reprographic applications where beam directional stability is vital, this quantity is specified on many data sheets. Beam-pointing stability is specified at ± 10 to 50 μrad/°C, with the explicit inclusion of temperature reflecting the crucial influence of temperature on alignment stability of the laser cavity. Stability of beam position and divergence may also be specified, with typical values ± 0.1 mm (at the output mirror) and ± 1 mrad.

Suitability for Use with Laser Accessories. He–Cd lasers often are used with external modulators to adjust output intensity for applications such as printing and writing.

Modelocking is preferable to Q switching for generating short pulses with high peak power. Energy-storage capacity of He–Cd lasers is too low for practical Q switching, but the 5-GHz linewidth is large enough to generate transform-limited modelocked pulses shorter than 1 ns. Modelocked pulses could be used for harmonic generation, although the 442-nm blue line is close to the short-wavelength limit for practical frequency doubling in crystals (the 325-nm wavelength is much too short). However, He–Cd lasers are rarely used in harmonic generation because their low average power makes harmonic-generation efficiency much lower than for higher-power lasers.

Operating Requirements

Input Power. During normal operation, a He–Cd laser emitting a few milliwatts in the blue draws about 165 W of electrical power, while one emitting 50 mW in the blue will require about 350 to 560 W. Similar input powers will produce roughly one-half as much light on the 325-nm ultraviolet line. Heating of the cadmium metal during tube warm-up to produce adequate metal vapor pressure may require an added 10 to 50 W of electrical power. Most models operate from a 115-V ac line, but some higher-power models may have 220-V operation as an option or requirement.

Cooling. Electrical input to a He–Cd laser is typically a few watts per centimeter of tube length, leading to heat-dissipation needs of 100 W and up from the tube. Convection air cooling is adequate for smaller tubes; higher-power models and coaxial tubes need forced-air cooling.

Consumables. He–Cd lasers are built as sealed units and require no consumables per se.

Required Accessories. Normally He–Cd lasers require no special accessories other than those needed for a specific application, such as an external modulator if the beam is to be modulated.

Operating Conditions and Temperature. Specified operating temperatures for He–Cd lasers are 10 to 30 or 40°C, with storage temperatures specified at 0 to 50°C, or in some cases −20 to 80°C. Specified maximum humidity during operation is 90 to 100 percent.

All He–Cd lasers must be warmed up from a cold start. About 5 minutes is needed to heat the cadmium metal to produce enough vapor for operation at a reasonable fraction of normal laser output, as shown in Fig. 9.4. Up to half an hour may be needed to fully stabilize operation. Such delays are avoided in some models by a standby mode, which keeps the tube heated and cadmium pressure up, without generating a beam, cutting warm-up time to a couple of minutes.

Mechanical Considerations. Standard commercial He–Cd lasers include a laser head and separate power supply. The laser heads are 45 to 130 cm long, with round or rectangular cross section that measures 10 to 15 cm in most models. Low-power heads weigh a few kilograms, while those with output in the 50-mW range weigh up to about

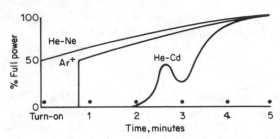

Figure 9.4 Types of power variations seen in He–Cd lasers during initial warm-up, as compared to those of He–Ne and argon ion lasers. (*Courtesy of Omnichrome.*)

16 kilograms (kg). The highest-power lasers, which are not in common use, may have heads that are heavier and larger.

Most power supplies are boxes 10 to 15 cm high, 30 to 50 cm across, and about as deep as wide. Moderate-power versions weigh 10 to 20 kg.

Safety. The main hazards posed by the 442-nm blue He–Cd line are to the retina of the eye. The eye appears to be more vulnerable to retinal damage from long-term exposure to the He–Cd line than to exposure to longer visible wavelengths, with the threshold perhaps an order of magnitude lower than for the 632.8-nm He–Ne line (Sliney and Wolbarsht, 1980, p. 136). Thus users should be particularly cautious, even though He–Cd power levels may appear low.

The 325-nm line lies in the ultraviolet-A region and presents different hazards. Some retinal damage has been produced by modelocked pulses at that wavelength (Sliney and Wolbarsht, 1980, p. 109), but standard measurements indicate that under 1 percent of the light at that wavelength entering the eye reaches the retina. The lens absorbs strongly at that wavelength, and there is some indication that long-term exposure to ultraviolet-A radiation could contribute to cataracts. Neither He–Cd line presents serious hazards to the skin or other parts of the body except at power densities higher than normally produced by He–Cd lasers.

He–Cd lasers which simultaneously emit red, green, and blue lines pose a special safety hazard because goggles are not readily available to block all three wavelengths at once. Users may be able to find goggles which block two wavelengths at once (e.g., the blue and green wavelengths with goggles sold for use with blue-green argon lasers). However, blocking all three is more difficult, and runs the risk of blocking so much of the visible spectrum that vision is obstructed to a dangerous degree.

The high voltages (one to several kilovolts) applied across the He–Cd discharge tube are a potential shock hazard, but the direct-current levels are low enough that they do not pose a severe electrocution hazard (although they can give one a nasty jolt). The high tube temperature needed to vaporize the metal presents a potential burn hazard, but in commercial lasers the hot tube is packaged safely out of reach.

Reliability and Maintenance

Lifetime. Much work has gone into extending He–Cd laser tube life (Dowley, 1982, 1987). Commercial blue and ultraviolet lasers are rated for 6000-hour operating lives. Blue and ultraviolet lasers often show essentially no degradation in output power after 7000 hours of operation. Multiline white-light He–Cd lasers have a rated life of 3000 hours. Manufacturers define tube failure as the point when it no longer meets published specifications, or when power drops below a certain fraction of the specified level. This may not be the point at which the customer can no longer use the tube.

Other components normally outlive the tube, so laser lifetime can be extended by simply replacing the tube.

Maintenance and Adjustments. Some models may require periodic adjustment and/or cleaning of optics in the laser cavity external to the tube, while other types require no active maintenance during tube lifetime.

Mechanical Durability. Like other glass tubes, He–Cd tubes are fragile, but they can be packaged to minimize that problem. One company specifies that its packaged lasers can survive a shock of 20 g's.

Failure Modes and Causes. Several factors contribute to degradation of He–Cd tubes:

- Discharge contamination
- Growth of cadmium deposits on the beam path, bore, or optical surfaces
- Depletion of helium gas by cathode sputtering, burial under condensing cadmium metal, and wall diffusion
- Depletion of the cadmium reservoir

Tube designers have found various mechanisms to reduce these problems and extend tube lifetime, but they remain limits on operation.

Possible Repairs. Because it is the tube that fails, the simplest way to fix a He–Cd laser is to replace the tube. Replacement tubes typically cost one-third to one-half the price of a complete new laser, and usually are installed at the factory.

Commercial Devices

Standard Configurations. The standard configuration for He–Cd lasers is a laser head with separate power supply. The laser usually is factory-built to emit in the blue or ultraviolet, but if the optics are external to the tube, they can be changed. Major manufacturers offer two parallel product lines which differ primarily in packaging—one intended for stand-alone use in the laboratory, the other for incorporation into larger laser-based systems by equipment manufacturers.

Options. He–Cd lasers are sold mostly as standard models, with users selecting a particular model to obtain the desired power level, wavelength, polarization, and cooling method. A few models have interchangeable optics for blue or ultraviolet wavelengths.

Pricing. Prices of both blue and ultraviolet He–Cd lasers start at a few thousand dollars and reach about $15,000 for powers of about 25 mW in the ultraviolet or 40 mW in the blue. White-light versions are priced at $24,000 to $31,000. Replacement tubes cost about half as much as a new laser for low-power models and somewhat less for higher-power lasers.

Suppliers. Over a thousand He–Cd lasers are sold around the world each year. There are three major suppliers of blue and ultraviolet lasers, two in the United States and one in Japan. A second Japanese company is the only one offering what it calls "red-green-blue" He–Cd lasers, which so far have not found wide acceptance. Despite the small number of manufacturers, there is active competition, both among companies manufacturing He–Cd lasers, and among manufacturers of He–Cd, argon, and He–Ne lasers.

Applications

Uses of helium-cadmium lasers fall into three major categories: information-handling systems, inspection and measurement, and research.

The largest numbers of He–Cd lasers are used in information handling because they offer both short wavelength and moderate output power. The blue and ultraviolet He–Cd lines have two important advantages over red lines of He–Ne lasers: they can focus on smaller spots than can the longer-wavelength lasers, and many light-sensitive materials have a much stronger response at shorter wavelengths. These combine to allow faster recording at higher densities. The blue and ultraviolet wavelengths also are somewhat shorter than the blue-green lines of argon ion lasers, an advantage that can be significant for some applications.

Printing has been another major information-handling application. The laser scans a photoconductive drum, writing images that are transferred to paper by a toner process similar to that used in copiers. He–Cd lasers are used mainly in high-speed systems which can print up to 100 pages per minute. (Semiconductor lasers are standard in the slower laser printers used with personal computers.) He–Cd lasers also are used in typesetting systems, where the tightly focused laser spot writes characters on light-sensitive paper with resolution so high that no spots are visible. He–Cd lasers also are used in systems which separate colors for full-color printing; they create the halftone dot patterns used to generate the images printed in different colors.

Optical recording is another important application. He–Cd lasers can write master disks for videodisks, audio compact disks, and compact-disk read-only memories (CD ROMs). Those masters are replicated by sophisticated pressing techniques. He–Cd lasers also can generate patterns for other applications.

He–Cd laser wavelengths are valuable in some measurement and inspection applications. He–Cd lasers can inspect printed-circuit boards and integrated circuits. They are used in scanning laser microscopy and for stepper alignment in integrated-circuit manufacture. They inspect U.S. paper currency because the special paper used has a unique fluorescence response at the He–Cd line. Other measurement applications include ultraviolet fluorescence studies, stereolithograpy, and flow cytometry.

Research applications cover a broad range, with many involving fluorescence induced by the 325-nm ultraviolet line. The 441.6-nm blue line can be used for short-wavelength holography.

Bibliography

William B. Bridges: "Ionized gas lasers," in Marvin J. Weber (ed.), *CRC Handbook of Laser Science & Technology*, vol. 2, *Gas Lasers*, CRC Press, Boca Raton, Fla., 1982, pp. 17–269 (review article; covers He–Cd as well as other ion lasers).

Mark W. Dowley: "Reliability and commercial lasers," *Applied Optics* 21:1971–1975, 1982 (short review).

Mark W. Dowley: "A modern helium cadmium laser—the evolution of a design," *Proceedings of the SPIE*, vol. 741, *Design of Optical Systems Incorporating Low Power Lasers*, SPIE, Bellingham, Wash., 1987, pp. 31–38 (short review).

K. Fujii, T. Takahashi, and Y. Asami: "Hollow-cathode type CW white light laser," *IEEE Journal of Quantum Electronics QE-11*(3):111–114, 1975 (research paper).

M. Otaka et al.: "He–Cd$^+$ whitelight laser by a novel tube structure." *IEEE Journal of Quantum Electronics QE-17*(3):414–417, 1981 (research paper).

Wakeo Sasaki, Hideshi Ueda, and Tatehisa Ohta: "Spectral quality of He–Zn II and He–Cd II hollow-cathode metal-vapor lasers in the magnetic fields." *IEEE Journal of Quantum Electronics QE-19*(8):1259–1269, 1983 (research paper).

Wolfgang K. Schuebel: "Transverse discharge slotted hollow-cathode laser." *IEEE Journal of Quantum Electronics QE-6*(9):574–575, 1970 (research paper).

William T. Silfvast, G. R. Fowles, and B. D. Hopkins: "Laser action in singly ionized Ge. Sn. Pb. In. Cd. and Zn." *Applied Physics Letters 8*:318–319, 1966 (first report of He–Cd).

David Sliney and Myron Wolbarsht: *Safety with Lasers and Other Optical Sources.* Plenum, New York, 1980 (safety reference).

S. C. Wang and R. D. Reid: "Parametric performance of a hollow-cathode white laser," paper WK3 presented at Conference on Lasers and Electro-Optics. Washington, D.C., June 1981 (research report).

Carbon Dioxide
Lasers

The carbon dioxide laser is one of the most versatile types on the market today. It emits infrared radiation between 9 and 11 micrometers (μm), either at a single line selected by the user or on the strongest lines in untuned cavities. It can produce continuous output powers ranging from well under 1 watt (W) for scientific applications to many kilowatts for materials working. It can generate pulses from the nanosecond to millisecond regimes. Custom-made CO_2 lasers have produced continuous beams of hundreds of kilowatts for military laser weapon research (Hecht, 1984) or nanosecond-long pulses of 40 kilojoules (kJ) for research in laser-induced nuclear fusion (Los Alamos National Laboratory, 1982).

This versatility comes from the fact that there are several distinct types of carbon dioxide lasers. While they share the same active medium, they have important differences in internal structure and, more important to the user, in functional characteristics. In theory, the structural variations could range over a nearly continuous spectrum, but manufacturers have settled on a few standard configurations which meet most user needs. Thus users see several distinct types, such as waveguide, low-power sealed-tube, high-power flowing-gas, and pulsed transversely excited CO_2 lasers. This chapter covers all these major types.

Internal Workings

The active medium in a CO_2 laser is a mixture of carbon dioxide, nitrogen, and (generally) helium. Each gas plays a distinct role.

Carbon dioxide is the light emitter. The CO_2 molecules are first excited so they vibrate in an asymmetrical stretching mode. The molecules then lose part of the excitation energy by dropping to one of two other, lower-energy vibrational states as shown in Fig. 10.1. These two decay paths are the two principal laser transitions: a shift to a symmetrical stretching mode accompanied by emission of a 10.6-μm photon, or a shift to a bending mode accompanied by emission of a 9.6-μm photon. Superposition of changes in the molecules' rotational states on the vibrational transitions yields large families of laser lines surrounding the 9.6- and 10.6-μm transitions. Once the molecules have emitted their laser photons, they continue to drop down the energy-level ladder until they reach the ground state.

The nitrogen molecules help to excite CO_2 to the upper laser level. The lowest vibrational state of N_2 is only 18 inverse centimeters lower in energy than the asymmetric stretching mode of CO_2, a difference that is less than one-tenth the mean thermal energy of room-temperature molecules, and hence insignificant from a practical standpoint. This lets the nitrogen molecules absorb energy and transfer it to the carbon dioxide molecules, thereby raising them to the upper laser level.

Carbon dioxide molecules can also reach the upper laser level in other ways. They can directly absorb energy from electrons inserted into the gas in a discharge or electron beam. An alternative way of producing the population inversion needed for laser operation is to rapidly expand hot, high-pressure laser gas into a cool near-vacuum; this is the basic principle behind the gas-dynamic carbon dioxide laser. In practice, the presence of N_2 significantly enhances laser operation, and that gas is almost always present in CO_2 lasers.

Helium plays a dual role. It serves as a buffer gas to aid in heat transfer and helps the CO_2 molecules drop from the lower laser levels to the ground state, thus maintaining the population inversion needed for laser operation.

The optimum composition and pressure for the gas in a CO_2 laser vary widely with the laser design. In a typical flowing-gas CO_2 laser,

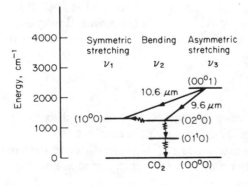

Figure 10.1 Energy-level structure in the CO_2 laser, showing the relevant vibrational modes of the CO_2 molecule. Numbers in parentheses indicate the excitation levels of the symmetric stretching, bending, and asymmetric stretching vibrational modes, respectively, of the molecule.

the total pressure might be around 15 torr [2000 pascals (Pa)], with 10 percent of the gas CO_2, 10 percent N_2, and the balance helium. In general, the concentrations of nitrogen and carbon dioxide are comparable, but much lower than that of helium. Low pressures are needed for continuous operation, but pulsed CO_2 lasers can be operated at pressures well above 1 atm.

The discharge breaks down some CO_2 into carbon monoxide and oxygen. In some cases, other gases may be added to the laser mixture to improve performance or prevent degradation of tube components by free oxygen. For example, hydrogen or water can be added to promote regeneration of CO_2 during operation of a sealed-tube laser, and sometimes carbon monoxide is added for similar reasons. It is even possible to operate pulsed lasers with a 50:50 mixture of air and carbon dioxide, albeit at reduced output power. However, too much water vapor can absorb enough infrared light to reduce laser gain.

Performance of the CO_2 laser depends on temperature because of its energy-level structure. As can be seen in Fig. 10.1, the lower laser levels are only about 1200 inverse centimeters above the ground state. As the gas is heated above room temperature, the populations of these levels increases. To keep gain at high levels, the temperature must be kept below about 150°C.

Internal Structure

The classification of carbon dioxide lasers into types is based on their internal structure. Several key parameters are involved, including gas pressure, gas flow, type of laser cavity, and excitation method. Many different combinations have been demonstrated in the laboratory, but only a few are in practical use.

Types of CO_2 Lasers

Sealed-tube lasers.   The classic sealed-tube CO_2 laser is a glass tube filled with CO_2, He, and N_2, with mirrors forming a resonant cavity, as shown in Fig. 10.2. Electrodes are placed near the two ends of the tube. It sounds simple to operate: just fill the tube with the proper gas mixture, seal it, and apply a high voltage to the electrodes to pass a discharge through the gas. However, life is not that simple; the electric discharge in the tube breaks down the CO_2, freeing oxygen which can corrode internal components. Storing CO_2 in a gas reservoir outside of the main discharge path does not solve the problem. Left by itself, a sealed CO_2 laser with an ordinary gas mixture would stop operating within a few minutes.

One solution is to add hydrogen or water to the gas mixture so that it can regenerate CO_2 by reaching with the carbon monoxide produced

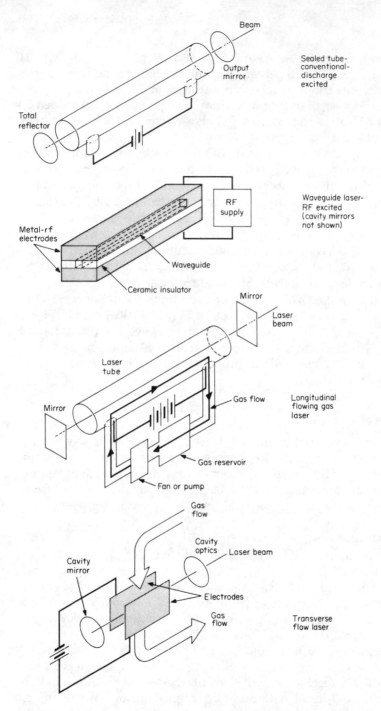

Figure 10.2   Major types of CO_2 lasers are (top to bottom) sealed-tube conventional laser with longitudinal dc discharge excitation, waveguide laser with radio-frequency excitation, longitudinal flowing-gas laser, and axial-flow laser. Slow and fast axial-flow lasers differ in gas speed through the tube; in the latter the laser runs cooler and more efficiently and thus can generate more power.

by the discharge. (However, as noted above, too much extra gas can reduce gain.) Alternatively, a 300°C nickel cathode can catalyze the recombination reaction. Such measures can be used to produce sealed CO_2 lasers which can operate for up to several thousand hours before their output seriously degrades. Tubes must then be refilled and cleaned or replaced.

In traditional sealed CO_2 lasers, a direct-current electric discharge passes between electrodes near the ends of the tube, transferring energy to the laser gas. The maximum output power possible with this longitudinal discharge is about 50 W per meter of cavity length, and maximum continuous-wave output is about 100 W. This approach has been most common for CO_2 lasers generating 25 to 100 W continuous-wave, or comparable average powers in pulsed operation.

A newer approach first used for waveguide CO_2 lasers (described below) is application of a radio-frequency discharge transverse to the tube axis. This design does not require standard high-voltage electrodes and offers some other advantages, including the ability to electronically control output at rates to 10 kilohertz (kHz), lower operating voltage, and potentially lower tube cost. On the other hand, radio-frequency power supplies are more complex and less efficient than dc supplies. Radio-frequency excitation has been growing in popularity for conventional sealed-tube CO_2 lasers. It can generate somewhat more power because it can excite a broader area than a dc discharge can, but it also works well at low powers.

All sealed-tube CO_2 lasers, including the waveguide types described below, are limited in output by the difficulty in removing waste heat if the laser gas is not flowing through the tube.

Waveguide lasers.   If the inner diameter of a sealed CO_2 laser is shrunk to a few millimeters and the tube is constructed in the form of a waveguide, the result is a "waveguide" laser similar to that shown in Fig. 10.2. The waveguide design limits diffraction losses that would otherwise impair operation of a narrow-tube laser. The tube normally is sealed, with a gas reservoir separate from the waveguide itself. Waveguide lasers may be excited by dc discharges or intense radio-frequency fields.

Waveguides may be made of metal, dielectric, or combinations of the two. For example, blocks of metal may form the top and bottom of the waveguide, while a dielectric ceramic forms the vertical walls. In such a laser, the radio-frequency field is applied between the two metal walls, while the ceramic serves as an insulator.

The waveguide laser is very attractive for powers on the low end of the CO_2 range, particularly under about 50 W. It provides a good-quality beam, can operate continuously or pulsed, and can be readily tuned to many discrete lines in the CO_2 spectrum. One obvious advantage is its compact size, fitting into a package similar in size of a helium-neon laser, but able to generate watts instead of milliwatts. (However, most of those small models require water or forced-air cooling.) The need for expensive ceramics keeps costs far above those for He–Ne tubes; one manufacturer reports that beryllium oxide waveguides cost about $1000 each when purchased in quantity.

Developers are working on folded waveguides and arrays in an effort to increase output power from compact waveguide lasers (Newman and Hart, 1987). One approach is to form a V-, Z-, or X-shaped resonator with waveguides connecting mirrors that bend the cavity. Another is to fabricate an array of parallel waveguides which will lase independently, and then have beams combined externally.

Longitudinal (or axial) slow-flowing-gas lasers. An obvious way to solve the problems of the sealed CO_2 laser is to flow the gas through the laser tube, as shown in Fig. 10.2. The oldest approach is to pass the gas through the length of the laser tube longitudinally or along the tube's axis; hence the name. Generally the electric discharge that excites the gas is also applied along the tube's axis. The gas pressure is low, as the name implies the gas flows slowly, and gas consumption can be reduced further by recycling options on many lasers.

Slow axial-flow CO_2 lasers produce continuous-wave output that is roughly linearly proportional to the tube length. Typical output limits are 80 W/m. The laser beam can be folded or bent with mirrors through multiple tube segments, avoiding the need for unwieldy packages, and the design is simple enough that it remains common for CO_2 lasers emitting average or continuous power of less than about 500 W. However, powers much higher than a kilowatt or two would require tubes that would be impractically long.

Cooling is by conduction of heat energy through the laser gas to the walls of the tube, where it can be removed by flowing water or other coolant around the tube. (This heat transfer is attributable to the large concentration of helium in the laser gas.) One major advantage of this approach is its operational simplicity, which enhances its reliability. It uses round laser tubes, which produce beams of good quality, important for many industrial applications. However, the round tube means that the gas is hottest in the center of the active region.

One recent advance is the arrangement of many thin tubes in parallel, in either a flat or annular array, as shown in Fig. 10.3. Generally the tubes are radio-frequency-excited. The flat or annular surfaces are cooled to keep the laser gas temperature low enough for efficient operation. The maximum power per channel is 50 W/m, which can be multiplied by the number of parallel channels, although at some sacrifice in beam quality.

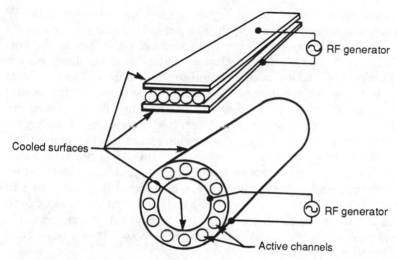

Figure 10.3 Slow axial-flow tubes can generate higher powers if several are arranged in parallel, either in a ring or on a flat surface, and cooled from both sides. (*Courtesy of Coherent Inc.*)

Fast axial-flow lasers. The efficiency of axial flow lasers can be increased dramatically by using a pump or turbine to move the gas rapidly through the discharge area. This design allows short resonators to produce relatively high powers; 800 W/m is a typical value of power per unit length. Excitation usually is with a longitudinal discharge, as in slow axial-flow lasers, but some fast axial-flow lasers are powered by radio-frequency discharges.

The principal advantage of the fast flow is that it cools the laser gas better. Slow-flow lasers rely on heat conduction to the walls of the discharge tube to remove the excess heat. (This is why so much helium is needed.) In fast-flow lasers, the gas moves so quickly through the discharge area that it does not become hot enough to become inefficient and the highest temperatures are at the downstream end of the tube. After leaving the discharge region, the gas is cooled in a heat exchanger. Most of the gas recirculates through the laser, but a small amount is replaced with fresh gas to maintain peak performance.

A typical design might include four to eight parallel flow tubes, optically in series, for each kilowatt of output. With its short resonator and small "footprint" of required floor space, the fast axial-flow laser has become the most common industrial CO_2 laser design for powers from about 500 W to 5 kW.

The fast-flow design increases system complexity and uses gas pumps which have suffered some reliability problems. It also reduces mode quality somewhat.

Transverse-flow lasers. Even higher powers, on the order of 10 kW per meter of active medium length, are possible if the gas flows in a direction perpendicular to the laser cavity axis, as shown at the bottom of Fig. 10.2. The electric discharge that powers the laser also is applied transverse to the laser axis and to the gas flow. The gas flows across a much wider region and does not need to travel as far as in a fast axial-flow laser, so it does not move as rapidly. This eases pumping requirements. The gas usually is recycled by passing it through a system which regenerates CO_2 and adds some fresh gas to the mixture.

The short cavity length of transverse-flow lasers is a mixed blessing. It is short enough to cause beam quality problems. Both beam mode structure and beam symmetry are considerably poorer than in fast or slow axial-flow lasers. Nonetheless, the design is the most common one for lasers in the 5-kW class to the highest power available from commercial lasers, 25 kW. A few higher-power lasers have been built for military research.

Gas-dynamic lasers. Transverse flow is also used in another type of high-power CO_2 laser, the gas-dynamic laser, shown in Fig. 10.4. In the gas-dynamic laser, the excitation energy comes from heat applied to the laser gas, which is initially at a pressure of several atmospheres. (Both the heat and some components of the laser gas may come from combustion of hydrocarbon fuels.) The hot gas is then expanded through a nozzle into a low-pressure chamber. The rapid cooling of the fast-moving gas produces a population inversion—i.e., more CO_2 molecules in the upper laser level than in the lower one. A laser beam is extracted from the gas by placing a pair of mirrors on opposite sides of the expansion chamber.

At the end of the 1960s, the gas-dynamic laser was an important breakthrough that made it possible for the first time to reach power levels of 100 kW or more. Such powers are required only for military applications, and because of their complexity and high power, gas-

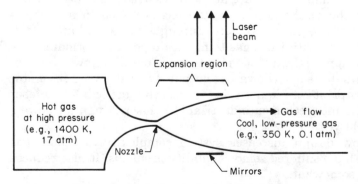

Figure 10.4 Basic structure of a gas-dynamic laser.

dynamic lasers have never entered the commercial world. It now looks as if they are impractical for military field use as well.

Transversely excited atmospheric (pressure). Discharge instabilities make it impossible to operate a CO_2 laser continuous-wave at gas pressures above about 100 torr (13.3 MPa). However, pulses lasting tens of nanoseconds to microseconds can be produced by passing electric pulses through the gas in a direction transverse to the laser cavity axis. Such lasers are called *transversely excited atmospheric (pressure)* (TEA) lasers because they operate near or above atmospheric pressure. They are available in versions with sealed tubes, slow or fast axial flow, or transverse gas flow, depending on power levels.

The prime attractions of TEA lasers are their generation of short, intense pulses and the extraction of high power per unit volume of laser gas. High-pressure operation also broadens emission lines, permitting the use of modelocking techniques to generate pulses lasting about 1 ns. At pressures around 10 atm, the broadening is sufficient to allow near-continuous tuning over most of the CO_2 wavelength range.

The basic TEA design is adaptable, and commercial models range from tabletop versions to dumpster-sized behemoths. The same basic design can be used with other laser gases, and some companies have offered "multigas" lasers adaptable for use as excimer, carbon monoxide, and chemical lasers by switching optics and gases. (However, the tube may require purging when switching between some pairs of gas mixtures.) Some low-power sealed lasers may be filled with special gas mixtures of controlled isotopic composition, giving more precise control over operating wavelengths because specific CO_2 vibrational lines shift slightly depending on isotopes in the gas mix.

Optics. Low- and moderate-power carbon dioxide lasers typically have a pair of laser mirrors at the ends of the cavity, one totally reflecting and one partly transparent. If the cavity is long and "folded," additional totally reflective mirrors direct the beam between segments of the cavity. The power levels are sufficiently high even in small CO_2 lasers that heat handling is an important criterion for mirror materials. Common choices include silicon with high-reflectivity coatings and metals such as molybdenum and copper. Because CO_2 gain is comparatively high, the fraction of the beam transmitted is higher than in low-gain lasers such as He–Ne.

One common variant on this design is to cut one of a pair of metal mirrors to allow some light to escape from the cavity to form the laser beam. Another variation is use of Brewster angle windows on the ends of the discharge tube, which polarize the beam and allow the use of external optics to select wavelength. In some cases, the optics may allow the laser tube to amplify a signal from an external laser.

The transmissive materials used as windows for the 10-μm output of carbon dioxide lasers are rather different from ordinary optical glass. Two common choices, gallium arsenide and germanium, are not transparent to visible light. Zinc selenide is transparent at visible wavelengths but has a strong orange color. Some of the materials which are most transparent at 10 μm are hygroscopic alkali halides such as sodium chloride, which absorb atmospheric moisture unless they are protected. Some infrared window materials are toxic.

Optics become a problem at high powers, where low absorption and high thermal capacity are critical. Copper mirrors have better thermal conductivity and thus can withstand higher powers than molybdenum, but copper is a softer metal, harder to polish, and more costly. At high power densities, mirrors may be actively cooled by flowing water or other coolant through the substrate.

Transmissive optics are more of a problem at very high powers, so engineers have developed "aerodynamic" windows. These resemble holes in the laser cavity, but they actually isolate the interior of the cavity by flowing gas past the hole so rapidly that air cannot enter and contaminate the laser gas. Aerodynamic windows are used in some commercial lasers emitting more than 5 kW continuously.

Cavity Structure and Length. The basic types of laser cavity structure have been described above. The cavities of some slow longitudinal-flow lasers can be very long, containing many separate segments connected optically in a folded configuration that lets the laser package be much smaller than the tube length. Cavity lengths for other types typically range from about 0.3 to 2 m. Folded-cavity waveguide lasers are in development.

Stable resonators are standard in most commercial CO_2 lasers. However, unstable resonators are used in the highest-power continuous-wave models with transverse gas flow.

Inherent Trade-offs

In addition to the usual trade-offs encountered with most types of laser, a few are specific to CO_2 lasers:

- Different CO_2 designs operate best in different power ranges, but the trade-offs can become complex if two or more types can produce the desired characteristics. Beam quality, laser reliability, cost, and size all can be factors.

- High powers trade off directly against single-line operation and beam quality. The highest powers are from multimode lasers designed to operate throughout the CO_2 emission spectrum. Usually their emission is dominated by the strongest lines. Only a few lines can deliver the highest possible power levels.

- Radio-frequency and dc excitation offer different advantages. Radio-frequency power supplies are more costly and complex and less efficient, but radio-frequency excitation allows faster control over changes in output and permits the use of simpler, less costly tubes. The two approaches also have different geometries, which may be attractive in different situations.

- In TEA lasers, generally the faster the repetition rate, the lower the energy and peak power per pulse, when the laser is operated at its peak ratings. The average power usually decreases above a certain repetition rate.

- In many types of CO_2 laser, cavity length increases with output power.

Beam Characteristics

Wavelength and Output Power. The nominal operating wavelength of the CO_2 laser is usually written as 10 μm, 10.6 μm, or 9 to 11 μm. The actual emission spectrum is complex. The CO_2 laser has two principal vibrational transitions, at 9.6 and 10.6 μm. Many closely spaced rotational transitions are superimposed on the vibrational transitions, producing a total of roughly 100 possible distinct emission lines, as shown in Fig. 10.5.

A diffraction grating or other tuning element can be put into the laser cavity to select a single line in the CO_2 spectrum, or the laser can operate untuned, emitting a higher total power on multiple lines. Because laser amplification builds up the power on the strongest emission line, multiline lasers are often said to operate at 10.6 μm, the strongest transition, although other wavelengths are also present.

Single-line output is not available or required on lasers designed to provide high power for materials working, but it is available on lower-

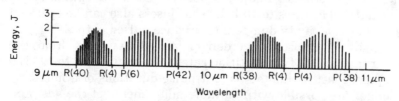

Figure 10.5 Spectrum of wavelengths produced by a transversely excited atmospheric (pressure) CO_2 laser, showing pulse energy emitted at each line in single-line operation by a commercial laser. R and P denote rotational sublevels for each of the two main vibrational transitions of the carbon dioxide laser. (*Courtesy of Lumonics Inc.*)

power continuous-wave lasers and on many TEA models. Single-wavelength output on the strongest CO_2 line is generally about half the multiline output. There is some additional loss of power in tuning to other lines, but not as much as might be expected. Depending on the laser model, some 60 to 80 lines can emit at least half as much power as the strongest wavelength in single-line operation. Increasing the gas pressure to about 10 atm broadens the individual lines enough to permit continuous tuning of TEA lasers across much of the CO_2 emission range. Because the laser transitions are vibrational, their wavelength can be changed by substituting different isotopes into the carbon dioxide molecule.

Maximum output (normally measured in multiline operation) depends on the type of CO_2 laser. Commercial waveguide lasers now produce up to about 30 W continuous-wave; higher powers have been obtained from folded waveguide lasers in the laboratory, but the effects of high power density on cavity optics may limit power from a single waveguide. Conventional sealed tube CO_2 lasers are available with powers of up to 200 W, although powers of 100 W or less are more typical. Slow axial flow lasers typically generate 50 to 500 W, with a few models at higher or lower powers. Commercial fast axial-flow lasers generate 500 W to 5 kW. Transverse flow lasers are available with powers ranging from 3 to 25 kW; higher powers can be produced, but there is no demand for it. Gas-dynamic lasers also operate in the multikilowatt range, but they are not offered commercially.

TEA lasers have average powers from under 1 W to over 100 W, which depend on repetition rate and pulse energy, as described below.

Temporal Characteristics of Output.   Except for TEA types, most CO_2 lasers normally produce a continuous beam. However, nominally continuous-wave lasers can be pulsed electrically to produce a power "spike" lasting 0.1 to 1 ms with a peak power 5 to 10 times the continuous-wave level. Such pulses are useful for many materials-working applications because they can break down the surface and start the cutting process. Radio-frequency-excited waveguide lasers can be modulated at rates to 10 kHz. CO_2 lasers also can be operated with Q switches or cavity dumpers to produce pulsed output.

Commercial TEA lasers can deliver multiline pulses from the millijoule range to 500 J at repetition rates from single-shot to 300 Hz. Shortest pulses last around 40 ns, but pulse durations of a microsecond or longer are possible with suitable adjustments of the gas mixture and electric discharge. There are typically two elements to the pulse, a short peak arising from direct electrical excitation of the gas, and a longer-duration falloff (of lower power) as a result of transfer of energy from excited N_2 to CO_2 molecules which then emit light.

Because of the high pressure of the gas in TEA lasers, the gain bandwidth is about 4 gigahertz (GHz), enough to allow the use of modelocking techniques to generate subnanosecond pulses. (The theoretical limit on modelocked pulse duration is that it should be inversely proportional to gain bandwidth; thus 4-GHz bandwidth theoretically permits generation of quarter-nanosecond pulses.) A certain jitter in pulse timing (generally around 20 to 30 ns) and a small misfire rate (generally well under 1 percent) are inherent in the electronics of TEA lasers. Pulses can be triggered remotely or internally in most models.

Efficiency. The overall efficiency of carbon dioxide lasers typically runs between 5 and 20 percent, not good when compared to other types of electrical equipment, but higher than most other lasers. Efficiencies are lower if operation is on a single line, if the electrical power supply is inefficient, or if the optics are particularly inefficient in extracting energy from the laser cavity. Users concerned with efficiency should make sure that it is defined explicitly—wall-plug efficiency is always lower than efficiency in converting energy deposited in the active medium into laser emission.

Spectral Bandwidth and Frequency Stability. The spectral bandwidth is irrelevant to people using carbon dioxide lasers as heat sources. With multiline output, the bandwidth is large—although limited to the 9- to 11-μm emission of the laser—and unspecified in manufacturers' literature. The natural bandwidth of a single emission line is on the order of 100 megahertz (MHz) at low gas pressures, but it broadens as pressure increases until eventually the lines overlap. At such high pressures, the laser operates only in pulsed mode. Sophisticated optical cavities can reduce the spectral bandwidth well below the natural linewidth of a single emission line.

Lasers tunable to individual transitions in the CO_2 emission band normally can operate stably on that transition at low pressures. Accessories can boost frequency stability to between one part in a million and one part in a billion for continuous-output lasers. Similar stabilities can be achieved with pulsed lasers operating at higher pressures if suitable optics are used.

Amplitude Noise and Pulse-to-Pulse Variations. Power levels of both pulsed and continuous-wave carbon dioxide lasers are subject to fluctuations which can be reduced by stabilization techniques. Generally materials-working lasers have fluctuations in continuous output limited to a few percent, with some models including active stabilization to provide even tighter control. Unstabilized waveguide lasers can have amplitude variations of ± 10 percent over a few hours and ± 4

percent on a scale of minutes, but better results can be obtained with stabilization. Pulse-to-pulse variations for repetitively pulsed TEA lasers are often specified at within a few percent, but variations could be larger for lasers producing high-energy pulses at low repetition rates. Grating-tuned lasers can be made more stable than multiline lasers; for example, specifications of one model claim power can stay constant to ± 0.25 percent over several hours.

Beam Polarization, Quality, and Modes. The design of the laser cavity and the resonator optics determine the modes in which a laser can oscillate, and these modes, in turn, play a major role in determining beam quality. There are four basic transverse modes of emission common in carbon dioxide lasers:

- *Multimode emission*, in which the laser cavity simultaneously supports many different modes of oscillation. This extracts the most output power from a laser cavity, but the beam tends to be large in diameter and rapidly diverging, with some unevenness in quality.

- *TEM_{00} emission*, in which the laser cavity allows oscillation only in its fundamental (TEM_{00}) mode—producing an intensity pattern with a peak in the center and a decline in intensity to the edges. The lower intensity at the edges limits the amount of energy that can be extracted from the laser cavity, but the resulting beam is of better quality than a multimode beam. Although there can be exceptions, a small-diameter laser operating in TEM_{00} mode typically will emit a quarter to a half as much power as it would in a multimode beam, and the beam diameter and divergence will be reduced proportionately.

- *Unstable-resonator emission*, in which the cavity optics produce a beam with a doughnutlike cross section. Despite the name, lasers with such cavities can operate stably. They are attractive for CO_2 lasers because they offer good beam quality while extracting more energy from the active medium than is possible with TEM_{00} mode emission, particularly for a large-diameter laser. Output power with an unstable resonator typically is one-half to two-thirds that of a multimode beam. The unstable-resonator design also permits use of totally reflecting metal mirrors which can handle high levels of optical power.

- *Waveguide lasers* have a mode structure determined by the nature of the waveguide rather than by the resonator mirrors. The lowest-order EH_{11} waveguide mode is functionally the same as the TEM_{00} mode of nonwaveguide lasers once the beam emerges from the laser. The laser cavity's small diameter allows efficient extraction of energy from the laser gas.

If a continuous-wave CO_2 laser is tuned to emit at a single wavelength, normally it will emit in a single longitudinal mode. This occurs even without internal line-narrowing optics because the long laser wavelength leads to spacing of longitudinal modes that is broader than the gain bandwidth on low-pressure CO_2 lines. A conventional stable-resonator CO_2 laser emitting a single line produces a TEM_{00} beam; unstable-resonator and waveguide CO_2 lasers also can be tuned to a single line.

Multiline CO_2 lasers normally emit randomly polarized light unless their tube designs incorporate Brewster angle windows. The optics used for grating tuning impose linear polarization on the beam (although the angle of polarization may change as wavelength is changed), so single-line lasers emit linearly polarized beams.

Coherence Length. Coherence length generally does not matter for most CO_2 laser applications. The broad emission bandwidth of multiline lasers leads to a very short coherence length. Single-line CO_2 lasers have coherence lengths on the order of a meter, or longer if line-narrowing accessories are used.

Beam Diameter and Divergence. The shape and length of the laser cavity and the nature of the resonator optics determine beam diameter and divergence. Thus these quantities differ among CO_2 laser types. Typical ranges are

Sealed-tube conventional	1- to 7-mm diameter, 2- to 6-mrad divergence
Waveguide	1- to 2-mm diameter, 3- to 10-mrad divergence
Slow axial-flow gas	3- to 15-mm diameter, 1- to 4-mrad divergence
Fast axial-flow gas	8- to 25-mm diameter, 1.5- to 5-mrad divergence
Transverse flow	13- to 50-mm diameter, 1- to 3-mrad divergence
TEA	4- to 200-mm diameter, 0.5- to 10-mrad divergence

In general, beam diameter and divergence are smaller for single-mode beams than for multimode output, except for waveguide lasers. Some large-diameter laser tubes produce beams with oval cross sections.

Stability of Beam Direction. Pointing stability is unmentioned in most data sheets, but one company specifies stability of ±0.15 mrad for its 150- to 1200-W axial-flow lasers, which are designed for materials-working applications.

Suitability for Use with Laser Accessories. Carbon dioxide lasers are versatile enough to be used with a number of accessories although few applications require them. Their wavelength is enough longer than those of other major lasers that they require special components. Important accessories for CO_2 lasers include

- Q switches to produce microsecond or nanosecond pulses from continuous-wave beams.

- Shutters or pulse choppers to produce longer pulses from continuous-wave beams.

- Modelockers to produce nanosecond pulses from lasers which operate at pressures high enough (near atmospheric pressure) to produce the required broadening of laser lines.

- Harmonic generators can produce integral multiples of the laser frequency (or equivalently reduce the wavelength by an integral factor) but have found little practical use.

- Optically pumped far-infrared lasers, in which the energy from a carbon dioxide laser is used to produce laser emission at a longer wavelength in the far-infrared or submillimeter region as described in Chap. 15.

Operating Requirements

Input Power. The input power for CO_2 lasers, like the output power, ranges over a few orders of magnitude. The smallest CO_2 lasers can plug into an ordinary wall socket; the largest consume prodigious amounts of power. dc discharge-driven lasers require higher continuous or pulsed voltages than do radio-frequency-excited lasers. Typical requirements depend on the type.

Waveguide lasers powered by a dc discharge generally draw 1 to 3 A from a 110-V ac source, or comparable power levels from sources drawing other voltages. That energy goes to produce a current of a few milliamperes at about 15 kV for a discharge along the length of the waveguide.

Sealed CO_2 lasers powered by a discharge require somewhat higher input power to produce a comparable output. Most draw 3 to 15 A from a 110-V ac source. A voltage on the order of 10 kV may drive several milliamperes of current through the gas.

Radio-frequency excitation of sealed conventional or waveguide CO_2 lasers generally requires about twice as much power as discharge excitation. The extra power is lost in the power supply, which generates strong radio signals. Most operate at frequencies near 27 or 40 MHz, reserved by the Federal Communications Commission (FCC) for instrumentation use. The use of other frequencies requires shielding to avoid exceeding FCC permitted levels of radiation.

Continuous-wave flowing-gas CO_2 lasers require peak electrical input capacity in the range of 10 to 20 times their optical output, and may require extra startup power. This energy goes to power vacuum pumps, cooling equipment, and other accessories as well as the laser tube itself. Some models can operate from single-phase 110-V ac supplies, but large versions require 440-V, three-phase sources. One 1.2-kW laser, for example, requires 22-kV · A capacity from a 440-V supply, which reduces to 15 kV · A after start-up.

TEA lasers have fairly complex power requirements because of the nature of their pulsed operation. Typically, some energy in the form of electrons or ultraviolet photons is discharged into the laser gas slightly before the main pulse to make it possible to obtain higher output power. Energy for the main pulse typically comes from a storage capacitor which accumulates a charge from a dc power supply operating at tens of kilovolts and tens of milliamperes. Some of the highest-power TEA lasers use more complex—and more expensive—pulse-generating electronics and may have overall voltages above 100 kV. Wall-current requirements for the charging power supply range from a couple of amperes at 110 V to 50 A of three-phase 220-V power.

Cooling. Forced-air cooling is used in some small CO_2 lasers. Higher-power lasers require water cooling, and the largest devices have sophisticated multicycle cooling systems. Typical water-flow requirements are 2 liters per minute (L/min) for a 150-W slow axial-flow laser and 20 to 40 L/min for a 1.5-kW fast axial-flow laser.

Consumables. The prime consumable material used with CO_2 lasers is the laser gas because the discharge breaks down CO_2 molecules. Even low-power models with sealed tubes require periodic gas replacement, on the order of every thousand to several thousand hours of operation. Higher-power lasers require continuous replenishment of the laser gas. Gas consumption is roughly proportional to power level for lasers of the same design. For example, one continuous-output 150-W laser each hour needs 40 L of helium, 4 L of CO_2, and 7 L of N_2, against hourly consumption of 425 L He, 54 L CO_2, and 127.5 L N_2 for a 1.2-kW laser. However, gas-regenerating catalysts can reduce this gas consumption by up to 90 percent.

Some TEA lasers need dry air for spark-gap switching. Note also that cooling water is needed for many CO_2 lasers.

Required Accessories. Carbon dioxide lasers are offered in configurations ranging from complete materials-working systems to "kits" which contain little beyond the laser cavity itself. While complete systems normally are almost ready to run after being hooked up to power, water, and gas supplies, the less-complete systems or kits may also need high-voltage supplies, cavity optics, vacuum pumps, and other equipment.

Operating Conditions and Temperature. Carbon dioxide lasers are designed to operate at room temperature and can function well in normally clean environments. Many models are intended for industrial use. Extremes of dirt and/or vibration in factories have caused problems in some cases, and such protective measures as air filters may be needed in particularly severe environments.

Mechanical Considerations. Waveguide CO_2 lasers are impressively compact devices, with heads as light as a few kilograms and power supplies not much heavier. The smallest heads are shorter than 40 cm, and have square cross sections only about 5 cm across. The compactness of these devices has led to their use in portable military range finders.

Some TEA and most sealed-off continuous CO_2 lasers are compact enough to fit on a laboratory bench; a few are very compact. Flowing-gas and the largest TEA lasers are massive devices that weigh in at a few hundred kilograms to a few tons and require at least a few square feet of floor space. The biggest commercial CO_2 laser, a 25-kW materials-working system, occupies 288 ft^2 (27 m^2) of floor space and weighs about 30,000 pounds (lb) [13,600 kilograms (kg)].

Safety. There are two significant hazards with CO_2 lasers: high voltages and the laser beam.

The high voltages needed to operate the laser are by far the more deadly hazard. In complete laser systems, these voltages generally are accessible only inside the laser cabinet, but in other cases (such as when an external high-voltage supply is used), external connections could carry dangerous voltages. Capacitors and other components in pulsed lasers can retain charges long after the laser has ceased operating; be certain that they have been discharged before getting near them.

The beams from virtually all CO_2 lasers fall under Class IV of the federal laser product standard. That means that the beams could, at least in theory, burn the skin or the surface of the eye (the 10-μm wavelength is not transmitted into the eye). Fortunately, the damage threshold for skin is above the level at which the body senses warmth from a continuous-wave beam, so normally a feeling of warmth would give adequate warning for a person to move out of the beam for moderate-power lasers (Sliney & Wolbarsht, 1980, p. 166). The 10-μm CO_2 wavelength is absorbed very strongly by tissue, so any damage is concentrated on the surface.

The CO_2 beam normally is invisible in air and when it is reflected off a surface, except at power densities high enough to significantly heat the surface or dust particles in air. Some CO_2 lasers are packaged with helium-neon lasers with beams aligned collinearly, so that the red He–Ne beam traces the path of the invisible infrared beam.

In standard industrial practice, the beam travels no more than a matter of centimeters between the focusing lens and the workpiece, reducing both optical losses and the risk of beam exposure. Many materials-working CO_2 lasers are shielded by Plexiglas enclosures which transmit visible light but absorb the infrared laser beam. The federal laser product code requires elaborate interlocks that make it difficult for operators to expose themselves to the beam or for unauthorized persons to operate the laser.

Special Operating Considerations. Optics which transmit the 10-μm CO_2 laser beam are made of materials quite different from optical glass. Many materials transparent at 10 μm do not transmit visible light. Unlike ordinary optical glass, many 10-μm window materials are chemically reactive, hygroscopic (water-absorbing), and/or toxic, although suitable coatings can reduce their sensitivity. Ordinary glass totally absorbs 10-μm energy and can be cut with a CO_2 laser.

Discharging capacitors in TEA lasers may produce audible "pops" in normal operation. Such pulsed lasers also can produce electromagnetic interference and may require shielding to avoid interfering with other equipment. Radio-frequency discharges normally operate at frequencies designated by the FCC for instrumentation, but they may cause interference with sensitive external equipment. Although commercial lasers generally incorporate shielding, users should check that it is sufficient.

Some lasers, particularly TEA types, are built for operation with several different gas mixtures. Thus a single laser cavity would function as a carbon dioxide laser when filled with a CO_2 gas mixture but could also function as other types—typically hydrogen fluoride, excimer, or carbon monoxide—when used with different gas mixtures and different optics. When gas mixtures are changed, the system should be purged before operation with the new mixture, and optics should be changed as required.

Reliability and Maintenance

Laser Lifetime. Different factors limit operating lifetimes of different types of CO_2 laser, including gas lifetime in sealed lasers and degradation of optics and power supplies in all types. Sealed continuous-output CO_2 lasers are designed to operate from one to several thousand hours on a single gas fill. TEA lasers can deliver millions of shots from a single gas fill. Output power drops gradually if the gas is not changed, but refilling (and in some cases cleaning) the tube should restore the laser to original output levels.

Lifetimes of flowing-gas CO_2 lasers normally are not specified, and in practice depend heavily on environmental factors, such as contamination of optics, which may require periodic replacement in materials-working applications. Some lasers can last a long time. One company reports that two 3-kW lasers have survived 14 years of continuous use.

Maintenance and Adjustments.   A few basic types of maintenance are common for CO_2 lasers:

- Replacement of optics subject to degradation
- Replacement of gas in sealed-tube lasers after 1000 hours or more of operation
- Replacement of high-voltage components such as spark gaps
- Lubrication and other upkeep of essential accessories such as vacuum pumps

Many CO_2 lasers designed for industrial applications have preventive maintenance schedules and come with elapsed-time meters that indicate when maintenance is due.

Mechanical Durability.   Military contractors have built compact CO_2 lasers designed for field use as range finders, although they were not deployed when this book was written. Otherwise, CO_2 lasers are generally designed for use in a fixed location, not to withstand extensive vibrations and shock.

Failure Modes and Possible Repairs.   Degradation of the optics, the power supply, or (in sealed-tube lasers) the laser gas can all cause failure in CO_2 lasers. Repair is generally by replacing the affected components.

Commercial Laser Systems

Standard Configurations.   Outside of safety standards set by the federal government, there are no formal or de facto standards on the configurations of carbon dioxide lasers sold in the United States. Users can buy anything from a bare-bones laser kit to a complete system including beam-direction optics. There are several key elements of a CO_2 laser system which generally are packaged in one or more units:

- The laser cavity itself, including the housing for the laser gas and the electrodes which excite the gas.
- The resonator or beam-forming optics.
- The power supply, which in pulsed lasers includes two segments, a high-voltage dc supply and a pulse-generating capacitor or network. In continuous-wave lasers only the dc supply is present.
- A control panel.
- Gas mixing and monitoring equipment.
- Cooling equipment, including (where needed) hookups for water cooling.
- Beam-directing optics, which take the beam from the laser to the place it is used. Typically these are included with laser systems sold for surgery or materials working, but not with other CO_2 lasers.

Small lasers intended for laboratory use and waveguide lasers may be packaged as integrated units containing laser cavity, resonator, and power supply. Some small models have been developed and packaged for use in portable military systems. Larger lasers typically are packaged in two or three units, a laser head, a power supply, and sometimes a separate control panel or remote control. A typical laboratory laser head contains the laser tube, optics, and other components needed on the laboratory bench; the power supply usually is mounted on wheels and rests on the floor. Hoses for cooling water run to the laser head. In many industrial lasers, the power supply and laser head are housed in a single large unit, with control panel mounted remotely. Industrial lasers may operate in an enclosure transparent to visible light but opaque to the CO_2 laser's 10-μm output.

In many TEA lasers, the laser head includes an energy-storage capacitor which is discharged across the laser cavity to produce a laser pulse. The dc power supply which charges the capacitor is mounted remotely.

Some companies offer a family of lasers based on a single modular design, with extra laser-cavity modules being added to increase output power. TEA lasers are often available with Brewster angle windows rather than resonator mirrors to allow them to be operated as amplifiers for the beam from a separate external CO_2 laser.

Options. A wide range of options is available for CO_2 lasers. Some are intended for specific applications (e.g., materials working), while others simply enhance the laser's versatility or operation. Some of the most important are listed below.

- Materials-handling systems (for moving parts under the laser beam in industrial applications)
- Grating-tuning optics which select one line in the CO_2 emission spectrum
- Cavity optics to select TEM_{00} mode output, or to operate the laser with an unstable resonator
- Beam-delivery systems for surgery and materials working
- External triggering of pulses
- Q switches to produce pulses of 1 μs or less from continuous-wave lasers
- Modelockers to generate nanosecond pulses from atmospheric-pressure CO_2 lasers
- Gas cleaning and recycling to reduce gas consumption in flowing-gas lasers
- Temperature stabilization of output
- Power meters for monitoring output power
- Electronics to generate pulses in the 0.1- to 1-ms range from continuous-wave lasers
- Computerized controllers
- Gas-mixing and flow-control systems
- Premixed laser gases
- Brewster angle windows to permit operation of the laser cavity as an amplifier or with external optics

Some of these accessories are available from laser manufacturers, while others are offered by companies specializing in optics or laser accessories. Optical components for application outside the laser cavity are generally offered by companies which specialize in optics.

Pricing. The prices of carbon dioxide lasers vary almost as widely as does their performance. The values listed below are representative, as of this writing, of those quoted for particular types, with lower-priced versions generally having lower output and fewer options than the higher-priced versions. With the exception of low-power models, CO_2 lasers are almost invariably purchased in small lots, and quantity discounts generally are not mentioned.

Waveguide	$4,500 to $18,000
Sealed tube	$2,000 to $40,000
TEA	$4,500 to $150,000
Slow axial flow	$10,000 to $50,000
Fast axial flow	$60,000 to $300,000
Transverse flow	$100,000 to $1 million

The highest-power and most-costly systems generally are built specifically for materials working and include some expensive accessories not normally found on CO_2 lasers.

Suppliers. Many companies offer CO_2 lasers, but most specialize in certain types of laser or certain applications. Some industry directories break down manufacturers of CO_2 lasers by types produced, which can be helpful in screening vendors. Most users should be able to find multiple vendors who can meet their needs, but they may have to search to identify those companies. Note also that two or more types of CO_2 laser (e.g., waveguides and conventional sealed tubes) in some cases can meet the same performance requirements.

Applications. Carbon dioxide lasers can serve many functions, from a comparatively inexpensive source of laser photons to an emitter of precise infrared wavelengths. This versatility has opened up many applications for CO_2 lasers in both industrial and laboratory environments. Some of the major applications of CO_2 lasers and their principal requirements are summarized briefly below.

- *Materials working:* CO_2 lasers are used for materials working, primarily cutting and welding of metals and nonmetals. Nonmetals such as plastics, ceramics, cloth, and rubber are much more efficient absorbers than metals at 10 μm, so lower laser powers are required. Despite reflectivities that can exceed 90 percent at 10 μm, CO_2 laser metal working is growing because no other commercially available laser can generate as intense continuous-wave output. Cutting and welding generally require continuous-wave powers of 50 W and up, with much higher powers needed for most metals except titanium, which absorbs more 10-μm radiation than other metals and is hard to cut with mechanical saws. CO_2 lasers are sometimes used for drilling, but that task is more often done with pulsed neodymium lasers, which produce high peak powers at about 1 μm.

- *Heat treating:* To alter the characteristics of metal surfaces, CO_2 lasers are used for heat treating, generally improving the durability of the metal. This requires a continuous beam, which for the sake of efficiency should contain a high power that can be spread over a large area.

- *Marking letters or symbols:* Pulsed TEA lasers can etch a stencil-like image onto a workpiece in a single (or sometimes multiple) shot, a noncontact technique for permanent marking that is used in a variety of industries. A pulsed laser can write alphanumeric characters as series of dots, but that is generally done with pulsed neodymium lasers.

- *Spectroscopy and photochemistry:* CO_2 lasers can generate narrow-band output in the 10-μm range for spectroscopic applications, and/or high powers for studies of laser-induced chemistry. TEA or waveguide lasers are the commonest types for these applications.

- *Range finding:* For military applications, use of waveguide CO_2 lasers as range finders is in the development and demonstration stages.

- *Laser radar or "lidar":* CO_2 lasers can be used in systems which generate 10-μm pulses and analyze the returns, much as an ordinary radar system in the microwave range. Alternatively, returns from the atmosphere can be analyzed spectroscopically to determine concentrations of various species in the atmosphere.

- *Surgery:* Surgery can be performed with 10-μm powers in the 50-W range because water in living tissue absorbs that wavelength strongly. Continuous-wave lasers are used, with waveguide lasers becoming increasingly popular because of their compact size, good beam quality, and low cost.

- *Generation of far-infrared or submillimeter radiation:* A CO_2 laser can be used to optically pump or energize another gas so that the second gas can emit a laser beam at longer wavelengths. The use of different gases makes it possible to generate many wavelengths in the far-infrared and submillimeter regions. Commercial far-infrared lasers are described in Chap. 15.

- *Plasma production:* An intense pulse from a TEA laser (often modelocked) can deposit energy quickly onto a small spot to produce a plasma. One specific application has been laser fusion research, although most work in that field is with shorter-wavelength lasers.

- *Directed-energy weapons:* Weapon experiments have been conducted with carbon dioxide lasers (Hecht, 1984). However, interest in CO_2 laser weapons has decreased in recent years.

Bibliography

Coherent Inc., Engineering Staff: *Lasers: Operation, Equipment, Application and Design*, McGraw-Hill, New York, 1980. (Despite the general title, this emphasizes materials working, mostly with CO_2 lasers.)

W. W. Duley: *Laser Processing and Analysis of Materials*, Plenum, New York, 1983 (overview of materials-working applications of lasers).

Jeff Hecht: *Beam Weapons: The Next Arms Race*, Plenum, New York, 1984 (covers laser weapons).

Peter Laakmann: "Sealed off low and medium power CO_2 lasers," *Lasers & Optronics* 8(3):35–41, March 1989 (review, emphasizing radio-frequency excitation and waveguide lasers).

Los Alamos National Laboratory: *Inertial Fusion Program Jan–Dec 1981*, Los Alamos National Laboratory, Los Alamos, N.M., Progress Report LA-9312-PR, 1982.

Leon A. Newman and Richard A. Hart: "Recent R&D advances in sealed-off CO_2 lasers," *Laser Focus/Electro-Optics* 23(6):80–91, June 1987.

Aris Papayoanou: "CO_2 lasers for tactical military systems," *Lasers & Applications 1*: 49–55, December 1982.

John F. Ready: *Industrial Applications of Lasers*, Academic Press, Orlando, Fla., 1978 (overview of industrial laser applications).

Michael W. Sasnett: "Comparing industrial CO_2 lasers," *Lasers & Applications* 3(9):85–90, 1984 (review article).

Michael W. Sasnett: "Kilowatt-class CO_2 lasers meet present and future industrial needs," *Laser Focus/Electro-Optics* 24(3):48–67, March 1988 (review of high-power CO_2 development).

David Sliney and Myron Wolbarsht: *Safety with Lasers and Other Optical Sources*, Plenum, New York, 1980 (comprehensive reference on laser safety).

Orazio Svelto: *Principles of Lasers*, 3d ed., Plenum, New York, 1989 (textbook description of lasers).

W. J. Witteman: *The CO_2 Laser*, Springer-Verlag, Berlin and New York, 1987 (review of CO_2 laser technology).

William L. Wolfe and George J. Zissis (eds.): *The Infrared Handbook Revised Edition*, Office of Naval Research, Arlington, Va. 1985 (covers infrared technology; available from Environmental Research Institute of Michigan, PO Box 8618, Ann Arbor, MI 48107).

Chemical Lasers

Chemical lasers are those in which a chemical reaction excites an atom or molecule to the upper laser level. J. C. Polanyi proposed the idea at a 1960 conference (Polanyi, 1961), but the first chemical laser did not operate until a few years later (Kasper and Pimentel, 1965). Chemical reactants can store energy quite efficiently and release it quite rapidly in a suitable chamber, so the prime attraction of chemical lasers has long been their potential to generate high powers. Much work in this area has been sponsored by military agencies seeking laser weapons (Hecht, 1984). Most chemical lasers operate on infrared vibrational transitions of diatomic molecules, notably hydrogen halides. Some chemical lasers produce excited atomic states; the most important laser generates excited iodine atoms.

Three types of chemical laser have attracted the most attention: hydrogen fluoride, deuterium fluoride, and iodine. Hydrogen fluoride emits at 2.6 to 3.3 micrometers (μm) in the infrared, where atmospheric absorption is strong, and at overtone wavelengths of 1.3 to 1.4 μm (Jeffers, 1989), where transmission is better. Substitution of deuterium (hydrogen-2) shifts emission to 3.5 to 4.2 μm, where the atmosphere is more transparent, at the cost of lower efficiency and extra expense for the rare isotope. Iodine atoms emit on a 1.3-μm electronic transition which can be readily pumped by excited molecular oxygen produced by a chemical reaction in a chemical oxygen-iodine laser (Shimizu et al., 1991).

Military researchers have spent hundreds of millions of dollars on building-sized demonstration lasers that have generated nominally continuous outputs to a couple of megawatts. Commercial hydrogen fluoride (HF) and deuterium fluoride (DF) lasers are much smaller-scale devices, generating much lower powers for materials studies, spectroscopy, chemistry, biomedical research, and other laboratory uses.

Internal Workings

Figure 11.1 shows how gases enter a chemical laser through a set of nozzles, which mix them together in a region where they react to form vibrationally excited HF or DF. The gas flows rapidly through the laser cavity, passing through an optical resonator with its axis perpendicular to the flow direction. The resonator extracts energy from the excited molecules by stimulated emission before the spent gas is removed by a vacuum pump.

Active Medium.   The active medium which generates a chemical laser beam differs from the input gases, which include two reactants and often other diluent or scavenger gases. Molecular hydrogen is the most common hydrogen source, although hydrocarbons are used sometimes. Molecular fluorine is extremely hazardous to handle, so most lasers generate fluorine internally from compounds such as SF_6 (sulfur hexafluoride) and NF_3 (nitrogen trifluoride). Total pressure in the laser is a few torr (a few hundred pascals). Gas flow is supersonic in high-energy military lasers, but subsonic in commercial lasers.

Energy Transfer.   In commercial continuous-wave HF lasers, an 8-kilovolt (kV) electric discharge breaks down SF_6, producing free fluorine. (Higher voltages are needed in pulsed lasers because SF_6 has very high resistance.) Oxygen is added to the discharge mixture to scavenge sulfur and improve discharge quality; helium is added as a diluent. That gas mixture then flows into the laser chamber where the fluorine atoms react with H_2 in a highly exothermic reaction which produces vibrationally excited HF.

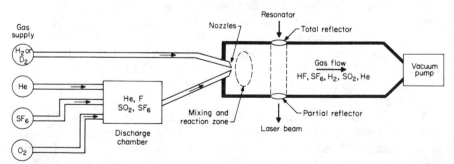

Figure 11.1   Basic elements of a commercial continuous-wave chemical laser include a gas supply, a discharge chamber that produces free fluorine, nozzles which mix the reactants, a mixing region, a laser resonator, and a vacuum pump to collect spent laser gas. Gas flow is from left to right; the laser beam is perpendicular to the gas flow.

Some custom-built laboratory chemical lasers rely on chain reactions involving H_2 and F_2, which start with a small quantity of atomic fluorine and proceed

$$H_2 + F \rightarrow HF^* + H$$

$$H + F_2 \rightarrow HF^* + F$$

$$H_2 + F \rightarrow HF^* + H$$

ad infinitum. An electric discharge may start the reaction.

HF or DF molecules have a series of vibrational energy levels, with spacing smaller at higher energies, as shown in Fig. 11.2. The initial energy of the excited molecules depends on the reactants. Laser transitions on the fundamental band of each molecule (2.6 to 3.3 μm for HF) occur when molecules drop one vibrational level. Molecules also can drop two levels at once (e.g., from level 3 to level 1) and emit on an "overtone" transition at a shorter wavelength. Laser gain is higher on fundamental bands, but HF lasers can operate on the 1.3- to 1.4-μm overtone bands if the cavity optics suppress lasing at 2.8 μm and are reflective at 1.3 μm (Jeffers, 1989). Output on the overtone bands can reach a large fraction of that on the fundamental bands for HF. The gain on DF overtone bands is low, and they have not lased.

The vibrational levels have rotational sublevels, so vibrational and rotational transitions occur simultaneously. This produces a series of lines for each vibrational transition. Emission tends to concentrate at the shorter wavelengths in the fundamental bands, but energy-level populations in pulsed lasers are time-dependent, so relative line

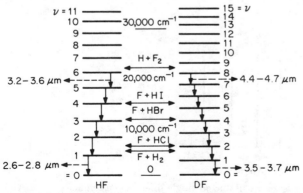

Figure 11.2 Vibrational energy levels of HF and DF, shown with energies of certain reactions which produce HF and DF. Only fundamental-band transitions are shown; rotational sublevels are not shown, but account for the ranges in wavelength. (*From Chester, 1976.*)

strengths can change during a pulse. The H_2 + F reaction used in commercial HF/DF lasers concentrates emission on the lower vibrational transitions, but the many rotational sublevels produce a rich line structure, shown in Fig. 11.3.

Internal Structure

Gas flow is a critical concern in continuous chemical lasers because even mixing of reactants is critical to laser performance. Nozzle design is crucial for good mixing. The gas flows continuously through the reaction zone and resonator optics to a vacuum pump. HF and DF lasers have similar designs but use different resonator optics because of their different wavelengths.

Low-power pulsed lasers can operate differently, with the cavity first filled with a mixture of H_2 and a fluorine source, and optionally preionized with ultraviolet light, before a strong transverse electric discharge initiates the chemical reaction. This approach is similar to the transverse excitation of high-pressure transversely excited atmospheric pressure (TEA) carbon dioxide lasers. Only a small fraction of the reactants are consumed, so the HF can be removed from the spent gas and the rest recycled through the laser, allowing operation for several hours on one gas fill with addition of small amounts of SF_6 and H_2.

Optics and Cavity Configuration.   The axis of a chemical laser cavity is transverse to the gas flow direction. Typically output is coupled from the laser cavity through Brewster angle windows of zinc selenide or calcium fluoride to a totally reflecting rear mirror and a partly trans-

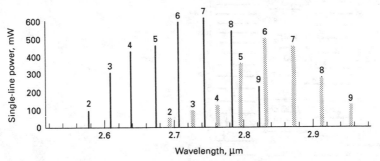

Figure 11.3   Relative power levels available on single lines from a commercial HF laser with a multiline cavity. Solid lines are rotational $1Pn$ sublevels of the $1 \rightarrow 0$ transition; shaded lines are rotational $2Pn$ sublevels of the $2 \rightarrow 1$ transition. The shortest-wavelength sublevel ($1P1$) of the $1 \rightarrow 0$ transition, at 2.55 μm, can be obtained with different cavity optics. (*Courtesy of Helios Inc.*)

parent output mirror. High-power lasers require special optics and typically couple light out through a hole in one mirror.

Gain is fairly high in chemical lasers, so a reasonable fraction of the intracavity beam emerges through the output mirror. Unstable resonators are common on commercial pulsed lasers because they offer good beam quality. However, low-power continuous-wave lasers use stable resonators because unstable resonators offer little beam improvement and reduce output power significantly. Both pulsed and continuous-wave lasers have cavities 0.5 to 1 meter (m) long.

Both pulsed and continuous-wave commercial chemical lasers can be tuned to single lines with gratings in their resonant cavities. Operation on the overtone bands requires optics which suppress reflection in the 2.6- to 2.8-μm range and reflect in the 1.3-μm region.

Variations and Types Covered.   Important chemical lasers are listed in Table 11.1, along with the reactions which power them. HF and DF are the most important, and the only two available commercially. Other hydrogen halides do not generate as much power as HF or DF; their wavelengths also shift when deuterium replaces normal hydrogen. HF and DF are functionally interchangeable in many lasers. Only HF generates usable output on its overtone bands, although overtone emission has been observed in the laboratory from other chemical lasers. This chapter concentrates on HF and DF lasers.

HF and DF are "direct" chemical lasers, in which a chemical reaction produces the light-emitting species. The chemically excited species also can transfer its excess energy to a light emitter; the chemical oxygen-iodine laser is an example of a "transfer" laser. O_2 excited by a chemical reaction transfers its energy to atomic iodine, which then emits at 1.3 μm. Japanese researchers have demonstrated chemical oxygen-iodine lasers with continuous outputs to 1 kW (Shimizu et al., 1991), but commercial versions are not available. Chemically excited

TABLE 11.1   **Major Chemical Lasers**

Laser	Typical reactions	Wavelength, μm
I	$O_2{}^* + I \rightarrow O_2 + I^*$ (transfer)	1.3
HF overtone	Same as HF	1.3–1.4
HF	$F + H_2 \rightarrow HF^* + H$	2.6–3.5
	$H + F_2 \rightarrow HF^* + F$	
HCl	$H + Cl_2 \rightarrow HCl^* + Cl$	3.5–4.1
DF	$F + D_2 \rightarrow DF^* + D$	3.5–4.1
	$D + F_2 \rightarrow DF^* + D$	
HBr	$H + Br_2 \rightarrow HBr^* + Br$	4.0–4.7
CO	$CS + O \rightarrow CO^* + S$	4.9–5.8
CO_2	$DF^* + CO_2 \rightarrow CO_2{}^* + DF$ (transfer)	10–11

DF can transfer energy to power a CO_2 laser emitting at 10 μm. Note that chemical reactions are not the only ways to excite some transitions; both iodine and carbon monoxide lasers can be excited electrically, as described in Chap. 16.

Beam Characteristics

Wavelength and Output Power. Commercial continuous-wave HF lasers have multiline output of a few watts to about 500 W between 2.5 and 3.0 μm. Spectral lines and relative intensities for one laser are shown in Fig. 11.3. With DF, multiline output at 3.4 to 4.0 μm is 60 to 80 percent of that from HF. Maximum single-line power in the fundamental band is up to 100 W for HF and 50 W for DF. Both multiline and single-line output powers can be adjusted over considerable ranges, with minimum powers 0.1 to 1 percent of maximum power. Major fundamental HF and DF lines are listed in Table 11.2. HF normally oscillates on the two lowest vibrational transitions (between levels 2 and 1 and between levels 1 and 0), and DF operates on the three lowest (including between levels 3 and 2).

Commercial pulsed chemical lasers produce much lower average power in multiline operation and do not generate true single-line output. Typical values are about 3 W for HF and 2 W for DF; highest powers are about 15 W. One model delivers 1-joule (J) multiline HF pulses at a 0.5-Hz repetition rate; others deliver 50-mJ pulses at 25 hertz (Hz) or 500 mJ at 5 Hz. Pulse durations are 50 to 400 nanoseconds (ns). DF pulse energies are a little over half those available from HF.

Overtone HF lasers produce continuous-wave multiline power of 25 W in the 1.3- to 1.4-μm region when molecules drop two vibrational energy levels. Table 11.3 lists single HF overtone lines from commercial lasers; outputs can reach 10 to 15 W.

Efficiency. Infrared output of commercial continuous chemical lasers is 1 to 2 percent of electrical input to the discharge; efficiency is reduced to 0.25 to 0.5 percent if power requirements of other components are included. Pulsed fast-discharge lasers are somewhat more efficient. Electrical efficiency of laboratory lasers can exceed 100 percent when a discharge triggers the highly exothermic chemical reaction, but in that case it is the chemical efficiency which is more meaningful, and that runs from 1 to 20 percent.

Beam Quality, Polarization, and Modes. Continuous-wave lasers oscillate in one to three transverse modes, with about 90 percent of output in TEM_{00} mode. Single-line lasers can oscillate in TEM_{00} mode if beam diameter and cavity aperture are limited. Unstable resonators

TABLE 11.2   Selected Major Laser Transitions on HF
and DF Fundamental Bands

	Wavelength, μm	
Transition	HF	DF
Vibrational State $1 \to 0$		
1P1	2.5507	3.4660
1P2	2.5787	3.4930
1P3	2.6084	3.5211
1P4	2.6397	3.5303
1P5	2.6727	3.5806
1P6	2.7074	3.6122
1P7	2.7440	3.6449
1P8	2.7826	3.6789
1P9	2.8231	3.7142
1P10	2.8656	3.7508
1P11	2.9103	3.7887
1P12	2.9573	3.8279
1P13	3.0065	3.8686
1P14	3.0583	3.9108
1P15	3.1125	3.9544
Vibrational State $2 \to 1$		
2P1	2.6668	3.5788
2P2	2.6963	3.6068
2P3	2.7275	3.6360
2P4	2.7604	3.6663
2P5	2.7952	3.6979
2P6	2.8319	3.7307
2P7	2.8705	3.7648
2P8	2.9112	3.8001
2P9	2.9539	3.8369
2P10	2.9989	3.8749
2P11	3.0462	3.9144
2P12	3.0958	3.9554
2P13	3.1480	3.9978
2P14	3.2027	4.0417
2P15	3.2602	4.0873
Vibrational State $3 \to 2$		
3P1		3.6966
3P2		3.7257
3P3		3.7560
3P4		3.7875
3P5		3.8203
3P6		3.8544
3P7		3.8899
3P8		3.9267
3P9		3.9650
3P10		4.0046

SOURCE: Helios Inc.

TABLE 11.3   HF Overtone Lines Observed from Commercial Chemical Lasers*

Transition	Wavelength, μm
20P1	1.2970
20P2	1.3045
20P3	1.3125
20P4	1.3212
20P5	1.3305
20P6	1.3404
20P7	1.3510
20P8	1.3622
20P9	1.3740
20P10	1.3866
20P11	1.3999
20P12	1.4139
20P13	1.4287
20R0	1.2839
20R1	1.2781
20R2	1.2729
20R3	1.2683

*All are on transitions from second to ground vibrational states.
SOURCE: Helios Inc.

can limit pulsed lasers to a single mode and are often required to achieve reasonable beam quality. Beam quality can become a serious problem at the extremely high powers of military chemical lasers.

Suitability for Use with Laser Accessories.   Chemical lasers do not store much energy in the laser resonator, but Q switching is possible at repetition rates of 1 MHz, producing peak powers 10 times the continuous-wave level. HF overtone bands are close to the second harmonic wavelength of HF, and harmonic generation has attracted little interest.

Operating Requirements

Input Power.   Major power consumption in continuous-wave chemical lasers is by the high-voltage discharge supply, which draws 2.6 to 24 kW, and the vacuum pump, which draws 5.6 to 15 kW. The smallest models can operate from a 110-V ac line, but 220-V operation is more common, and the largest models can require three-phase 440-V supplies. Pulsed models require as little as 3 amperes (A) at 110 V, not counting vacuum pump.

Cooling.   Pulsed HF and DF lasers can be air-cooled, but continuous-wave lasers require water cooling of the laser head and (in many cases) of the vacuum pump at 0.5 to 4 gallons per minute (gal/min) [1.9 to 15 liters per minute (L/min)].

Consumables. Continuous-wave chemical lasers operate in an open cycle, consuming SF_6, O_2, He, and H_2 or D_2. Gas consumption per minute at maximum power is:

Gas	Standard liters	Standard cubic feet
SF_6	0.45–70	0.016–2.5
O_2	0.94–28	0.033–1
He	5.7–40	0.2–1.4
H_2	4.8–110	0.17–4
D_2	4.8–85	0.17–3

Different lasers require different ratios of the component gases. SF_6 costs several cents per standard liter, and this can be the single largest operating cost for an HF laser. Deuterium costs about a dollar per standard liter, so DF lasers cost much more to operate than HF lasers.

Some pulsed lasers can be operated with gas recirculators that remove the small amount of HF or DF produced during laser operation and recycle the rest of the gas. This reduces gas consumption to about 1 percent the level of an open-cycle laser.

Required Accessories. A vacuum pump is needed to remove laser gas; recirculators may be used with pulsed lasers. An exhaust system, scrubber, and/or waste disposal system are needed to handle HF and SO_2 produced during laser operation.

Mechanical Considerations. The smallest chemical lasers are pulsed types weighing under 40 kilograms (kg) and packaged in a single case as small as 0.7 by 0.5 by 0.3 m. Continuous-wave chemical lasers are larger, including separate laser head, control console, vacuum pump, and (in larger models) a high-voltage supply and ballast systems. Combined weights range from 270 to 2550 kg.

Safety. Chemical laser wavelengths cannot penetrate the eye, but power densities above 10 W/cm^2 can burn the cornea before the ocular blink reflex can protect it. The wavelengths are strongly absorbed by human skin, so high-power beams can cause skin burns.

Two input gases, SF_6 and H_2 (or, equivalently, D_2), present moderate hazards. H_2 is a fire hazard, and H_2-air mixtures can explode if H_2 concentration exceeds 4%. SF_6 is a health hazard above 1000 parts per million (ppm), but other fluorine donors present considerably worse hazards. F_2 is both dangerous to handle and toxic above 0.1 ppm. NF_3

is somewhat more stable and less toxic than F_2 (10 ppm is hazardous), but it is still much more dangerous than the comparatively inert SF_6.

HF and SO_2 in the waste gas also are health hazards, so laser exhaust must be directed outdoors and scrubbed. American National Standards Institute (ANSI) standards limit HF exposure to no more than 3 ppm, and SO_2 is harmful in similar concentrations (Weast, 1981). HF and most fluorine donors (except SF_6) are highly corrosive.

The high voltages in discharges pose serious electrocution hazards.

Reliability and Maintenance

HF and DF chemical lasers are laboratory instruments which handle extremely corrosive gases, and are not designed to meet industrial standards for reliability and maintenance. Nonetheless, some commercial models are used 8 to 16 hours per day, and have operated for up to 5000 hours altogether. Regular maintenance, such as cleaning optics, changing vacuum-pump oil, cleaning the fuel injector, and replacing scrubber material, is required about every 50 hours; discharge tubes should be rebuilt after 250 to 500 hours of operation.

Commercial Devices

Standard Configurations. Pulsed chemical lasers are about the same size as TEA CO_2 or excimer lasers, with most major components except the vacuum pump (and sometimes the power supply) in a single case. Continuous-wave chemical lasers have a bench-mountable laser head, with separate vacuum pump, control console, and (in higher-power models) power supply and ballast. Production is small, so manufacturers can readily accommodate special requests.

Pricing. Prices of complete pulsed chemical laser systems start in the $40,000 range. Complete continuous-wave lasers run $50,000 to $200,000, including vacuum pump, scrubber, and interconnections.

Suppliers. Only one company specializes in commercial chemical lasers, but a couple of others produce a few chemical lasers. High-energy military demonstration lasers are custom-built by large aerospace contractors.

Special Notes. Most chemical laser development has long been sponsored by military agencies seeking laser weapons, which allow publication of basic research but classify details on high-energy lasers. Except for efforts to develop visible chemical lasers (Woodward et al., 1990), little has been published recently on chemical lasers.

Applications. Chemical lasers are used exclusively in research. Many small commercial lasers are used in military programs. They also are used as sources of midinfrared light for chemistry, materials research, spectroscopy, and biomedicine. HF wavelengths may be attractive for bone surgery (Izatt et al., 1990). Gas-handling requirements continue to make chemical lasers more cumbersome to use than CO_2, but chemical laser specialists believe that a viable commercial market would justify the investment in engineering needed to make chemical lasers considerably more practical.

Bibliography

T. Y. Chang: "Vibrational transition lasers," in Marvin J. Weber (ed.), *CRC Handbook of Laser Science & Technology*, vol. 2, *Gas Lasers*, CRC Press, Boca Raton, Fla., 1982, pp. 313–409 (review article listing all vibrational transition lasers, including chemical types).

Arthur N. Chester: "Chemical lasers," in E. R. Pike (ed.), *High-Power Gas Lasers 1975*, Institute of Physics, Bristol and London, 1976, pp. 313–409 (useful review article despite its age).

R. W. F. Gross and J. F. Botts (eds.): *Handbook of Chemical Lasers*, Wiley, New York, 1976 (reference, useful despite its age).

Jeff Hecht: *Beam Weapons: The Next Arms Race*, Plenum, New York, 1984 (overview of laser weapon development).

Joseph A. Izatt et al.: "Ablation of calcified biological tissue using pulsed hydrogen fluoride laser radiation," *IEEE Journal of Quantum Electronics* 26(12):2261–2270, 1990 (research report).

William Q. Jeffers: "Short wavelength chemical lasers," *Journal of Aircraft* 27(1):64–66, 1989 (research report).

J. V. V. Kasper and G. C. Pimentel: "HCl chemical laser," *Physical Review Letters 14*: 352, 1965 (first chemical laser).

J. Munch et al.: "Frequency stability and stabilization of a chemical laser," *IEEE Journal of Quantum Electronics .QE-14*(1):17–28, 1978 (research report).

John C. Polanyi: "Proposal for an infrared maser dependent on vibrational excitation," *Journal of Chemical Physics 34*:347, 1961 (first proposal).

K. Shimizu et al.: "High-power stable chemical oxygen iodine laser," *Journal of Applied Physics 69*(1):79–83, 1991 (research report).

Robert Weast (ed.): *CRC Handbook of Chemistry & Physics*, CRC Press, Boca Raton, Fla., 1981 (lists chemical safety hazards).

J. R. Woodward et al.: "A chemically driven visible laser transition using fast near-resonant energy transfer," *IEEE Journal of Quantum Electronics* 26(9):1574–1587, 1990 (research report).

Copper and Gold Vapor Lasers

Copper and gold vapor lasers are the most important members of a family of neutral metal vapor lasers which emit in or near the visible region. They are unusual in their high power and high efficiency in this region, and in that their normal operation is at repetition rates of several kilohertz (the internal physics prevent continuous-wave emission). Commercial copper vapor lasers can emit over 100 watts in the green and yellow; gold vapor lasers can generate several watts in the red. These high average powers make copper and gold vapor lasers attractive for applications including pumping dye lasers, detection of fingerprints and trace evidence, and certain biomedical procedures.

The copper vapor laser was first demonstrated in 1966 by W. T. Walter, N. Solimene, M. Piltch, and Gordon Gould at TRG Inc. (Walter et al., 1966). The laser's high efficiency was recognized even then, but the need to heat the tube to 1500 to 1800°C to get high enough copper vapor pressures presented severe technical difficulties. In the early 1970s, Soviet workers showed that waste heat from the laser discharge could heat the tube, raising overall efficiency (Isaev et al., 1972). Others studied the use of copper halides, which do not require such high temperatures (Chen et al., 1973).

Commercial copper vapor lasers based on discharge heating have been developed partly as an outgrowth of technology developed for laser isotope separation at the Lawrence Livermore National Laboratory. Copper vapor lasers were chosen as the best candidate for pumping high-power dye lasers tuned to selectively excite uranium-235 (^{235}U) in uranium vapor, and to purify the fissionable plutonium-239 (^{239}Pu) isotope. This led to a large investment in copper vapor laser technology, including construction of a laser with average power of 6000 W, which Livermore calls "the highest average power visible laser system operating in the world today" (Lawrence Livermore National Laboratory, 1989). Some of that technology is applicable at the lower powers of commercial lasers.

Other neutral metal vapor lasers use technology similar to that of copper vapor. The only one commercially available is the gold vapor laser, which is of interest because its 628-nm red output is at a wavelength useful for photoradiation or photodynamic therapy of cancer (McCaughan, 1983).

Internal Workings

Neutral metal vapor lasers operate in ways similar to each other, but quite different from ionized metal vapor lasers such as helium-cadmium, described in Chap. 9. In neutral metal vapor lasers, a fast electric discharge directly excites metal atoms, producing a population inversion and laser emission. The laser pulse terminates when the lower laser level fills, but laser action can occur again after slower processes depopulate that level. High repetition rates permit high average power.

Active Medium.   Vaporized copper and neon form the active medium in commercial copper vapor lasers. Pieces of copper metal are put into the discharge tube, where they are heated to about 1500°C to obtain the required 0.1 torr [13 pascals (Pa)] of metal vapor. Adding an inert gas improves discharge quality enough that sufficient power can be coupled into the active medium; it also may help depopulate the lower state, although electron quenching appears much more effective. Neon at about 25 torr (3300 Pa) is the normal choice, but helium and argon also can be used.

Waste heat from the discharge keeps the inner part of the laser tube hot, but the copper does condense on outer, cooler portions of the tube, so extra metal must be added periodically. Similar processes occur with other neutral metal vapor lasers, although some of the details such as operating temperature differ. Lead vapor lasers, for example, operate at about 900 to 1100°C, and gold vapor must be heated to about 1550 to 1850°C.

The use of copper halides and other compounds to provide copper vapor has been demonstrated in the laboratory. This approach requires lower temperatures to produce copper vapor, but that advantage is more than offset by practical problems in operating the laser, and the technology has not come into commercial use. However, plans for a do-it-yourself version have been published in *Scientific American* (Walker, 1990). That design is hardly suitable for a casual "amateur scientist," but it could be useful for a laboratory with a limited budget and access to glass-blowing equipment and suitable power supplies.

Energy Transfer. Energy is delivered to a copper vapor laser by a pulsed high-voltage discharge passing longitudinally along the length of the plasma tube. Electron collisions raise copper atoms to one of the two upper levels shown in Fig. 12.1. At vapor densities of 10^{11} atoms per cubic centimeter, copper atoms remain in the upper levels only about 10 nanoseconds (ns) before dropping back to the ground state, making laser action impossible. Raising copper density to about 5×10^{13} per cubic centimeter (cm^{-3}) increases effective lifetime of the upper laser level (relative to the ground state) to 10 milliseconds (ms) because of radiation trapping, although the actual time an individual atom remains excited is unchanged. That is sufficient to permit laser emission.

Copper vapor has two main laser transitions between separate pairs of states, one at 510.6 nanometers (nm) in the green, and one at 578.2 nm in the yellow. Both terminate in lower levels that are not thermally populated, but are metastable. Relaxation times from these lower levels are tens to hundreds of microseconds, depending on operating conditions, long enough that the populations build up and ter-

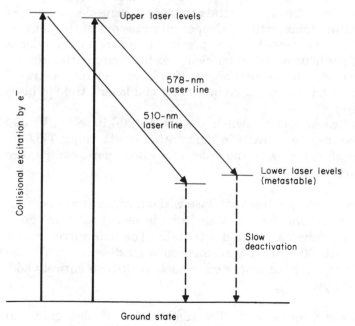

Figure 12.1 Energy levels and laser transitions in copper vapor.

minate laser emission in under 100 ns. Laser action cannot resume until the lower laser level is depopulated.

The self-terminating nature of the laser makes fast discharge rise time crucial. Operation at high average powers requires high repetition-rate discharges, typically 4 to 12 kilohertz (kHz) in commercial copper vapor lasers, although experimental lasers have passed the 100-kHz mark (Grove, 1982). Energy-transfer processes in other neutral metal vapor lasers are qualitatively similar to those in copper vapor, although different wavelengths and different numbers of emission lines are involved. All share the self-terminating nature that dictates repetitively pulsed operation.

Internal Structure

The high gain of copper vapor, 10 to 30 percent per centimeter, makes optical requirements for laser oscillation easy to meet. The major design challenges come in sustaining the required high temperatures and vapor pressure, and in delivering fast current pulses at a high repetition rate.

The discharge is contained in ceramic alumina tubes 1 to 8 cm in diameter. External insulation helps the tubes maintain enough waste heat to vaporize metal from pieces of copper put inside, typically near the ends of the bore. Refractory metal electrodes are needed because of the high operating temperatures. Copper migrates out of the hot portion of the tube to condense in cooler portions at a rate on the order of 0.01 gram (g) per hour of operation. Schemes for recirculating the escaping copper back into the discharge region have been demonstrated in the laboratory but are not used in commercial lasers. Gold diffuses somewhat faster.

Thyratrons are needed to switch the 10- to 20-kilovolt (kV) discharge pulses at repetition rates in the multikilohertz range. This, together with peak currents on the order of a kiloampere, can require special charging circuits.

Optics.   The copper vapor laser can lase without any mirrors because of its high gain. In practice, the laser tube is sealed with flat glass windows, and the mirrors mounted externally. The rear mirror is a total reflector, with 90 percent transmission a good selection for the output-coupling mirror. An uncoated window makes an entirely adequate output coupler.

Cavity Length and Configuration.   The combination of high gain and short pulse length poses imposes some constraints on design of cavities for copper vapor lasers. The pulse duration allows the light to make only a few round trips of the laser cavity. In 10 ns, light travels 3 meters (m), and a typical copper vapor laser cavity is on the order of 1 m long. Because the light makes so few round trips of the cavity, beam divergence is comparatively large with stable resonators.

Beam quality can be improved by the use of unstable resonators. A confocal cavity, with a large concave mirror and a small convex one along the cavity axis, produces higher power along the laser axis at a small sacrifice in average power and efficiency. Placing the smaller mirror off-axis can increase both average and on-axis power (Webb, 1987).

Output power can be raised above the levels available from a single oscillator by using external power amplifiers. Average powers above 5 kW have been produced by a linear chain of amplifiers following a copper vapor oscillator; each amplifier can generate 200 W when used as an oscillator.

Variations and Types Covered

This chapter concentrates on the copper vapor laser, which because of its wavelength and high power has attracted the most attention of any neutral metal vapor laser. Gold vapor lasers also are mentioned because they have been available commercially. Structurally they are similar to each other and the other neutral metal vapor lasers listed in Table 12.1. That table lists major lines of selected elements; many weaker and less efficient lines are listed in more extensive data tabulations (see, e.g., Davis, 1982).

TABLE 12.1   Wavelengths of Major Neutral Metal Vapor Laser Lines, with Relative Powers That Might Be Expected from Devices of Comparable Scale*

Element	Wavelength, nm	Relative power	Remarks
Copper	511, 578	1	
Gold	628	0.1–0.3	
	312	Low	Secondary line
Barium	1130	Low	Ba liquid a problem
	1500	0.3–0.5	Ba liquid a problem
Lead	722.9	0.2–0.3	1000–1100°C temperature
Manganese	534	0.2–0.3	Mn vapor a problem
	1290		Mn vapor a problem
Calcium	852.4	—	
	866.2	—	

*For copper vapor and the 628-nm gold line, values are for commercial devices; other results are from laboratory experiments. Laser action has been demonstrated experimentally on many other lines.

Beam Characteristics

Wavelength and Output Power.   Copper vapor lasers emit at two wavelengths, 510.6 and 578.2 nm. Both lines are produced simultaneously, with relative intensity a function of operating temperature as shown in Fig. 12.2. When the operating temperature is picked to give peak output power, about two-thirds of the power is at 510.6 nm. If only a single line is desired, the two wavelengths can be separated in the output beam. Average power ranges from a few watts to 100 W for commercial copper vapor lasers. Amplifier chains of copper vapor lasers also are available; their total output typically scales as the product of the number of lasers times the power the amplifier can generate when operated as an oscillator.

Commercial gold vapor lasers produce up to several watts of average power at 628 nm. Wavelengths of other metal vapors and their relative power levels are listed in Table 12.1; some lines are shown in Fig. 12.3.

Efficiency.   Overall wall-plug efficiency is 0.2 to 1 percent for commercial copper vapor lasers. Sometimes higher efficiency figures may be quoted when the measurement compares laser output with energy stored in the discharge circuit. Gold vapor lasers have 0.1 to 0.2 percent overall efficiency.

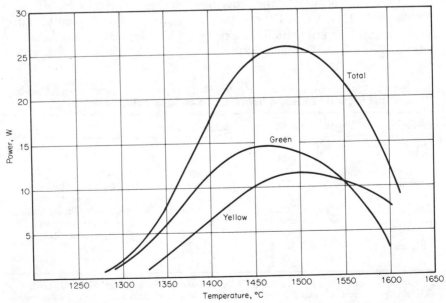

Figure 12.2 Relative strengths of copper vapor lines as a function of temperature. (*Courtesy of Oxford Lasers Ltd.*)

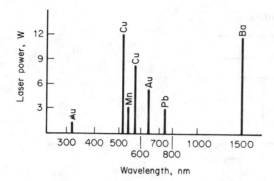

Figure 12.3 Wavelengths and relative intensities of some major neutral metal vapor laser lines. (*Courtesy of Quentron Optics Pty. Ltd.*)

Temporal Characteristics. Average output power of commercial copper vapor lasers generating up to 25 W peaks at repetition rates of 4 to 15 kHz. Higher-power lasers generate peak power over a more limited range of repetition rates. Optimum repetition rate depends on tube construction, and higher rates are possible with certain tube designs.

Individual pulses last from 8 to 80 ns. The shorter pulses represent light extracted in essentially a single pass of the laser cavity; the longer pulses are generated in multiple passes. Their shapes differ, as shown in Fig. 12.4. Special cavity designs are used to generate the shorter pulses. Pulse energies range from 1 to 20 millijoules (mJ), with peak powers from tens to hundreds of kilowatts.

Gold vapor lasers have similar pulse length and can operate at similar repetition rates, but output power appears to be highest and repetition rates are somewhat higher than copper.

Both gold and copper vapors have repetition rates sufficiently high that the beams appear continuous to the human eye.

Spectral Bandwidth. A single line from a copper vapor laser has a spectral bandwidth of 6 to 9 GHz. Typically, the green line has a bandwidth of 6 GHz and the yellow line, 9 GHz. The bandwidths differ because the two lines have different hyperfine structures, which spread the yellow line more than the green one. Gold vapor lasers, with a less complex hyperfine structure, have smaller linewidths.

Amplitude Noise and Stability. Instantaneous power levels vary widely because of the repetitive pulsing. Pulse-to-pulse amplitude variations often are unspecified, but one company reports amplitude stability of ±1 percent root mean square for 90 percent of shots. Pulse shape stays reasonably constant if operating conditions stay unchanged, but the shape does vary with changes in repetition rate, gas pressure, resonator temperature, or other factors. Pulse-timing jitter is specified at 1 to 2 ns.

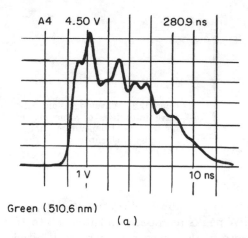

Green (510.6 nm)

(a)

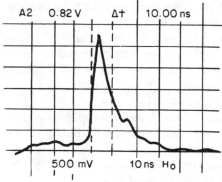

Timescale: 10 ns/div

(b)

Figure 12.4 Pulse shapes of copper vapor lasers operated on the green line in (a) long- and (b) short-pulse mode. Time scale is 10 ns per division. (*Courtesy of Oxford Lasers.*)

Beam Quality, Polarization, and Modes. Stable resonators produce a high-order multimode beam with uniform intensity across the beam. Unstable resonators can produce a doughnut-shaped mode with much lower divergence, but more recent designs (with an off-axis mirror) can produce a gaussian mode with a divergence three to four times the diffraction limit. Either type of unstable resonator can concentrate about 30 to 40 percent of the laser power in a spot within 0.2 milliradian (mrad) of the axis, with the off-axis design slightly more efficient (Webb, 1987). Normally the output is unpolarized, but intracavity polarization is possible if desired.

Beam Diameter and Divergence. Copper vapor lasers have beams 2 to 8 cm in diameter. With stable resonators, beam divergences are 3 to 5 mrad. An unstable resonator can produce a mode with 0.1- to 0.5-mrad divergence, but it may not contain all the laser power.

Suitability for Use with Laser Accessories. The 578-nm line of copper vapor can be frequency-doubled well in nonlinear crystals. A new material, β-barium borate, is an efficient harmonic generator for the 510.6-nm line. Little attention has been paid to modelocking, presumably because of the ease of modelocking argon ion and dye lasers. The short energy storage times in the laser medium make Q switching impractical.

Operating Requirements

Input Power. Operation of a copper vapor laser requires a few kilowatts to about 16 kW of electrical input. The power may be drawn from a 220-V, single-phase ac source for smaller models; larger ones may require three-phase sources of 220 or 415 V.

Cooling. Many commercial copper vapor lasers are water-cooled, with flow rates of 1 to 8 liters (L) of tap water per minute. Forced-air cooling is used for some small models, such as a 10-W laser which dissipates about 2 kW of heat.

Consumables. Copper slowly migrates out of the discharge region, and new metal must be loaded into the laser tube after a few hundred hours of operation. The same phenomenon occurs with gold, which also must be added to the laser tube periodically. The cost of the added gold corresponds to about $20 per 8-hour day of operation, but the gold is not lost—just deposited elsewhere in the tube where it can be recovered.

Buffer gas normally flows slowly through the laser to purge contaminants at a rate of around 1 L/h.

Operating Conditions, Temperature, and Warm-up. Copper vapor lasers are designed for operation in normal laboratory conditions and temperatures. Half an hour to an hour is needed to heat the laser tube to the 1500°C temperature needed to produce enough copper vapor for laser operation. This is an unusually long time by laser standards, comparable only to that for lasers with special frequency-stabilization equipment, and much longer than that for ionized metal vapor lasers such as He-Cd, which operate at lower temperatures. Warm-up times needed to reach the 1550 to 1850°C operating temperature of gold vapor lasers are comparable to those for copper vapor.

Mechanical Considerations. Copper vapor lasers are large, with laser heads typically 1.5 to 2.5 m long, generally built to be put on an optical bench. The power supply and controls are normally in a separate housing placed on the floor, but in some cases the power supply, controls, and laser head may be combined in a single floor-mounted unit.

Safety.   The green and yellow lines of copper vapor and the red line of gold vapor are potential hazards to the retina, because the rest of the eye transmits visible light well. Average power levels are high enough to pose serious dangers to the eye. Copper vapor power levels are high enough that they could cause skin burns if the beam were tightly focused, but such lesions are far more likely to heal than retinal damage. Because the two copper vapor wavelengths are widely separated, users should make sure that their safety goggles provide adequate protection at *both* wavelengths, not just one.

The high-voltage, high-current discharges which drive metal vapor lasers can electrocute anyone who comes in contact with them. The repetitively pulsed operation of the laser makes it likely that some components may retain a charge when the laser is turned off, so users and service personnel should be sure all voltages have been discharged before working inside the laser.

The high operating temperature of the laser tube poses thermal hazards—do not touch it unless it has had plenty of time to cool off.

Reliability and Maintenance

Operating conditions of a copper vapor laser are difficult, so much of the research and development devoted to copper vapor lasers has been aimed at improving reliability. Great strides have been made in building practical lasers, but the "hands-off" operating time of copper vapor lasers remains smaller than that of many other types.

Lifetime.   Some experimental copper vapor lasers have operated for up to a few thousand hours without replenishment of the metal (Grove, 1982), but to sustain maximum output power commercial models require addition of metal every few hundred hours. With regular addition of metal, commercial copper vapor laser tubes have rated lifetimes of 1000 to 3000 hours of operation. Typical manufacturer warranties are 1 year on the laser tube and 500 or 1000 operating hours to 1 year on the thyratron.

Gold vapor lasers are generally less reliable than copper vapor because they must operate at higher temperatures.

Maintenance and Adjustments.   The most frequent maintenance needed in metal vapor lasers is addition of more metal. Typically, copper vapor lasers require more metal every 100 to 300 hours. The lasers are designed to facilitate the addition of metal. One manufacturer claims a user can fill its lasers with 18-gauge (American Wire Gauge) household copper wire in 15 minutes, without realigning the laser cavity. (Cavity alignment is not crucial because the laser medium has high gain.)

Switching thyratrons are the components that require most frequent replacement. Some lasers require periodic replacement of plasma tubes.

Commercial Devices

Standard Configurations. Most neutral metal vapor lasers are packaged for general laboratory use, with separate laser head and power supply. Copper vapor lasers are available in oscillator-amplifier configurations.

Copper and gold vapor lasers are similar in design, and copper can lase in some gold vapor tubes, if operating temperature is reduced. Operation of gold vapor in copper tubes is more difficult because it requires higher temperatures than usually used in copper lasers. When other neutral metal vapor lasers are available, they normally are modified versions of copper or gold lasers.

Options. Common options include choice of stable or unstable resonator, choice of green or yellow copper vapor line, short-pulse operation, output stabilization, and polarizing optics. Lasers can be triggered externally for applications such as high-speed photography or pumping of dye-laser amplifiers. Because production quantities are small, manufacturers are willing to build customized lasers for special requirements.

Pricing. As of this writing, commercial copper and gold vapor lasers carry prices of about $40,000 to $200,000, depending on packaging and power levels. Replacement plasma tubes run about $600 to $10,000, with prices based on exchange of the old tube for a new one.

Suppliers. Only a handful of companies manufacture copper vapor lasers, reflecting both the small market and the need for specialized technology. Product lines differ significantly, and suppliers are based both in the United States and overseas. The use of copper vapor lasers in isotope separation of uranium and plutonium has led to stringent export controls, especially on higher-power models.

Applications

The biggest single application for copper vapor lasers has been in pumping dye lasers. The high average power of copper vapor makes it possible to obtain higher wavelength-tunable dye output than when pumping with an argon laser, although the dye output—like the copper vapor—is repetitively pulsed. The use of the visible wavelength for pumping helps extend dye lifetime, although it limits the choice of dyes to those with emission wavelengths longer than the pump line. This set of advantages is particularly important for applications in photochemistry, specifically for the separation of isotopes of uranium and plutonium. The atomic vapor laser isotope separation (AVLIS) program at the Lawrence Livermore National Laboratory may lead to the first large-scale application of laser photochemistry. The same 6000-W copper vapor laser is used in both uranium and plutonium experiments. The uranium program is intended to increase the concentration of ^{235}U from the low natural levels to the higher levels needed to power light-water nuclear reactors. The plutonium program is intended to purify plutonium for use in nuclear weapons, removing ^{238}Pu, ^{240}Pu, and ^{241}Pu, leaving only ^{239}Pu.

A far more innocuous use of copper vapor pumped dye lasers is the amplification of ultrashort pulses. Colliding-pulse modelocked dye lasers can generate pulses on the order of 100 femtoseconds (100×10^{-15} seconds) long, but only at low energies. Copper vapor pumped dye amplifiers can boost the powers of these pulses while retaining their short duration (Knox et al., 1984; Knox, 1988), giving enough energy for many experiments.

Another application that is moving from the research stage to practical applications is the detection of trace evidence for law enforcement. The best-known example is detection of latent fingerprints by illuminating with the green copper vapor line and looking for fluorescence produced at longer wavelengths by substances in the fingerprints. Early work was done with argon lasers, but copper vapor is attracting attention because of its higher average power. The technique has also been extended to detection of other trace evidence, such as bloodstains, and examination of documents to check for forgeries. Most work has been done with the green copper vapor wavelength, but prospects for using the yellow line are being studied also.

A number of other potential applications of copper vapor lasers also are being studied. These include

- Pulsed photography and holography, taking advantage of the short pulse length of the repetitively pulsed lasers. Coherence lengths appear adequate for pulsed holography when emission is restricted to a single line.

- Measurement of combustion processes, using the short pulse length to separate fluorescence of signals from noise.

- Transmission of light underwater at the 511-nm line, which is close to the wavelength where water is most transparent.

- Color displays.

- Semiconductor processing and research.

- Biomedicine. The 578-nm copper line appears useful in treating "portwine" birthmarks. Such skin discolorations have been treated with argon lasers, but results have been inconsistent and results hard to predict.

- High-speed flash photography, in place of strobes. Copper vapor lasers offer comparable kilohertz repetition rates, but the lasers' tens of nanosecond pulse lengths are much shorter, avoiding image blurring for very rapid events.

Interest in the gold vapor laser has centered on biomedical applications of its red line. The 628-nm wavelength matches an absorption peak of hemoporphyrin derivative, a substance selectively absorbed by cancer cells which breaks down when it absorbs light, producing byproducts which kill the cancer cells. Photoradiation therapy is based on this effect (McCaughan, 1983). However, recently copper vapor pumped dye lasers have overtaken gold vapor lasers for this application. One advantage of the dye laser is that its wavelength can be tuned to match the absorption peak. Gold vapor lasers also suffer from poorer reliability and lower overall efficiency than does the copper vapor pumped dye combination, which can convert 30 percent of the copper vapor energy into dye output.

Bibliography

C. J. Chen, N. M. Nerheim, and G. R. Russell: "Double-discharge copper vapor laser with copper chloride as a lasant," *Applied Physics Letters* 23:514–515, 1973.

Christopher C. Davis: "Neutral gas lasers," in Marvin Weber (ed.), *CRC Handbook of Laser Science & Technology.* vol. II *Gas Lasers.* CRC Press, Boca Raton, Fla., 1982, pp. 3–168 (reference review; see especially pp. 134–139).

Robert E. Grove: "Copper vapor lasers come of age," *Laser Focus* 18(7):45–50, July 1982.

A. A. Isaev, M. A. Kazaryan, and G. G. Petrash: "Effective pulsed copper vapor laser with high average generation power," *JETP Letters* 16:27–29, 1972.

Wayne H. Knox et al.: "Amplified femtosecond optical pulses and continuum generation at 5-kHz repetition rate," *Optics Letters* 9:552–554, December 1984 (research report).

Wayne H. Knox: "Femtosecond optical pulse amplification," *IEEE Journal of Quantum Electronics* QE-24(2):388–397, 1988 (review paper).

Lawrence Livermore National Laboratory: "Uranium AVLIS program" and "Special Isotope Separation Program," *Energy and Technology Review*, pp. 22–23, 24–25, July–August 1989 (short status reports).

James S. McCaughan, Jr.: "Progress in photoradiation therapy of cancer following administration of HpD," *Laser Focus* 19(5):48–56, May 1983 (review of medical research).

G. G. Petrash (ed.): *Metal Vapor and Metal Halide Vapor Lasers*, Nova Science Publishers, Commack, N.Y., 1989 (review of Soviet research with copper vapor, barium, gold, bismuth, and copper halide lasers; translation of a 1987 Russian-language original).

Jearl Walker: "The Amateur Scientist: A homemade copper chloride laser emits powerful bursts of green and yellow light," *Scientific American 262*(4):114–117, April 1990 (how-to article).

W. T. Walter et al.: "Efficient pulsed gas discharge lasers," *IEEE Journal of Quantum Electronics QE-4*:474–479, 1966 (research report on first copper vapor laser).

Colin E. Webb: "Copper and gold vapor lasers: Recent advances and applications," *Proceedings of the International Conference on Lasers '87*, Society for Optical and Quantum Electronics, McLean, Va., 1987, pp. 276–284 (short review).

Colin E. Webb: "Copper vapor pumping of dye lasers," in Frank Duarte (ed.), *High Power Dye Lasers* Springer-Verlag, New York and Berlin, in press (review article).

13

Excimer Lasers

The term *excimer laser* does not describe a single device, but rather a family of lasers with similar output characteristics. All emit powerful pulses lasting nanoseconds or tens of nanoseconds at wavelengths in or near the ultraviolet. Most commercial excimer lasers can be operated with different gas mixtures to produce different output wavelengths. The technology is relatively new; while commercial devices for scientific applications are fairly well developed, those for medical and industrial applications are still evolving.

The term *excimer* originated as a contraction of "excited dimer," a description of a molecule consisting of two identical atoms which exists only in an excited state; examples include He_2 and Xe_2. It now is used in a broader sense for any diatomic molecule (and sometimes for triatomic types) in which the component atoms are bound in the excited state, but not in the ground state. That property makes them good laser materials with similar output characteristics. The most important excimer molecules are rare gas halides, compounds such as argon fluoride, krypton fluoride, xenon fluoride, and xenon chloride, which do not occur in nature, but which can be produced by passing an electric discharge through a suitable gas mixture. There are also other types of lasing excimers. Many of these molecules are so similar that they can be made to lase in the same device, which has come to be called an excimer laser. This chapter will use the terminology in the broadest commercial sense, even though some of the species which can be made to lase in excimer laser devices are not really excimers.

Although the excimer laser is one of the newest types on the commercial market, it has gained widespread acceptance in the research laboratory and is attracting increasing interest from potential industrial users. Its quick transition from laboratory demonstration to commercial product is due partially to a coincidence: the same excitation scheme can be used for excimer lasers as for transversely excited atmospheric pressure (TEA) carbon dioxide lasers. A group at the Naval Research Laboratory in 1976 found that they could build a "quick and dirty" excimer laser by using suitable optics and gases in a commercial TEA CO_2 laser. Lifetime was initially a problem because of the highly corrosive halogens in the excimer gas mixture, but laser makers soon designed versions specifically for excimer operation. Excimer lasers have become standard laboratory tools, and a significant fraction are being sold for industrial and medical applications.

Introduction and Description

Active Medium. Excimer lasers contain a mixture of gases at total pressure usually well below 5 atmospheres (atm). The bulk of the mixture, 90 to 99%, is a buffer gas which mediates energy transfer, usually helium or neon. The buffer gas does not become part of the light-emitting species. The rare gas that does combine to form excimer molecules in rare-gas-halide lasers is present in much smaller concentrations, typically 1 to 9%. The halogen donor concentration is usually 0.1 to 0.2%; it may be a diatomic halogen such as molecular fluorine, or a halogen-containing molecule such as hydrogen chloride. (F_2 is diluted with helium or neon, usually to a 5% concentration, for safety and handling.)

The optimum gas mixture is a complex function of operating conditions and gas kinetics and differs among laser models and excimer molecules. For example, argon concentration in argon fluoride lasers typically is much higher than xenon concentration in xenon fluoride lasers. Gas mixtures vary with time because halogen concentrations gradually decline regardless of whether the laser is used. High halogen concentrations extend gas lifetime between halogen replenishments; lower concentrations give more uniform beams. Manufacturers normally recommend gas mixtures for their products, which users should follow unless they are familiar with excimer gases.

Energy Transfer. The detailed kinetics of energy transfer in excimer lasers are complex and took years to unravel. The complex interactions among atomic and molecular species in the laser gas, the excitation energy, the walls of the laser tube, and the electrodes can all affect the operating parameters of the laser. The details are far beyond the scope of this book, but it is possible to give a brief overview.

In almost all commercial excimer lasers, energy is deposited in the laser gas by an electric discharge. The only commercial alternative is pumping with an electron beam. Up to 5 percent of the discharge energy can be converted into laser energy. Electron beams can deposit more energy in the gas and may offer slightly higher conversion efficiency. However, electron-beam generators are large, complex, expensive, inefficient, and limited in repetition rate. With a commercial krypton fluoride laser, overall wall-plug efficiencies to about 2.5 percent are possible, much higher than would be possible with electron-beam pumping. A third alternative, microwave excitation, has only recently been introduced in a small waveguide excimer laser. The developers chose microwave excitation because it has little tendency to form arcs or other discharge inhomogeneities, allowing uniform excitation of a small volume of high-pressure laser gas. Microwave excitation can produce pulses longer than 200 nanoseconds (ns) at repetition rates to about 2000 hertz (Hz), but typical pulse energies are under 1 millijoule (mJ).

To improve energy-transfer dynamics, and avoid discharge arcing, the laser gas is normally "preionized" before the excitation pulse. In commercial excimer lasers, preionization is normally by a pulse of ultraviolet light, but an electric discharge pulse, a pulse of x rays, or a shot from an electron beam can also be used. The electrons in the excitation pulse then produce the excited rare gas atoms and halogens that react to form excited diatomic molecules or excimers, such as xenon fluoride. (The excited state is often indicated by following the chemical formula with an asterisk, for example, XeF*.)

Excimer molecules are peculiar things. When electronically excited, the two component atoms attract each other to form a stable molecule. However, in the ground state the two atoms are mutually repulsive or in some cases weakly bound. Thus, as shown in Fig. 13.1, when an excimer drops from the excited state to the ground state, the force between the two atoms changes from attraction to repulsion, and the molecule breaks up.

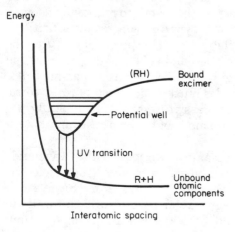

Energy

(RH) — Bound excimer

Potential well

UV transition

R+H — Unbound atomic components

Interatomic spacing

Figure 13.1 Internal energy of excimer molecules as a function of interatomic spacing for excited and ground states. The potential well makes the excited state stable, but the two atoms repel each other at all interatomic spacings in the ground state, so it is unstable. Transitions can occur over a range of energies.

This unusual energy-level structure makes excimers very good laser media. Because the ground state essentially does not exist, there is a population inversion as long as there are molecules in the excited state. Thus if excimers exist, so do the right conditions for lasing. The kinetics also lead to very high gain on excimer laser transitions.

Excited-state lifetimes on the order of 10 ns set the time scale for excimer laser pulses. Components which drive the discharge have switching times on the same scale. With conventional preionization and pulse generation technologies, the pulse length is limited to a few tens of nanoseconds by discharge instabilities, although the internal kinetics can allow pulse durations of hundreds of nanoseconds. Avalanche discharges allow pulses to 150 ns, and microwave excitation of waveguide lasers can generate 200-ns pulses. Pulses lasting 250 ns can be produced using a magnetic "spiker-sustainer" approach in which two separate discharge circuits drive the laser (McKee et al., 1989). One produces a short high-voltage low-energy spike which breaks down the laser gas; the other is a lower-voltage, impedance-matched current source which can sustain the discharge for a longer time.

Internal Structure. A conventional excimer laser contains a sealed tube filled with laser gas through which an excitation pulse is passed. In discharge-driven lasers, the electric pulse is perpendicular to the laser beam axis. In an electron-beam-pumped laser, the electrons are more likely to be directed along the beam axis. Normally some of the laser gas lies in a reservoir outside of the excitation region, and the gas may be circulated through the laser for operation at high powers or high repetition rates.

The laser cavity can be sealed and repeatedly refilled, often with different laser gas mixtures. This is needed because the laser gas degrades during use; new gas must be flowed into the active zone between shots. The gas fill degrades during use, lasting thousands to millions of shots in commercial lasers, depending on gas type and operating conditions. (Longer gas-fill lifetimes are possible with gas reprocessing.) After the gas is spent, the cavity must be emptied and refilled, with purging and passification required when changing between different gas mixtures. The laser cavity, optics, and electrodes must be designed to resist corrosion by halogens in the laser gas. Passive components typically are coated with Teflon, and electrodes are made of halogen-resistant materials such as solid nickel or brass.

An alternative that recently became available is the waveguide excimer laser. The excimer gas is housed in a small-bore dielectric waveguide, on the order of a millimeter across. A microwave discharge is applied across the waveguide; for best results the walls in contact with the electrodes should be less than 1 millimeter (mm) apart. Repetition rates to 8 kHz have been demonstrated in xenon chloride without flowing the gas (Christensen, 1987), but commercial models flow the gas slowly through the waveguide.

Optics. Excimer lasers have such high internal gain that they are virtually superradiant, and this is a dominant factor in determining the design of their optical cavities. As in most other gas lasers, the rear cavity mirror is highly reflective. However, the high gain eliminates the need for a reflective coating on the output mirror. The normal reflectivity of about 4 to 10 percent from an uncoated optical surface provides enough feedback for laser operation; some ArF lasers use sapphire optics for about 18 percent feedback. A diffraction grating can be inserted in the cavity to allow wavelength tuning of output. In some models, gate valves can be shut to isolate the cavity optics from the gas mixture so that the optics can be changed without contaminating the gas mixture.

Quartz or fused-silica optics may be listed as standard, but they are used only in lasers which operate only with XeCl; they are etched by fluorine in other gas mixtures. The most common optical material is magnesium fluoride; calcium fluoride also is used. Both are more transparent than quartz at short wavelengths. Optics affixed directly to the laser cavity are exposed directly to the laser gas, which can cause chemical etching with ArF or dust deposition with XeCl. Some lasers put windows in contact with the laser gas and mount cavity optics externally, to allow servicing of windows in contact with the gas without affecting alignment. Reflective coatings for the rear of the laser cavity normally are deposited on the outside of the window to avoid damage from the laser gas.

Air strongly absorbs short-wavelength ultraviolet light. Wavelengths shorter than about 200 nm are known as the "vacuum ultraviolet" because atmospheric absorption is so strong that they propagate well only in a vacuum. ArF beams at 193 nm are attenuated about 10 percent per meter in air and produce large amounts of ozone, but they can propagate in helium. Wavelengths between 200 and 300 nm can travel over laboratory scales, but virtually no solar radiation at wavelengths shorter than 290 nm can reach the earth's surface.

Cavity Length and Configuration. Excimer laser cavities generally are somewhat under a meter long. The standard stable resonator generates a fairly large, divergent beam because of the high gain of excimers. Optional unstable resonator optics give smaller and more uniform beams, but those qualities may come at the cost of reduced output power. Another possible configuration is the master oscillator–power amplifier (MOPA) arrangement shown in Fig. 13.2 (often called

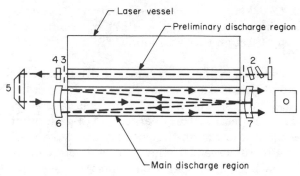

Figure 13.2 Narrow-line output of a master oscillator (top) is amplified in an unstable-resonator amplifier (bottom), extracting energy from most of the unstable-resonator cavity. Components are (1) master oscillator rear reflector, (2) etalons, (3) apertures, (4) master oscillator output coupler, (5) prism, (6) rear unstable-resonator optic, (7) front unstable-resonator optic. (*Courtesy of Lumonics Inc.*)

injection locking), in which an unstable-resonator amplifier amplifies the output of a narrow-line oscillator. Alternatively, a power oscillator–power amplifier (POPA) configuration can create high energy beams with low divergence.

Inherent Trade-Offs. Like other repetitively pulsed lasers, excimers produce less energy per pulse once the repetition rate exceeds a certain level. At low repetition rates, average power increases linearly with the number of pulses per second. However, above a certain point, the pulse energy decreases with repetition rate unless discharge power is increased. At somewhat higher levels, the average power actually declines as repetition rate increases, because pulse energy is decreasing faster than the number of pulses per second is increasing.

Three factors combine to limit the average laser power: the energy density limit of the discharge, the flow of gas available between the electrodes, and the maximum electrical power the high-voltage supply can deliver to the laser. Modern commercial lasers can operate at repetition rates above 300 Hz, with limitations becoming important at higher repetition rates. Figure 13.3 shows how the pulse energy and average power change with repetition rate for different gases in the same laser tube. Note that the scales are different.

No one type of excimer laser is ideal. Krypton fluoride has the highest pulse energy and average power, but its operating life is shorter than that of xenon chloride, and its 249-nm wavelength is not transmitted well by air over long distances. XeCl has a longer operating life and its 308-nm output is better suited for atmospheric transmission,

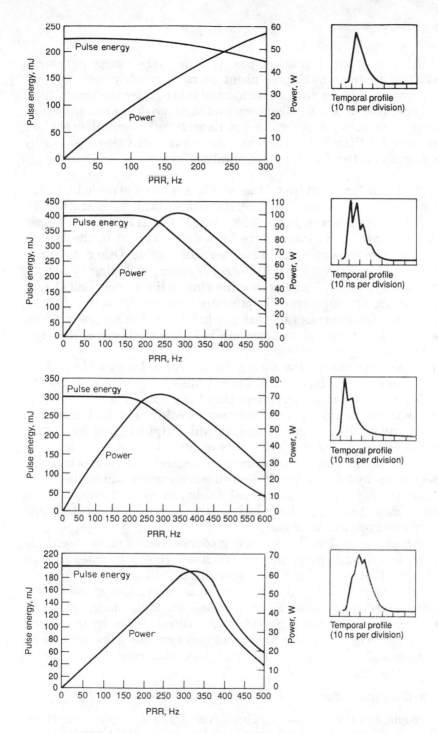

Figure 13.3 Pulse energy, average power, and pulse shape for different laser gases in the same excimer laser. Pulse energies normally are constant at low repetition rates, but eventually drop at higher levels. Each gas has a slightly different pulse shape. Note that scales for pulse energy, repetition rate, and average power are different for each gas. (*Courtesy of Lumonics Inc.*)

but it is only two-thirds as powerful as KrF—except for x-ray preionized lasers, which have about twice as much power on XeCl lines than on KrF. In practice, industrial users prefer the longer-lived XeCl over KrF unless the shorter wavelength produces better results. The other principal excimers, argon fluoride and xenon fluoride, are less powerful than XeCl, but have their own attractions, primarily wavelengths better for certain applications.

Variations and Types Covered.   Excimer lasers are a generic type, and commercial excimer lasers can operate with several gas mixtures. The best known are the rare gas halides: ArF, KrCl, KrF, XeCl, and XeF. Other rare-gas-halide lasers have been demonstrated in the laboratory, but they have not proved commercially viable. Other gases can also be used in some commercial excimer lasers, including F_2 and N_2 in the ultraviolet, N_2^+ and atomic fluorine in the visible, and CO_2 in the infrared, although special optics are required for some of them. This chapter concentrates on rare gas halides and other gases which are used in commercial excimer lasers but not otherwise common; thus N_2 and CO_2 are excluded.

The rare gas halides are not the only excimer lasers to have been demonstrated in the laboratory. There is also a family of mercury halide lasers with similar properties that have ground states in which the atoms are bound together only very weakly. The best known of these is the mercury bromide laser, in which HgBr formed by dissociation of $HgBr_2$ at about 160°C emits at 502 to 504 nm in the blue-green. No commercial versions are on the market, but it has received careful scrutiny because its blue-green wavelength could be useful for communicating with submerged submarines (Burnham and Schmitschek, 1981). Problems include the elevated temperature and the chemistry of the laser gas.

There are several variables among commercial excimer lasers, including wavelength, pulse energy, maximum repetition rate, average power, gas lifetime, and pulse duration. Table 13.1 gives a sampling of these characteristics for selected commercial lasers. Commercial models themselves also differ. Some models offer high-energy pulses at low repetition rate, others provide high average power by delivering moderate-energy pulses at a high repetition rate, and a few try to combine both high-energy pulses and high repetition rate.

Beam Characteristics

Wavelength and Output Power.   The wavelengths of major excimer laser gases together with pulse energy and average output power from a sampling of 1990-vintage commercial models are shown in Table 13.1.

TABLE 13.1 Pulse Energy, Average Power, and Repetition Rate for Representative Commercial Excimer Lasers*

Laser and gas	F2	ArF	KrCl	KrF	XeCl	XeF
Wavelength, nm	157	193	222	249	308	350
Lambda Physik LPX105						
Pulse energy, mJ	—	125	—	225	150	75
Average power, W	—	4	—	10	6	3
Repetition rate, Hz	—	50	—	50	50	50
Lambda Physik LPF205 (Vacuum Ultraviolet Optics Only)						
Pulse energy, mJ	60	100	—	—	—	—
Average power, W	3	5	—	—	—	—
Repetition rate, Hz	50	50	—	—	—	—
Lambda Physik LPX315i						
Pulse energy, mJ	—	500	—	800	600	400
Average power, W	—	45	—	100	75	45
Repetition rate, Hz	—	150	—	150	150	150
Lumonics Inc. Index-210 (Industrial)						
Pulse energy, mJ	—	100	—	250	150	—
Average power, W	—	30	—	75	45	—
Repetition rate, Hz	—	300	—	300	300	—
Lumonics Inc. Excimer-600						
Pulse energy, mJ	—	225	—	400	300	250
Average power, W	—	55	—	100	70	60
Repetition rate, Hz	—	350	—	500	600	600
Questek Inc. 2580vβ						
Pulse energy, mJ	—	300	—	500	400	300
Average power, W	—	40	—	100	60	50
Repetition rate, Hz	—	500	—	500	500	500
Questek Inc. 2920						
Pulse energy, mJ	—	700	—	900	500	400
Average power, W	—	5.6	—	7.2	4	3.2
Repetition rate, Hz	—	10	—	10	10	10
Siemens 2020						
Pulse energy, mJ	—	—	—	2000	—	—
Average power, W	—	—	—	40	—	—
Repetition rate, Hz	—	—	—	40	—	—

*All figures are maximums stated by manufacturers on data sheets or in industry directories. It may not be possible to realize all three at once. The lasers listed are representative of those available in 1990; each company offer other lasers, and other companies also produce excimer lasers.

Other ultraviolet and visible wavelengths can be produced by Raman shifting the standard excimer lines.

As a repetitively pulsed laser, the excimer laser has an average output power (in watts) that is the product of the pulse energy (in joules) times the number of pulses per second (hertz). On the strongest lines—ArF, KrF, XeCl, and XeF—typical average powers range from under a watt to 100 W or more. However, lasers with similar average power may have quite different output characteristics. A 10-W average power may be produced by generating 100 pulses of 100 millijoules (mJ) each, or ten 1-J pulses.

Many laser manufacturers list maximum values for average power, pulse energy, and repetition rate on their data sheets. However, a quick comparison often shows that the maximum average power is *not* the product of maximum repetition rate times peak pulse energy. Typically, the pulse energy is highest at low repetition rates, than starts dropping beyond a certain repetition rate. The maximum average power normally is reached at a repetition rate slightly above the point at which pulse energy starts dropping, as shown in Fig. 13.3. Further increases in repetition rate cause average power to decrease, because the pulse energy drops faster than the number of pulses increases. By adjusting the gas mixture, some laser manufacturers can avoid the drop in pulse energy at high repetition rates, but only by trading off with lower energies at low repetition rates. One manufacturer uses three different gas mixtures in a 0- to 500-Hz laser for XeCl and KrF: one for 0 to 200 Hz, a second for 200 to 350 Hz, and a third for 350 to 500 Hz.

As Table 13.1 indicates, the different excimer gases generate different power levels when operated in the same laser. The ratios of powers generated by different gases are not constant because the "best" discharge conditions differ among gases. Most lasers are designed to operate with the three most powerful excimers, ArF, KrF, and XeCl. Many also are specified for operation with XeF lines, and some can operate with F_2. Some models are optimized for vacuum-ultraviolet F_2 and ArF lines. The 222-nm KrCl laser is considerably weaker than the others, producing pulse energies and average powers only about a tenth those from the most powerful excimer, KrF.

Some applications require a specific wavelength, limiting choice to a single excimer. However, others may require only high-energy ultraviolet pulses. In that case, the choice is often between the 249-nm KrF laser and the 308-nm XeCl laser. KrF generally is more powerful, but XeCl gas fills last longer because they do not contain highly corrosive fluorine, and the longer wavelength is easier to handle optically. Those trade-offs should be factored into any consideration of excimer types. (Some brochures for industrial excimer lasers do not prominently specify the active medium; one way to solve the mystery is to look carefully at the warning label required by federal laser product standards, which includes a description of the laser and its highest power level.)

Research laboratories are working on excimer lasers with pulse energies and average powers much higher than are available in commercial products. High-energy single-pulse KrF lasers are being developed for research on inertial confinement fusion. The largest of these is the Aurora laser at the Los Alamos National Laboratory, designed to deliver 5-ns 5-kJ pulses to targets (Cartwright et al., 1990). Developers of military laser weapons have sought to build excimer lasers with multikilowatt average powers; they have studied XeF lasers because the better beam propagation at the longer wavelength offsets the lower power of KrF. However, few details have emerged on progress.

Efficiency.   Typical wall-plug efficiency for a discharge-driven commercial KrF laser is 1.5 to 2 percent. What is called "electrical" or "discharge" efficiency—the fraction of energy stored in the discharge circuit that emerges in the laser beam—is typically 2.5 to 3 percent, and can reach 5 percent for KrF.

Even higher efficiency can be calculated from the fraction of energy deposited in the laser gas, ignoring discharge energy not deposited in the gas. This is often done in the scientific literature. Typical figures are in the 4 to 5 percent range for commercial discharge-driven KrF. Electron-beam excitation gives higher efficiency for deposited energy, but those benefits are more than offset by the losses in generating the electron beam.

Other excimer gases are less efficient than KrF, with the relative efficiency roughly proportional to relative output power under the same operating conditions. For example, a laser with 5 percent discharge efficiency with KrF has 3.5 percent discharge efficiency with XeCl.

Temporal Characteristics.   Pulses from commercial discharge-excited excimer lasers typically last from a few nanoseconds to a few tens of nanoseconds. Pulse lengths can vary significantly among different gases used in the same laser, as indicated in Table 13.2. Individual pulses often have structure within them, which can differ among gases, as shown in Fig. 13.3.

Overcoming the discharge instabilities which limit pulse length to tens of nanoseconds allows pulse lengths to reach hundreds of nanoseconds. The spiker-sustainer approach can generate 200-ns pulses, and repetition rates to 220 Hz have been demonstrated with XeCl

TABLE 13.2   Differences in Nominal Pulse
Lengths for Different Gases in Two Multigas
Excimer Lasers

	Pulse length, ns	
Gas	Laser 1	Laser 2
F_2	—	12
ArF	8–10	23
KrCl	5–8	—
KrF	12–16	34
XeCl	8–12	28
XeF	14–18	30

(McKee et al., 1989). That technology has yet to make its way into standard commercial products. Electron-beam pumping can stretch pulses to the microsecond range, but only the cost of low repetition rates and efficiency.

Repetition rates vary considerably. A few lasers are designed for slow pulsing, at about one pulse per second, but rates of tens to several hundred pulses per second are common for commercial excimer lasers. Repetition rates over 3 kHz have been demonstrated with an experimental XeCl system with rapidly flowing gas, but average power dropped above 2.5 kHz, and at 4 kHz pulsing was irregular (Takagi et al., 1990). For excimer lasers using commercial discharge excitation technology, the average power drops as repetition rate increases above several hundred hertz.

Microwave excitation of a waveguide laser allows excimer pulses to reach about 250 ns, the length of which can be controlled electrically. Repetition rates can reach 2500 Hz in commercial versions, but pulse energies are well below 1 mJ, so average powers are low.

Extremely short pulses, lasting only tens of femtoseconds, have been generated by using excimer lasers to amplify ultrashort pulses from dye lasers. Normally, however, excimer lasers generate pulses of at least several nanoseconds.

Spectral Bandwidth.   The laser transitions in excimer lasers are broad, due to the lack of well-defined energy levels in the ground state, as shown in Fig. 13.1. Thus laser emission takes place over a range of wavelengths. Typical linewidth of an excimer laser is about 0.3 nm, with output wavelength tunable across about 3 nm with some loss of output energy. Standard commercial lasers produce broadband emission on the laser transition, but the user can add line-narrowing optics or grating tuning within the broad laser gain band. Injection locking to a narrowband source also can limit bandwidth.

Amplitude Noise and Power Stability. Pulse-to-pulse variations can be significant in excimer lasers, with specified values averaging ±3 percent for KrF, XeCl, and XeF. The figure depends on the laser gas, and is larger for the weaker laser lines. One supplier lists pulse stabilities of ±3 percent for KrF, XeCl, and XeF, ±5 percent for ArF, and ±15 percent for F_2 operated in the same laser. Some companies offer pulse-stabilization options to reduce variation on strong lines.

In the long term, output energy inevitably declines over hundreds of thousands of shots if the gas mixture is not cleaned and makeup gas added. Generally the decline is smooth and gradual. One way to delay such declines is to provide excess initial capacity in the laser and raise the discharge voltage to automatically maintain constant output power until the excess capacity is exhausted. Functionally, both gas cleaning and replacement and voltage adjustment serve as controls to stabilize energy.

Beam Quality, Polarization, and Modes. Stable-resonator excimer lasers have multimode output and poor beam quality because of their high gain. The highly multimode output (one estimate is 2500 modes) is desirable for applications such as microlithography for the fabrication of integrated circuits, where the speckle produced by lasers with better beam quality introduces undesired patterns on the surface.

Higher-quality beams are possible if unstable resonators are used, or with long laser pulses (which give light time to make many round trips of the cavity). Beams normally have flat or gaussian profiles (often flat in one direction and gaussian in the other), but some have doughnut-profile beams. Normally output is unpolarized, but polarizing optics can be added.

Coherence Length. The broad emission bandwidth of excimer lasers leads to limited coherence lengths. For a typical laser with 0.3-nm linewidth at 300 nm, corresponding to 10^{12}-Hz bandwidth, calculated coherence length is about 0.3 mm.

Beam Diameter and Divergence. Conventional excimer lasers with stable-resonator optics typically have oblong beams roughly 10 by 20 mm across at the output window. Although there are minor variations among different models and gases, all such discharge-driven excimer lasers are similar enough that beam dimensions almost always fall within a factor of 3 of those values. Waveguide lasers have much smaller beams, about 0.5 by 2 mm, reflecting their smaller internal diameters.

Stable resonator excimers have typical beam divergences of 1 × 3 or 2 × 3 milliradians (mrad), although some models have larger divergences. Waveguide excimers have divergences as large as 2×10 mrad.

Unstable resonators produce beams about the same size as stable resonators when used with conventional discharge-driven excimer lasers. However, their divergence is much smaller, typically 0.2×0.2 mrad, and their near-field energy is only about 40 percent that of a stable-resonator beam.

Stability of Beam Direction. The broad divergence and multimode output of most excimer lasers indicate that some variation in beam direction might be expected, but it is typically much smaller than divergence. However, published specifications do not list this quantity.

Suitability for Use with Laser Accessories. Because of their ultraviolet wavelength, excimers are not suitable for use with conventional frequency-doubling crystals. However, harmonics can be generated by passing excimer laser pulses through gases. This technique has been used to generate coherent radiation at wavelengths as short as 35.5 nm, the seventh harmonic of KrF (Bokor et al., 1983). Although this opens up the extreme ultraviolet to laboratory study, at this writing applications remain in the research stage.

Raman interactions in molecular gases such as hydrogen can produce smaller frequency shifts when a tightly focused excimer laser beam is passed through a suitable gas cell. The Raman shift can be toward either longer or shorter wavelengths, but generally the longer wavelengths are generated more efficiently. The wavelengths and relative output powers produced by a commercial Raman shifter are shown in Fig. 13.4; qualitatively similar shifts would be produced by other such devices.

Q switching of excimer lasers is impractical because of the high gain. Both active and passive modelocking have been demonstrated in the laboratory, but they are not used in practice.

Operating Requirements

Input Power. Small excimer-laser power supplies draw on the order of 10 A from 110-V, single-phase sources. Higher-power models require 208- or 380-V, three-phase service, and draw currents of 20 to as much as 100 A. Specifications for some of the highest-power lasers do not mention input power requirements. All discharge-driven excimer lasers require high internal voltages, typically in the range of tens of kilovolts.

Cooling. Many excimer lasers require tap-water cooling at rates of 2 to 10 liters per minute (L/min), with the largest reaching nearly 60

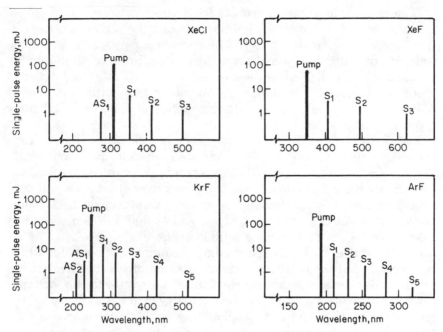

Figure 13.4 Typical output characteristics of a Raman-shifting cell for operation at 1-Hz repetition rate and optimum hydrogen pressure. S_n indicates Stokes-shifted line n; AS_n indicates anti-Stokes-shifted line n. The pumping laser is identified in the upper right corner of each small graph. (*Courtesy of Lambda Physik.*)

L/min. Others, operating at lower pulse energies or repetition rates, are air-cooled.

Consumables. The prime consumable in excimer lasers is the laser gas, which can account for a large part of operating costs. The gas deteriorates both during laser operation and with time, and eventually must be pumped out of the laser cavity and replaced. The number of shots possible varies among laser gases, as shown in Table 13.3 for one commercial laser. Gas lifetime also differs among different laser models.

TABLE 13.3 Excimer Gas Lifetimes in a Multigas Laser without Special Cleaning Equipment or Gas Replenishment*

Laser	Wavelength, nm	Life (10^6 shots)*
F_2	157	0.05
ArF	193	0.4
KrF	248	1.0
XeCl	308	10.0
XeF	351	2.0

*Representative data; actual life depends on operating conditions.

Generally excimer lasers operate from three separate gas supplies: the inert buffer gas (typically helium or neon), the active rare gas (argon, krypton, or xenon), and the halogen being used. Typically the halogens are supplied as hydrogen chloride or fluorine, diluted to 5% by the addition of helium or neon to render the gas safer and easier to store and handle. Neon buffer gas increases laser efficiency, but it costs more than helium. Premixed supplies of the combined laser gases may also be used.

The cost of gas to fill an excimer laser is significant. Recent estimates give costs as $25 to $35 for ArF, $35 to $45 for KrF, and $60 to $80 for XeCl (Jursich et al., 1989). Xenon is the most costly gas, but fortunately little is needed. Electrical power may also be a significant operating cost, but it normally disappears into overhead. Gas purchases come out of the user's budget, and at 200 pulses a second, it takes only 1.4 hours to reach a million shots.

Gas lifetime is normally specified as the number of shots for which a laser can deliver at least a certain fraction of its specified output power under certain operating conditions. This reflects the fact that discharge-produced impurities are a primary factor that degrades the gas. However, the laser gas also degrades as the halogen reacts with the walls and other internal components of the laser. These reactions occur even when the laser is off or operating at a low repetition rate, limiting the static or shelf life of the gas fill. Figure 13.5 shows how the two effects combine to reduce laser pulse energy. In practice, estimates of gas lifetime exclude this static degradation and assume that the laser is operating almost continually at a high repetition rate. Lifetime is very sensitive to operating conditions such as pulse energy, repetition rate, and halogen concentrations, so be cautious in making comparisons.

Gas lifetime can be extended by purifying it to remove impurities, by partial gas replacement, and by adding small volumes of halogen-rich gas. Cryogenic gas processors can trap less-volatile impurities, increasing gas life by a factor of 2 for XeCl to a factor of 10 for ArF. However, they cost around $10,000 and consume liquid nitrogen.

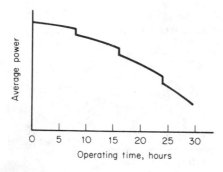

Figure 13.5 Average power fall-off for an excimer laser operated only during an 8-hour workday but left off overnight, plotted against *operating* (not total) hours. The scale of the overnight drop is roughly that for KrF and XeF: there would be a smaller drop with XeCl, but a larger one with ArF. (*Courtesy of Questek Inc.*)

Waveguide excimer lasers are not sealed, but pass a slow flow of the laser gas (XeCl is the standard type, made up of He, Xe, and HCl) through the tube. Gas flow rates are 10 to 20 cubic centimeters (cm^3) at standard temperature and pressure per minute.

For heavy industrial use, cavity optics and high-voltage components become significant "consumables" because they are put under heavy stress. Their costs often exceed those of the laser gas except in scientific applications requiring low laser utilization.

Required Accessories.   A vacuum pump is needed to remove spent laser gas, but this is often included with the laser. Equipment also is needed to handle the spent gas, which cannot be exhausted directly to the air because of its halogen content.

Vacuum systems which exclude air are needed for the propagation of vacuum ultraviolet beams—the 157-nm output of F_2 and sometimes the 193 -nm of ArF. The ArF wavelength can travel short distances through helium, argon, or nitrogen as well as through vacuum.

Operating Conditions and Temperature.   Commercial excimer lasers are designed for a normal laboratory or light industrial environment, and work well there.

The gas cavity of an excimer laser requires passivation to limit reactions involving halogens in the gas mixture. It is particularly important to thoroughly purge the old mixture from the cavity when switching to a different gas. Long-term operation with a chlorine mixture may impair operation with fluorine mixtures.

Mechanical Considerations.   Excimer lasers come in a wide range of shapes and sizes. Waveguide excimers are compact, with a laser head as small as 0.34 by 0.23 by 0.2 meter (m) with a separate controller power supply that measures 0.48 by 0.43 by 0.32 m and fits in a standard U.S. 19-inch (in) (0.48-m) instrumentation rack. Conventional laboratory lasers also have separate laser head and power supply, but the heads are considerably larger, with a typical one measuring 1.45 by 0.8 by 0.47 m and weighing 200 kilograms (kg). The vacuum pump and controls also may be housed separately.

High-power industrial excimer lasers can be considerably larger. One model is packaged as a single unit which measures 1.9 by 1.1 by 2.1 m and weighs 2600 kg. Others come packaged in two units, a laser head which rests on a massive base.

Safety.   There are four potential hazards from excimer lasers: high voltages, x rays used for preionization in certain models, ultraviolet light, and the laser gases themselves.

Discharge-driven excimer lasers apply tens of kilovolts across the laser gas. These high voltages are stored and switched within the laser and power supply, and care should be taken that all stored voltages have been dissipated before poking around inside the laser.

If x rays are used to preionize the laser gas, they could pose a radiological hazard. Such products should come with proper shielding and warning labels to allow safe operation.

Short, powerful ultraviolet pulses are hazardous to the eye, and ultraviolet exposure also can harm the skin. Pulses with wavelengths longer than about 315 nm can cause retinal damage, although exposure limits are not well quantified. Shorter ultraviolet wavelengths do not reach the retina but are absorbed in the cornea, where they can cause a painful but temporary sunburnlike effect called "snow blindness" or "welder's flash." Ultraviolet light also can present dangers to the lens of the eye. Skin exposure can lead to the same effects as overexposure to the sun's ultraviolet rays (Sliney and Wolbarsht, 1980).

The rare gases used in excimer lasers are relatively innocuous, although high-pressure cylinders pose some hazards and any inert gas can suffocate by displacing oxygen. Halogens are bad news. Safety standards limit fluorine content to 0.1 part per million (ppm) and chlorine content to 1 ppm in air. Both the laser mixture and waste gas from the laser present hazards. Pure F_2 is so corrosive that it presents fire and explosion hazards; to avoid these dangers, F_2 for excimer lasers normally is diluted to 5 percent with neon or helium before the gas cylinders are shipped. Users should be familiar with halogen safety (Dietz and Bradford, 1990).

Special Considerations.   The high-voltage discharges in excimer lasers generate strong electromagnetic interference (EMI) which can affect surrounding equipment. Newer lasers must limit emission to meet requirements set by the Federal Communications Commission (FCC), but older models may not meet these standards.

Many excimer lasers incorporate computer control or are designed to interface with personal computers. Equipment built into the laser or controller is shielded, and fiber-optic cables, which are impervious to EMI, often carry control signals between separate elements of an excimer laser, or within the laser package.

Operation of pulsed excimers can generate an audible "ping" at each pulse. The noise may be annoying, but it does not indicate a problem.

Reliability and Maintenance

Lifetime.   Excimer laser makers traditionally have not quoted lifetimes for scientific lasers, although they do list gas-fill lifetimes. They have collected more data on component and system reliability with the growth of their installed base and increasing industrial interest in excimers, and data on mean time between failures usually are available for industrial lasers. Many uncertainties remain because designs are continually evolving and operating conditions strongly affect lifetimes.

Most failures are of individual components rather than an entire laser. One manufacturer assigned the following lower limits to compo nent lifetimes based on its testing and experience (Andrellos et al., 1986):

Electrodes	5×10^8 (except 1×10^8 for ArF)
Preionizer	5×10^8 (except 1×10^8 for ArF)
Optics	$>5 \times 10^8$
Capacitors	$>10^9$
Thyratron	$>10^9$
High-voltage power supply	$>10^9$
Magnetic switch	5×10^9

The output windows on the laser cavity are exposed to corrosion and contaminants from the laser gas, as well as intense ultraviolet light. The damage depends on the laser gas and operating conditions. ArF tends to corrode output windows, while XeCl leaves a fine dust. The gas mixtures also can deposit material on inside surfaces. In many cases, the windows can be removed and cleaned. The short-wavelength ultraviolet output of ArF and KrF lasers can form color-center defects in windows; annealing for several hours at 300°C can remove color centers in MgF_2 optics.

Maintenance and Adjustments Needed.   Excimer lasers require several types of periodic maintenance at intervals usually specified by the number of shots. The most frequent maintenance is changing of the laser gas.

Output windows require periodic cleaning to remove contaminants and corrosion from surfaces exposed to the laser gas. This is not easy because cleaning interior optics requires breaking and later re-forming gas seals that prevent leakage around the windows. If air enters the laser cavity, the interior must be passivated with the gas mixture to be used. To simplify optics maintenance, some tubes include gate valves, which isolate the optics from the rest of the laser tube. The user can shut the valve to seal the laser tube before opening the end to clean or remove the optics. This exposes only a small part of the gas plumbing to the outside world, and can preserve a partly used gas mixture.

Typical intervals for routine maintenance recommended by one excimer manufacturer are

- Clean optics after 10^8 shots.
- Change halogen filter after 5×10^8 shots.
- Change optics after 5×10^8 shots.
- Replace electrodes and preionizer after 5×10^8 shots.
- Replace thyratron and high-voltage capacitor at major overhaul of laser after $\geq 10^9$ shots.

Mechanical Durability.   Excimer lasers are not designed to resist heavy shocks and should be given the usual care accorded to laboratory equipment.

Failure Modes and Causes.   The corrosive halogens used in excimer lasers still present some problems, although the use of nickel electrodes and Teflon-coated cavities prevents the severe corrosion problems that ate away early excimer lasers. Nonetheless, exposed electrodes and windows still suffer from corrosion and deposition of contaminants. The degradation of windows and other optics can be accelerated by the powerful ultraviolet radiation produced by excimer lasers. High-voltage switching electronics are subject to their own failure mechanisms.

One of the commonest problems is operator error, accidentally allowing laboratory air into the inside of the laser cavity. Reactions of air with the laser gas or contaminants within the laser cavity can cause problems and require repassification of the inside of the laser with fresh laser gas, or possibly cleaning of the cavity and optics.

Possible Repairs.   Failures in high-voltage switching electronics or in optics require replacement of the damaged components. Corrosion and contamination problems require cleaning or replacement of the affected components.

Commercial Devices

Standard Configurations.   Conventional discharge-driven excimer lasers typically are available in a variety of models based on a few common designs. Most are designed to operate on the major excimer transitions: ArF, KrF, XeCl, and XeF. Some can operate on the 157-nm F_2 band, and a few are specified for the 222-nm KrCl transition, which is losing popularity because of its limited performance. In the past, some manufacturers offered operation of excimer lasers in two other modes, as N_2^+ lasers at 428 nm, and as atomic fluorine lasers at 624 to 780 nm, but today those possible operating modes are ignored.

Major excimer manufacturers assemble their product lines from a number of common components, so they can meet a broad range of user needs without designing many different complete lasers. For example, several models may share the same laser cavity but include power supplies which generate pulses at different rates, giving different repetition rates and average powers. The same laser cavity and power supply may be packaged in different ways for industrial and scientific users. Some lasers are built especially for operation at the two shortest excimer transitions, the 157-nm F_2 line and the 193-nm ArF line, with power supplies as well as optics optimized for those gases.

Some companies offer both "industrial" and "scientific" product lines. The output characteristics may be similar, but the industrial models typically are designed to facilitate use and service, while the scientific models offer flexibility and performance. Industrial versions may include monitoring or active control of pulse energy and specially mounted optics. Typically they are built to operate on only one transition, and in some cases they may be incorporated into complete systems designed for a specific application. Such special-purpose systems mark the high end of the excimer price-performance range.

Most excimer lasers have pushed to higher repetition rates as the technology improved, but those designs require high-voltage thyratron switches, which are costly and have limited lifetime. An alternative is a "switchless" approach, which uses a high-voltage thryistor to control a pulse-forming network. However, thus far the circuit is limited to a 10-Hz repetition rate.

Waveguide excimer lasers have recently become available. These compact, inexpensive devices generate low-energy pulses at a high repetition rate, made possible by microwave excitation of the excimer gas. Unlike other excimer lasers, the pulse length can be altered electronically.

Options.   A variety of options are common on excimer lasers:

- Computer controls and interfaces.
- Stabilization of pulse energies.
- Integral helium-neon laser for beam alignment with visible light.
- Fiber-optic beam coupling.
- Resonator extensions to decrease pulse width by reducing the number of round-trip oscillations. (A 30 percent reduction in pulse width decreases pulse energy by the same amount.)
- Line-narrowing optics.

- Intracavity grating for tuning.
- Oscillator-amplifier configurations, including injection locking of a slave amplifier to a master oscillator.
- Unstable-resonator optics.
- Choice of less-expensive quartz optics or MgF_2 or CaF_2 optics which are more transparent at shorter ultraviolet wavelengths.
- Gas purification and/or replenishment systems to extend gas-fill lifetime.
- Polarizing optics.

Special Notes. Most excimer lasers allow internal or external triggering of pulses. Repetition-rate limitations often are imposed by the design of the charging electronics and/or constraints of the power supply.

Excimer lasers are slowly emerging from the laboratory and moving into real-world applications such as laser marking and semiconductor processing. Most manufacturers are very interested in industrial applications, and are adapting their designs toward more "industrial" environments.

Pricing. Current (1990) prices for excimer lasers start at around $14,000 for the lowest-priced waveguide laser. Conventional lasers typically run from $40,000 to over $100,000 for standard scientific and industrial models. The highest-power lasers, industrial systems, and some customized models can cost hundreds of thousands of dollars.

Suppliers. A few companies have dominated the excimer market for several years, but new companies have appeared recently. Some build their own excimer lasers for specialized applications such as medicine; others offer novel designs. The market is competitive.

Applications

The ability to generate high-energy ultraviolet pulses at sufficiently high repetition rates for average powers that can exceed 100 W has created a steady demand for excimer lasers in many fields. No other commercial laser can generate such high average power at such a short wavelength. Most excimer applications fall into three broad categories: scientific research, industrial applications, and medicine.

Scientific research has long accounted for most excimer laser sales, although industrial sales are rising. Major scientific uses include

- Pumping of tunable dye lasers
- Ultraviolet spectroscopy

- Nonlinear spectroscopy
- Photoexcitation and photochemistry
- Remote sensing
- Pollution monitoring
- Materials research
- Extreme-ultraviolet interactions
- Fabrication of high-temperature superconductors
- Plasma research
- Raman shifting
- Laser-induced nuclear fusion

Industrial applications of excimer lasers have increased slowly but steadily for several years, but have yet to find wide use, partly because many potential users worry that excimers may not be ready for regular use as practical tools. Many applications remain at the development stage; others are in limited use:

- Semiconductor fabrication systems, including pattern generators, wafer steppers, and mask aligners
- Marking of ceramics, glass, plastics, and metals
- Repair of integrated circuits
- Annealing and doping of semiconductors
- Precision micromachining and drilling
- Thin-film deposition
- Surface treatment of materials
- Chemical vapor deposition

Medical applications have attracted much interest, although they, too, have yet to gain widespread acceptance. Some are in clinical testing; others are in early development. Potential medical applications include

- Sculpting of the surface of the cornea to correct refractive errors in vision, reducing the need for corrective lenses
- Laser angioplasty, to remove plaque from clogged arteries by delivering laser pulses through an optical fiber threaded through the artery
- Microsurgery

Bibliography

Jack Andrellos, Mary Essary, and Herbert Pummer: "High average power commercial excimer lasers," paper presented at *SPIE Fiber Laser*, September 18, 1986, reprinted by Lambda Physik, Acton, Mass. (describes reliability issues).

J. Bokor et al.: "Generation of 35.5-nm coherent radiation," *Optics Letters* 8(4):217–219, 1983 (research report).

Ralph Burnham and Erhard J. Schmitschek: "High-power blue-green lasers," *Laser Focus* 17(6):54–66, June 1981 (review of early work in mercury halide lasers).

D. C. Cartwright et al., "KrF lasers for inertial confinement fusion," paper CWG2 in *Technical Digest, Conference on Lasers and Electro-Optics*, Baltimore, May 21–25, 1990, Optical Society of America, Washington, D.C., 1990 (short review).

C. P. Christensen et al.: "High repetition rate XeCl waveguide laser without gas flow," *Optics Letters* 12(3):169–171, March 1987 (research report).

Al Dietz and Elaine Bradford: "Safe handling of excimer laser gases," *1990 Photonics Design and Application Handbook*, Laurin Publishing, Pittsfield, Mass., 1990 (safety guidelines).

Gregory Jursich et al.: "Chemistry studies improve excimer gas lifetimes," *Laser Focus World* 25(6):93–100, June 1989 (overview of excimer-gas chemistry).

T. Lehecka et al.: "Gas lasers," in Kai Cheng (ed.), *Handbook of Microwave and Optical Components*, vol. 3, *Optical Components*, Wiley-Interscience, New York, 1990; see pp. 563–585 for review of excimer lasers.

Terrence J. McKee, Gary Boyd, and Thomas A. Znotins: "A high-power, long pulse excimer laser," *IEEE Photonics Technology Letters* 1(3):59–61, 1989 (research report).

Elmar Müller-Horsche, Peter Oesterlin, and Dirk Basting: "Excimer lasers: maturity and beyond," *Lasers & Optronics* 9(3):39–44, March 1990 (overview).

Charles K. Rhodes (ed.): *Excimer Lasers*, 2d ed., Springer-Verlag, Berlin and New York, (collection of review articles.

David Sliney and Myron Wolbarsht: *Safety with Lasers and Other Optical Sources*, Plenum, New York, 1980 (comprehensive handbook on optical safety).

S. Takagi et al.: "Laser power saturation in ultra-high repetition rate operation of a XeCl excimer laser," paper CThB3 in *Technical Digest, Conference on Lasers and Electro-Optics*, Baltimore, May 21–25, 1990, Optical Society of America, Washington, D.C., 1990 (research report).

J. F. Young et al.: "Microwave excitation of excimer lasers," *Laser Focus* 18(4):63–67, April 1982 (short review).

Nitrogen Lasers

First demonstrated nearly three decades ago (Heard, 1963), the 337-nanometer (nm) molecular nitrogen laser has been available commercially since 1972. One of the first ultraviolet lasers on the market, it was an early pump source for pulsed dye lasers in research laboratories. Excimer lasers and third- and fourth-harmonic output from neodymium lasers can outperform nitrogen, and have pushed it into the background. However, nitrogen lasers remain a viable source of pulsed near-ultraviolet light for a limited range of applications.

Nitrogen lasers offer low-cost, simple operation and short, high-power ultraviolet pulses. Atmospheric pressure nitrogen lasers can produce nanosecond or subnanosecond pulses without the need for complex modelocking schemes. The 337-nm N_2 wavelength is an excellent pump for dye lasers and can generate fluorescence from many materials. On the other hand, nitrogen is inefficient and low in power, with pulse energy limited to about 10 millijoules (mJ) and average power to a few hundred milliwatts in practical designs. Its main uses are for applications which require limited power, or laboratories which cannot afford higher-performance excimer or Nd–YAG lasers.

Internal Workings and Structure

The active medium in a nitrogen laser is nitrogen gas at pressures between 20 torr [2700 pascals (Pa)] and 1 atmosphere (atm). The gas usually flows through the laser, but sealed nitrogen lasers can operate at low repetition rates. A fast high-voltage discharge populates the upper laser level, an excited electronic state with 40-ns lifetime, which emits at 337.1 nm when it drops to the lower laser level shown in Fig. 14.1. The transition is a vibronic one, in which both electronic and vibrational energy levels change, making it broadband by laser standards. However, the N_2 laser is monochromatic compared to ultraviolet lamps.

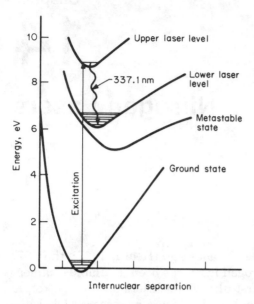

Figure 14.1 Energy levels in neutral nitrogen molecules involved in the 337.1-nm laser transition. The broad curves represent electronic energy levels, with the lines in the potential wells representing vibrational energy sublevels. The laser transition is a vibronic one with changes in both electronic and vibrational energy levels.

Nitrogen laser kinetics are not favorable. The lower level has a 10-microsecond (µs) decay time, much longer than the upper level, and drops to a metastable state with a lifetime of many seconds. That would make laser action impossible except that a fast electric discharge can populate the upper N_2 level very efficiently. The result is a transitory population inversion which generates laser pulses that terminate in nanoseconds or less as the lower state fills.

These features combine to give N_2 lasers very high gains during its short pulses; 50 decibels per meter (dB/m) or more are typical. This avoids the need for a high-quality resonator. Virtually all nitrogen lasers can operate in superradiant mode without cavity mirrors, but output can be more than doubled with a simple cavity with 100 percent reflective rear mirror and 4 percent reflective front mirror. Such a cavity greatly reduces beam divergence. Cavities normally are 15 to 50 cm long.

Nitrogen lasers are so simple to build that they were once featured in the "Amateur scientist" column of *Scientific American* (Strong, 1974), but they are far from trivial. The greatest challenge is the need to deliver a pulse of 15 to 40 kilovolts (kV) with rise time under 10 ns. Important factors include discharge circuitry, channels, electrodes, and discharge configuration.

Wavelength and Output Power. Standard commercial nitrogen lasers emit at 337.1 nm, where neutral molecular nitrogen has many closely spaced lines. Additional N_2 lines at 357.6 nm and in the near-infrared (Davis and Rhodes, 1982) do not operate in standard lasers.

Commercial N_2 lasers have peak powers from tens of kilowatts to about 2.5 megawatts (MW) at 337.1 nm. Specified pulse energies range from tens of microjoules to 9 mJ. Average powers range from about a milliwatt to hundreds of milliwatts because of the limited repetition rates and pulse energies.

Efficiency. Wall-plug efficiency of N_2 lasers typically is 0.11 percent or less if vacuum pumps are required or laser power is low. This low efficiency limits N_2 lasers to applications which require modest power and energy.

Temporal Characteristics. Internal kinetics limit nitrogen lasers to pulsed operation, with pulse length dependent on both gas pressure and the discharge circuit. Pulse lengths range from about 300 picoseconds (ps) at atmospheric pressure to about 10 ns at 20 torr (2700 Pa). Repetition rates range from 10 to 100 hertz (Hz) in commercial models; 1000 Hz has been achieved in the laboratory.

Spectral Bandwidth. Typical bandwidth is 0.1 nm, spread over multiple lines in the 337.1-nm vibronic band.

Amplitude Noise. Pulse-to-pulse variations in output power and energy are ±3 to ±5 percent for commercial nitrogen lasers.

Beam Quality, Polarization, and Modes. Like other lasers with high gain and without true resonators, N_2 lasers have poor beam quality, with output unpolarized and in multiple transverse and longitudinal modes.

Coherence Length. Coherence length is on the order of 1 mm.

Beam Diameter and Divergence. Low-power nitrogen lasers may have beams as small as a millimeter or two across. Higher-power models typically have oval or rectangular beams up to 6 by 32 mm across. Beams typically are oval or rectangular in cross section because of discharge cavity shape. Full-angle divergence ranges from 1 by 2 milliradians (mrad) to 6 by 14 mrad.

Suitability for Use with Laser Accessories. Nitrogen lasers cannot be Q-switched or modelocked, and their wavelength is too short for harmonic generation in nonlinear crystals.

Operating Requirements

Input Power. Small commercial N_2 lasers which do not need vacuum pumps can operate from 12-volt (V) batteries, or draw only modest currents from a 110-V ac line, but those voltages are stepped up to much higher levels in the laser discharge. High-power N_2 lasers with vacuum pumps may draw 20 amperes (A) at 110 V, or draw a smaller amperage at 220 V.

Cooling. Forced-air or convection cooling is standard on nitrogen lasers.

Consumables. Most nitrogen lasers require a continuous supply of nitrogen gas, which is exhausted into the air. Gas consumption ranges from 0.1 to 40 liters per minute (L/min). Standard-purity nitrogen is adequate because N_2 lasers are not very sensitive to impurities; some commercial models can operate with air, which is 78% nitrogen.

In some nitrogen lasers, the gas is sealed in a closed cavity, which is replaced after tens of millions of shots or more. Replacement is needed because discharge-induced breakdown of the N_2 can cause pulse amplitude and beam direction to vary.

Required Accessories. Flowing-gas nitrogen lasers require a vacuum pump if operated below atmospheric pressure.

Operating Conditions and Temperature. Most nitrogen lasers are designed for room-temperature operation in standard laboratory environments. Lower-power, compact models operate under a broader range of conditions; some can operate from batteries.

Mechanical Considerations. The smallest sealed nitrogen lasers measure 24.8 by 11.7 by 6.9 cm, weigh only 2.3 kilograms (kg), and have an internal power supply. The smallest flowing-gas lasers are 41 by 23 by 23 cm, not counting gas supply. The largest flowing-gas types are 140 by 50 by 26 cm, weigh about 80 kg, and may require an external vacuum pump as well as a gas supply.

Safety. Most nitrogen lasers fall under Class IIIb of the federal laser product code. The 337.1-nm N_2 wavelength falls in the ultraviolet-A region, where absorption in the lens provides some protection for the retina, but not enough to prevent damage from high-power pulses. Corneal damage also is possible (Sliney and Wolbarsht, 1980). Eye hazards in the near-ultraviolet have not been fully quantified, so caution is the best policy.

The high voltages needed to drive a nitrogen laser pose serious electrical hazards. Users should ensure that all components have been discharged fully before touching the power supply.

Reliability and Maintenance

Lifetime. Like most laboratory instruments, the lifetimes of N_2 lasers are not normally specified. The shortest-lived components usually are the high-voltage switches, with lifetimes of tens to hundreds of millions of pulses. Switching electronics may be packaged in disposable units with sealed nitrogen gas cells, which have comparable lifetimes.

Maintenance and Adjustments Needed. Sealed-tube lasers require periodic replacement of the gas module. Optics in N_2 lasers should be cleaned to prevent damage from exposure of dirty areas to high ultraviolet powers.

Commercial Devices

Standard Configurations. Many nitrogen lasers are small, inexpensive systems used either to pump tunable dye lasers or excite fluorescence. They appeal to users who do not need—or cannot afford—the higher pulse energies and repetition rates available from excimer lasers. Both sealed-tube and flowing-gas models are available, with the sealed versions offering advantages for use in the field or in instruments. Sealed-tube units tend to be the most compact. Some N_2 lasers are packaged with dye lasers as dedicated pump sources.

Larger nitrogen lasers are flowing-gas systems for laboratory use. They are supplied in bulky boxes which contain all the equipment needed to operate the laser except for a vacuum pump, where necessary. Figure 14.2 compares one of these lasers with a sealed-tube nitrogen laser.

Pricing. List prices of complete nitrogen lasers run from a few thousand dollars to over $20,000. Kits for building small nitrogen lasers may be available at even lower prices.

Suppliers. Several companies offer nitrogen lasers, which probably says more about the ease of entry into the market than about demand for nitrogen lasers; sales volume is modest. Most specialize in sealed or flowing-gas lasers.

Applications

The major application of nitrogen lasers is pumping pulsed dye lasers. Simple, compact, and inexpensive, nitrogen lasers are valuable pump sources that put dye lasers within the reach of limited budgets, such as college student laboratories.

Like other ultraviolet sources, nitrogen lasers can be used to study and

Figure 14.2   A pair of commercial nitrogen lasers. *Top*: A 1.2-m-long version delivers several-millijoule pulses. (*Courtesy of Laser Photonics*.) *Bottom*: A modular sealed laser about 25 cm long produces 0.1-mJ pulses. (*Courtesy of Laser Science Inc.*)

measure fluorescence effects. Their short pulses make them valuable for time-resolved measurements and for Raman scattering observations. Nitrogen lasers also can measure particle size by light scattering.

The high peak power of a focused beam allows the use of nitrogen laser pulses in cell surgery and microcutting. They also can cut thin films for making and correcting semiconductor wafers, and ablate tiny samples.

Bibliography

Robert S. Davis and Charles K. Rhodes: "Electronic transition lasers," in Marvin J. Weber (ed.), *CRC Handbook of Laser Science and Technology*, vol. 2, *Gas Lasers*, CRC Press, Boca Raton, Fla., 1982, pp. 273–312 (review of gas lasers).

H. G. Heard: "Ultraviolet gas laser at room temperature," *Nature 200*:667, November 16, 1963 (research report).

David Sliney and Myron Wolbarsht: *Safety with Lasers and Other Optical Sources*, Plenum, New York, 1980 (reference on laser safety).

C. L. Strong: "The amateur scientist [column]: An unusual kind of gas laser that puts out pulses in the ultraviolet," *Scientific American 230*(6):122–127, 1974.

Far-Infrared Gas Lasers

The "far-infrared" is the part of the electromagnetic spectrum loosely defined as extending from about 10 micrometers (μm) to 1 millimeter (mm). Neither endpoint is firmly or formally defined, and the longer-wavelength part of that region, beyond about 40 μm, is sometimes called the *submillimeter* region.

The haziness of the definition reflects the fact that the far-infrared had been little explored until recent years, when far-infrared laser sources became available. In these lasers, molecular gases emit at wavelengths of 10 μm to about 2 mm, on vibrational-rotational transitions at the short-wavelength end, and on purely rotational transitions at the longer wavelengths. Laser action has been demonstrated experimentally on well over 1500 far-infrared lines. Not all of those lines have been reported from commercial lasers, partly because manufacturers have used only a limited number of gases.

In practice, two commercial lasers which emit at the short-wavelength end of the far-infrared are not usually labeled "far-infrared lasers." These are the carbon dioxide laser (Chap. 10) and the nitrous oxide (N_2O) laser (Chap. 16), both of which can be excited electrically to emit in the 10-μm region. Most other far-infrared lasers require optical excitation with either a CO_2 or N_2O laser, although a handful emitting at longer wavelengths can be excited electrically.

Commercial far-infrared lasers emit at wavelengths from about 28 μm to 2 mm. A single laser cavity can house many different gases, and typically each gas can emit at a number of separate lines. So far most of the applications of far-infrared lasers have been in research, such as diagnostics of fusion plasmas, submillimeter astronomy, and studies of semiconductor materials. Potential applications in nondestructive testing and measurement are in development.

Common materials behave much differently in the far-infrared than at shorter infrared wavelengths and in the visible. Air effectively blocks long-distance transmission between about 30 μm and roughly 500 μm, and even at longer and shorter wavelengths transmission is limited. Most common infrared window materials are opaque at wavelengths longer than about 40 μm, and few transmission curves extend beyond 50 or 100 μm. Silicon, germanium, crystalline quartz, and diamond are among the few solids transparent beyond 50 μm. Plastics such as polyethylene also can be useful in the far-infrared. Fine metal meshes can serve as far-infrared reflectors because at long enough wavelengths their surfaces appear continuous. These differences in material response help make far-infrared lasers valuable for some applications but also make it important that users avoid unwarranted assumptions about far-infrared optics.

Internal Workings

Active Media. Many different molecular gases can be active media in far-infrared lasers, if they possess a permanent dipole moment. The most important molecules in commercial far-infrared lasers are alcohols and other carbon compounds. Minor changes to the molecules, such as isotopic substitution of deuterium (hydrogen-2) for normal hydrogen can give new laser lines. A sampling of major lines of commercial far-infrared lasers is given in Table 15.1. Tabulations of lines seen experimentally are much more extensive (see, e.g., Button et al., 1984; Knight, 1982a, 1982b).

In commercial far-infrared lasers, the gases are put into the laser tube at 30 to 300 millitorr [4 to 40 pascals (Pa)]. The gas may be flowing, or sealed into the tube. A sealed tube can be pumped out and a new gas inserted when the user wants another set of laser lines or when the gas has degraded.

TABLE 15.1   Major Far-infrared Laser Lines Available from Commercial Lasers

Wavelength, μm	Gas	Wavelength, μm	Gas
41.0	CD_3OD	255	CD_3OD
46.7	CH_3OD	375	$C_2H_2F_2$
57.0	CH_3OD	433	$HCOOH$
70.6	CH_3OH	460	CD_3I
96.5	CH_3OH	496.1	CH_3F
118.8	CH_3OH	570.5	CH_3OH
148.5	CH_3NH_2	699.5	CH_3OH
163.0	CH_3OH	764.1	$C_2H_2F_2$
184	CD_3OD	890.0	$C_2H_2F_2$
198.0	CH_3NH_2	1020.0	$C_2H_2F_2$
229.1	CD_3OD	1222.0	$C^{13}H_3F$

SOURCE: From tabulations by Laser Photonics Inc. and MPB Technologies Inc.

Not all far-infrared laser transitions demonstrated in the laboratory can be used in commercial far-infrared lasers. The most important group of these lines is the family of vibrational-rotational lines between about 10 and 25 μm, which some specialists call the *mid-infrared*, produced by molecules such as ammonia and carbon tetrafluoride (see Harrison and Gupta, 1983). Special optics are needed because the emission lines are close to the standard 10-μm pump wavelength, so these molecules could not be expected to lase well in a device designed for operation throughout the far-infrared.

Energy Transfer. Optical pumping is preferred because it permits precise selection of the initial excited state, important in producing a population inversion and laser emission in the far-infrared. Pumping is normally with an external CO_2 or N_2O laser, tuned to emit a narrow frequency band on a single line in the 10-μm region. This narrowband emission excites a specific vibrational state of the far-infrared laser molecule, generating a population inversion on a rotational or vibrational-rotational transition.

The processes involved in far-infrared laser action can be seen by studying Fig. 15.1, which shows some of the energy levels in a polar

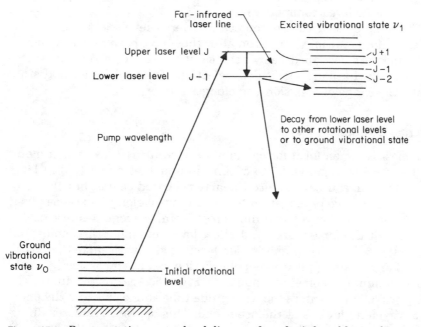

Figure 15.1 Representative energy-level diagram for a far-infrared laser, showing how optical pumping raises a molecule to an excited vibrational state, where laser action in the far-infrared takes place on a transition between two vibrational levels. The laser levels are pulled out of the series of other rotational levels for clarity; their spacing actually is similar to that of other levels.

molecule. Two vibrational levels are shown, the low-lying or ground state ν_0 and the adjacent excited level ν_1. Each vibrational state is associated with a large number of rotational levels (indicated by a J number). Far-infrared laser emission normally takes place between adjacent rotational levels of the same vibrational state.

It is hard to produce a selective population inversion on such a transition because the separation of rotational levels is less than thermal energy kT. Most excitation techniques tend to populate all the levels, but optical pumping can provide selective excitation if the pump frequency is within about 50 megahertz (MHz) of a molecular absorption frequency. The normal pump sources are carbon dioxide lasers and others (including isotopically substituted gases) emitting in the 10-μm range, with grating tuning of the output wavelength needed to precisely match emission and absorption frequencies. The selective excitation raises the far-infrared laser species to an excited vibrational level, and the laser transition occurs when the molecule drops from one rotational state of that level to a lower one.

A few far-infrared lasers can be excited by electric discharges. Two important examples are water vapor and hydrogen cyanide (HCN). These discharge-excited lasers can generate higher powers than optically pumped lasers, but they are not as versatile or as easy to use, and are not offered commercially.

The complex structure of molecular energy levels lets many molecules emit on multiple lines even after they have been excited to a specific state. As described below, the cavity is tuned to select a single wavelength. In many molecules, selection of the initial excited state limits the range of emission wavelengths.

Internal Structure

The most common form for commercial far-infrared lasers is a metal or dielectric waveguide, 25 to 50 mm in diameter and 1 to 3 m long. Dielectric waveguides produce linearly polarized beams, but they suffer from excessive propagation losses if the wavelength exceeds about 2 percent of the waveguide diameter. Metal waveguides are usable throughout the far-infrared and allow lower pumping thresholds, but they tend to produce higher-order laser modes with mixed polarization. (For certain applications polarization can be important.)

The pumping geometry, shown in Fig. 15.2, is longitudinal. The pump wavelength enters the waveguide tube at one end, and the emission wavelength exits at the other end. This approach imposes different requirements on the optical elements at each end of the laser. At the pump end, the vacuum window must transmit 10-μm pump radiation; zinc selenide is a common choice because it is also transparent

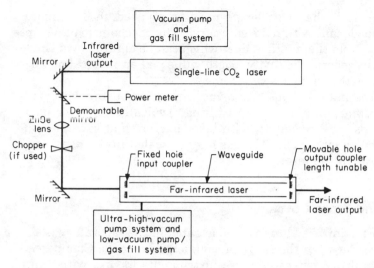

Figure 15.2   Typical arrangement for pumping a far-infrared laser optically. (*Adapted from diagram by MPB Technologies, Inc.*)

in the visible, aiding alignment. The mirror at that end of the cavity must transmit the pump beam but let little of the circulating power out of the cavity. The usual approach is to make a hole 1 or 2 mm in diameter in the center of the mirror, which admits a tightly focused pump beam, but represents only a tiny fraction of the mirror area.

At the output end, the vacuum window should transmit well throughout most of the far-infrared—a criterion that limits selection to materials such as pure silicon and special grades of crystalline quartz, which absorb strongly at 10 μm. The cavity output coupler is often a mirror with a central hole of 3 to 10 mm diameter, small enough to avoid excessive cavity losses. Other possibilities include partly transparent meshes, and hybrid mesh or hybrid hole designs.

Both mirrors must reflect at the 10-μm pump wavelength and throughout the far-infrared region. Reflection at the pump wavelength is important because the laser gas is tenuous and only weakly absorbing at the pump wavelength, making long path lengths essential for efficient energy transfer.

The two mirrors together with the waveguide tube define the resonant cavity. One mirror is fixed, but the other must be linearly movable with a micrometer or piezoelectric positioner. Adjustability of cavity length over a distance small compared to the tube length (typically a few millimeters) is needed to match cavity length to an integral multiple of half the emission wavelength. In shorter-wavelength lasers, the gain curve of the medium normally spans two or more longitudinal cavity modes, but in the far-infrared, the gain curve is much narrower than the cavity mode spacing. Thus the cavity length must be adjusted until a longitudinal cavity mode falls within the gain curve. This mechanism also helps select one line from the multiple lines which can be emitted by a single gas, because in general the cavity length will equal an integral number of half-wavelengths only for a single line.

The laser resonator is housed in a vacuum-tight housing with facilities for vacuum pumping and gas filling. Typically an outer glass tube surrounds the waveguide, with laser mirrors and vacuum windows mounted on either end. The gas may be sealed in the housing or may flow through the laser.

Variations and Types Covered

The range of far-infrared lasers demonstrated in the laboratory is significantly broader than the range of commercial models. Commercial models are built as general-purpose instruments for use with many gases and at many wavelengths. Experimental versions are often custom-built for use with a specific gas or in a specific wavelength region, and thus can be optimized for performance under those conditions. With the demand for far-infrared lasers limited, manufacturers simply cannot justify design of different models for each laser line or material—and researchers seeking to study the far-infrared could not justify buying half a dozen or more different instruments to cover that spectral region. However, some commercial models can be optimized for a specific wavelength range by changing the output coupler and/or waveguide as required.

The distribution of lines from commercial far-infrared lasers is adequate for most purposes, but it is hard to find many lines between 10 and about 40 μm. In part this reflects the scarcity of known lines in part of this region, but it also is related to the practical problems when pump and emission wavelengths are closely spaced. The problems include the need for different optics at such short wavelengths, and the difficulty in separating pump and output lines, mentioned earlier.

This chapter concentrates on commercial far-infrared lasers and their characteristics. Information on experimental lasers can be obtained from references cited in tabulations of far-infrared laser lines (e.g., Button et al., 1984; Knight, 1982a, 1982b; Leheck et al., 1990) or from recent issues of scholarly journals. Emphasis here is on waveguide lasers, in which the radiation is guided by a waveguide structure and does not obey the laws of free-space propagation. Such waveguide lasers tend to be more compact and to have much higher quantum efficiency than conventional open-resonator lasers.

Beam Characteristics

Wavelength and Output Power. Table 15.1 lists some of the important wavelengths available from commercial far-infrared lasers. Output powers range from under 1 milliwatt (mW) to more than 100 mW continuous-wave when pumping with single-line powers of 10 to 100 W. Peak powers with a pulsed pump laser are about five times the continuous-wave level, and peak powers about a thousand times the continuous-wave level are possible with a Q-switched pump laser. Megawatt pulses in the far-infrared can be produced by pumping with a transversely excited atmospheric pressure (TEA) CO_2 laser.

Efficiency. Maximum theoretical quantum efficiency of a far-infrared laser is given by

$$\eta_{max} = \frac{1}{2} \frac{\nu_{FIR}}{\nu_{pump}}$$

where ν_{FIR} is the far-infrared frequency and ν_{pump} is the pump laser frequency. Typical far-infrared lasers have quantum efficiencies of a few percent of this maximum value, but efficiencies to about 30 percent have been demonstrated experimentally. Overall energy efficiency is low because the emitted photons have much lower energy than the pump photons—1 to 25 percent for 1-mm to 40-μm emission pumped by a 10-μm laser.

Temporal Characteristics. Depending on the pump-beam characteristics, output from far-infrared lasers can be steady continuous-wave, chopped continuous-wave, or pulsed. Pulsed or chopped output is needed when using pyroelectric detectors, which detect changes in power rather than steady power. The pump beam need not be chopped to generate a chopped output in the far-infrared. Instead, part or all of a steady continuous-wave far-infrared output beam can be chopped. This approach is valuable for applications that require a continuous-wave beam, because it allows monitoring of a portion of the far-infrared beam that is split from the main beam, then chopped.

Spectral Bandwidth. Bandwidth of a far-infrared laser with tuned cavity can be on the order of a few megahertz, with long-term (tens of seconds) frequency jitter as low as a few tens of kilohertz. This corresponds to long-term frequency stability of better than 1 part in 10^8, and stabilities of 4 parts in 10^{12} have been demonstrated experimentally over 50-millisecond (ms) intervals. Normally, far-infrared lasers emit only a single wavelength in a single longitudinal mode, but it is at least theoretically possible to have simultaneous oscillation at two different wavelengths.

Amplitude Stability.   If input pump power and wavelength are stable, the amplitude of a far-infrared laser can remain stable to within a few percent for at least a 30-minute interval. Frequency stabilization of the *pump* laser is important to stability of far-infrared amplitude, because if the pump wavelength drifts away from the molecular absorption line, the pump-beam energy will not be absorbed, and far-infrared output will be reduced or nonexistent. Stabilization systems can be used with far-infrared lasers.

Beam Quality, Polarization, and Modes.   Waveguide far-infrared lasers operate in a single longitudinal mode. Dielectric waveguides can operate in the lowest-loss EH_{11} waveguide mode, which couples well into the gaussian free-space TEM_{00} mode. Dielectric waveguides produce linearly polarized beams, but cylindrical metal waveguides tend to produce beams of mixed polarization.

Specifications normally do not give beam divergence, which is large by laser standards because of the long wavelength, and which varies with wavelength. In the far-infrared, even a diffraction-limited beam diverges rapidly. For a 10-mm output aperture, the diffraction limit ($1.2 \ \lambda/D$, where λ is wavelength and D is output aperture) is 120 milliradians (mrad) at 1-mm wavelength and about 5 mrad at 40 μm. Diffraction-limited beam divergence is even larger if the output-coupling aperture is smaller. Divergence can be reduced by using whole-area couplers which couple radiation out of the waveguide across its entire cross-sectional area.

Operating Requirements

Input Power.   Far-infrared lasers require optical pumping with average or continuous power of a few watts to 100 W in the 10-μm region. The pump wavelength must be selected to match a molecular absorption line of the laser gas. Electricity is not needed per se to produce the laser beam, but it is required to operate controls, vacuum pumps, stabilization circuitry, and the pump laser.

Cooling.   Far-infrared lasers may be air- or water-cooled. Water cooling is advantageous at higher pump powers (to 200 W), where it improves amplitude stability and helps dissipate surplus pump energy. Many pump lasers require water cooling.

Consumables. The gases used in far-infrared lasers are consumables. They degrade gradually, with useful lifetimes measured in hours to weeks, although overall gas consumption is small because the tube operates at low pressure. Gases also must be changed to shift between wavelengths not available from the same materials.

Required Accessories. A single-line pump laser and a high-vacuum pump (to change gases) are needed to operate a far-infrared laser. The usual pump laser is a conventional or waveguide CO_2 laser which can be tuned to single emission lines and tuned off the peaks of those lines to precisely match absorption lines. Isotopic variants of CO_2, such as $^{13}CO_2$, may be used to generate additional pump lines (Davis et al., 1981; Wood et al., 1980). Single-line N_2O lasers offer more pump lines near 10 μm.

The vacuum pump is essential to exhaust old gas from the laser, especially when changing gases. In general, the lower the pressure, the better.

Operating Conditions and Temperature. Far-infrared lasers are designed for normal laboratory conditions. The dependence of output power on cavity length makes the laser sensitive to thermal gradients, but the use of Invar rods to stabilize the cavity length minimizes this problem. Although no special warm-up is needed for the far-infrared laser, some warm-up may be needed to assure stable single-line output from the pump laser.

Mechanical Considerations. Far-infrared lasers are small in cross section, but their cavities are 1 to 3 m long, making them cumbersome; weights are 50 kg and up. Bench space is required for the pump laser and any stabilization equipment used.

Safety. The hazards of far-infrared wavelengths longer than the 10-μm output of CO_2 lasers have not been well quantified. The best guideline is to use the same safety precautions and eye protection that would be used if the pump laser were being used by itself.

Reliability and Maintenance

Far-infrared lasers typically carry a 1-year warranty and can operate in most places where standard CO_2 lasers can function.

Particular attention must be paid to cleanliness and stability of the laser tubes. Cleanliness of the laser tube is important because operating gas pressures are low, and some of the gases are reactive. A major concern is ways in which impurities can react with some of the laser gases. For example, hydrogen-exchange reactions between water and deuterated methanol can convert the deuterated compound to normal methanol, producing a new set of possible emission lines. To limit contamination problems, the laser tube must be pumped down to pressures on the order of 10^{-5} torr (10^{-3} Pa) when changing gases. Some laser vacuum pumps can reduce pressure to about 10^{-7} torr (10^{-5} Pa).

A few gases used in far-infrared lasers can react with materials that may be used in the laser or vacuum pump. For example, formic acid can attack copper gaskets in the laser, forming compounds that must be cleaned out of the tube.

Thermal and mechanical stability are vital because of the need to match cavity length to oscillating wavelength. Changes in the distance between the mirrors can weaken oscillation at the desired wavelength. To provide the required stability, the laser tube is normally mounted in a housing with thick Invar rods running along its length.

Optical alignment of the flat mirrors so they are parallel with respect to each other is not as critical as for shorter-wavelength lasers, although it is still important. Minor misalignments can reduce output power and cause the laser to oscillate in higher-order modes. If ZnSe optics are used in the laser, the mirrors can be aligned visually with a helium-neon laser. However, some far-infrared lasers are designed for only factory alignment.

Commercial Devices

Standard Configurations. Commercial far-infrared lasers are built for general-purpose laboratory use throughout the far-infrared. The laser heads themselves are similar in general size and appearance—1 to 3 m long, with much smaller cross section, in a housing designed for use on an optical table. The heads may be sold alone, or as part of a complete system including pump laser, stabilization equipment, and controls. The pump laser normally fits on the optical bench, along with the output-monitoring equipment used to stabilize output wavelength of the pump laser. Power supply for the pump laser, operator controls, and other auxiliary equipment may be mounted on an equipment rack or separate console. A sampling of commercial far-infrared lasers and systems is shown in Fig. 15.3.

Options. In some senses, everything beyond the basic far-infrared laser head is optional. Typical options listed by manufacturers include

- Output-stabilization systems which control pump wavelength
- Selection of output coupler and waveguide configuration
- Special couplers for wavelengths below about 50 μm

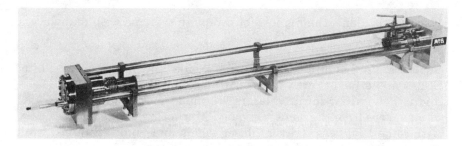

Figure 15.3 Two commercial far-infrared lasers showing (*top*) one simple structure (*Courtesy of MPB Technologies Inc.*), and (*bottom*) a system including computer controls. (*Courtesy of Laser Photonics Inc.*)

- Remote tuning of the far-infrared cavity length
- Detection systems
- Vacuum pumps and gas-fill systems
- Two-wavelength or dual-cavity options for heterodyning or interferometry
- Systems to stabilize cavity length
- Choice of pump laser integrated with far-infrared laser
- Types of pump laser, including CO_2, isotopically substituted CO_2, or N_2O
- Cooling equipment

Pricing. As of this writing, prices of far-infrared lasers run from slightly over $10,000 to more than $70,000. The lowest prices are for laser heads alone, the highest for complete systems with many accessories.

Suppliers. A handful of companies inside and outside the United States make far-infrared lasers. All of them also make pump lasers, and in general far-infrared lasers are made specifically to mate with pump lasers from the same company. Marketing emphasis does vary, with some suppliers stressing laser heads and pump modules, while others emphasize complete systems.

Makers of far-infrared lasers also can supply optics and detectors for the far-infrared. Outside of pyroelectric detectors, and front-surfaced mirrors, few optical components used at wavelengths beyond 50 μm are used at much shorter wavelengths, and such specialized far-infrared components are hard to obtain from suppliers of conventional visible and infrared optics.

Applications

Most far-infrared lasers are used in research and development. In large part, this reflects the recent development of the technology and the design of most far-infrared laser systems for laboratory use. Major research applications—in addition to the study of the far infrared in its own right—concentrate on measurement. They include

- Diagnostic measurements of plasmas, particularly for nuclear fusion research
- Studies of semiconductor materials
- Molecular spectroscopy
- Atmospheric spectroscopy
- Heterodyne sources for far-infrared astronomy
- Far-infrared imaging and radar
- Other measurement technology

Some research may eventually lead to practical applications.

Bibliography

Kenneth J. Button (ed.): *Infrared and Millimeter Waves*, vol. 1, *Sources of Radiation*, Academic Press, New York, 1979 (collected review papers).
Kenneth J. Button (ed.): *Infrared and Millimeter Waves*, vol. 7, *Coherent Sources and Applications, Part II*, Academic Press, New York, 1983 (collected review papers).
Kenneth J. Button, M. Inguscio, and F. Strumia (eds.): *Reviews of Infrared and Submillimeter Waves*, vol. 2, *Optically Pumped Far-Infrared Lasers*, Plenum, New York, 1984 (collected review papers).

Paul D. Coleman: "Far infrared lasers, introduction," in Marvin J. Weber (ed.), *CRC Handbook of Laser Science & Technology*, vol. 2, *Gas Lasers*, CRC Press, Boca Raton, Fla., 1982, pp. 411–419 (short review).

B. W. Davis et al.: "New FIR laser lines from an optically pumped far-infrared laser with isotopic $^{13}C^{16}O_2$ pumping," *Optics Communications* 37(4):303–305, May 15, 1981.

R. G. Harrison and P. K. Gupta: "Optically pumped mid-infrared molecular gas lasers," in Kenneth J. Button (ed.), *Infrared and Millimeter Waves*, vol. 7, *Coherent Sources and Applications, Part II*, Academic Press, New York, 1983 (review paper).

T. Lehecka et al.: "Gas lasers," in Kai Chang (ed.), *Handbook of Microwave and Optical Components*, vol. 3, *Optical Components*, Wiley-Interscience, New York, 1990, pp. 481–563 (review of progress in far-infrared lasers).

D. J. E. Knight: "Tables of CW gas laser emissions," in Marvin J. Weber (ed.), *CRC Handbook of Laser Science & Technology*, vol. 2., *Gas Lasers*, CRC Press, Boca Raton, Fla., 1982a (table of laser lines).

D. J. E. Knight: "Ordered list of far-infrared laser lines," NPL Report QU 45, National Physical Laboratory, Teddington, Middlesex, U.K., 1982b (table of laser lines, updated periodically).

Mary S. Tobin: "A review of optically pumped NMMW [near-millimeter-wave] lasers," *Proceedings of the IEEE* 73(1):61–85, 1985 (review paper).

R. A. Wood et al.: "Application of an isotopically enriched $^{13}C^{16}O_2$ laser to an optically pumped far-infrared laser," *Optics Letters* 5(4):153–154, 1980.

Other Commercial Gas Lasers

The commercially important gas lasers described in Chaps. 7 to 15 are only a few of the gas lasers that have been demonstrated in the laboratory. Most of the gas lasers listed in extensive tabulations based on research literature (e.g., Beck et al., 1980; Weber, 1982) are of purely academic interest. However, a few fall into an intermediate category, lasers which are available commercially, but only on a small scale, often with little marketing support.

This chapter covers five lasers that fall into that category:

- Carbon monoxide lasers at 5 to 6.5 micrometers (μm)
- Nitrous oxide (N_2O) lasers at 10 to 11 μm
- Xenon helium lasers at 2 to 4 μm
- Xenon lasers near 0.5 μm
- Iodine lasers at 1.3 μm

Not covered are a number of types that are minor members of families described in other chapters. Examples include the helium-selenium laser, a member of the metal vapor ion laser family described in Chap. 9, and the molecular fluorine lasers which are described with rare-gas-halide excimers in Chap. 13. Also not covered are isotopic variations, in which rare isotopes are substituted for common ones in order to obtain different wavelengths from gas molecules, particularly carbon dioxide.

Carbon Monoxide Lasers

Similar in many ways to the CO_2 laser, the CO laser was discovered in 1964 at Bell Labs by C. Kumar N. Patel (Patel and Kerl, 1964), who earlier had demonstrated the first CO_2 laser. Like CO_2, CO normally is excited with an electric discharge and emits on many vibrational-rotational lines. CO emission is at shorter wavelengths, with most lines between 5 and 6 μm.

The carbon monoxide laser has a potential for high power and high efficiency that led to study of it as part of the laser weapon program. By 1975, discharge efficiency as high as 63 percent had been measured, and researchers were talking about theoretical efficiency levels in the 80 percent range (Bhaumik, 1976). However, two serious practical problems limited interest in CO lasers. The strongest CO emission lines coincide with bands of high atmospheric absorption, leading to atmospheric transmission problems too serious for use in ground-based laser weapons. In addition, the gain of CO lasers drops off rapidly with temperature, and highly efficient operation is possible only at cryogenic temperatures.

Although these problems damped CO laser development, the CO laser remains the best source in the 5- to 6.5-μm region. It has found a variety of applications in spectroscopy. It also has been studied for potential medical applications.

Typically, commercial carbon monoxide lasers are modified versions of continuous-wave carbon dioxide lasers. CO can operate in both conventional and waveguide lasers and in sealed tubes or with flowing gas. Changes for CO operation include substitution of optics for use at 5.0 to 6.5 μm, insertion of the proper gas mix, and often addition of cooling equipment. One air-cooled waveguide model can generate up to 2 watts (W) continuous-wave in multiline operation and over 500 milliwatts (mW) on a single line with grating tuning. Water cooling allows operation at higher powers between -10 and $-15°C$. A laser head less than 0.75 meter (m) long can generate 8 W without cryogenic coolants, although it does require other active cooling. Efficiency and output power are higher when the laser is cooled to about $-70°C$.

With nonselective resonators, CO lasers typically oscillate on several lines, as shown in Fig. 16.1. For most applications, they are grating-tuned to operate on any of a few dozen lines between about 5.2 and 6.5 μm; Fig. 16.2 shows typical output on lines between 5.2 and 5.9 μm from a waveguide CO laser.

A few transversely excited atmospheric pressure (TEA) CO_2 lasers can be modified for use with CO. The output is multipeaked pulses lasting a few microseconds. The output is at multiple wavelengths, with complex energy-coupling effects causing the emission wavelength to shift during the pulse.

The gas mixture in CO lasers contains CO, nitrogen, helium, and usually xenon at pressures around 35 torr [4500 pascals (Pa)] in continuous-wave versions, or higher in TEA lasers. Air can be added

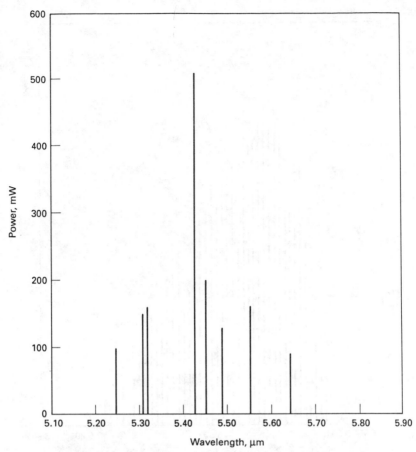

Figure 16.1 Typical output from a continuous-wave CO laser with nonselective resonator. (*Courtesy of Laser Photonics Inc.*)

to flowing-gas CO lasers. Operation of flowing-gas CO lasers poses potential safety hazards because their exhaust contains the odorless but deadly carbon monoxide. CO concentrations of 50 parts per million (ppm) are considered unhealthy, and higher concentrations can cause unconsciousness and death, making proper exhaust systems crucial.

Nitrous Oxide Lasers

Nitrous oxide (N_2O) is a linear triatomic molecule similar in structure to CO_2. The nitrous oxide laser was discovered shortly after the CO laser (Patel, 1965). Like CO_2, the N_2O laser can emit on up to 100 rotational sublevels of a vibrational transition near 10 μm. The 10.65-

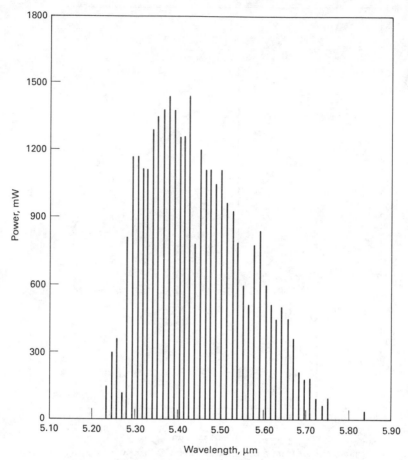

Figure 16.2 Typical output from a continuous-wave CO laser with line-tunable optics (with output on only one line at a time). (*Courtesy of Laser Photonics Inc.*)

μm center wavelength of the N_2O vibrational transition is slightly longer than the 10.4-μm center wavelength of the corresponding CO_2 transition. The N_2O laser lines in Fig. 16.3 lie between 10.3 and 11.1 μm, filling in some of the holes in the CO_2 line spectrum. As with CO_2 and CO, excitation is by passing an electric discharge through the gas.

Only minor modifications are needed to operate an N_2O laser in some CO_2 laser tubes. In line-tunable continuous-wave lasers, single-line N_2O powers can reach 10 W; in pulsed TEA lasers pulse powers can reach 30 W. These output powers are much lower than available from CO_2, about one-seventh the power level from continuous-wave line-tunable models, and a smaller fraction of the peak output of TEA pulsed lasers.

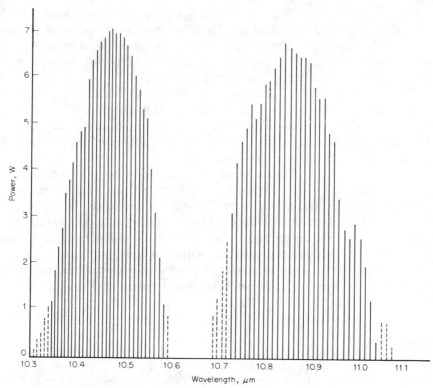

Figure 16.3 Line spectrum of N_2O laser for continuous-wave and chopped operation. The dashed lines can only be produced when laser output is chopped; all other lines indicate output power available on that line in continuous-wave operation. The gas mix is N_2O, 12%; N_2, 16%; CO, 9%; He, 63%. (*Courtesy of Laser Photonics Inc.*)

Because CO_2 lasers can produce much higher output power in the same wavelength region, there is little call for N_2O lasers except in applications which require precise wavelengths not available from CO_2. There are several such applications, including lidar experiments in the atmosphere, spectroscopy, frequency mixing, clock applications (Thomas et al., 1980), and pumping of far-infrared lasers (see Chap. 15).

The active medium in a typical continuous-wave N_2O laser is a mixture of about 12% N_2O, 16% N_2, 9% CO, and 63% He, at a total pressure of around 10 torr (1300 Pa). Normally this mixture flows through the laser and is exhausted through a vacuum pump. As with the CO laser, this exhaust gas is both potentially lethal and odorless, so care must be taken to vent the spent gas outside.

Xenon-Helium Lasers

Xenon has a number of laser lines between 2 and 4 μm in the near-infrared. The 2.026-μm line was among the early lasers to be demonstrated (Patel et al., 1962). The stronger 3.506-μm line was reported less than a year later (Bridges, 1963). Other lines are weaker and were discovered later. The laser emission can come from pure xenon, but addition of helium at pressures tens to hundreds of times that of the xenon raises output power significantly.

Gain is high on the 3.506-μm line, about 2.5 per centimeter in a 0.75 mm bore, and this has been called a high-power laser in pulsed operation (Davis, 1982). However, it suffers from power saturation that limits continuous-wave output and has attracted only minimal commercial development or interest.

Only one company makes commercial Xe–He lasers, which can operate continuous-wave, chopped, or pulsed at rates to 1 kilohertz (kHz). Output is tunable between 2.6 and 4.0 μm, with the 3.506-μm line the strongest wavelength. Maximum output is about 200 milliwatts (mW). Commercial lasers have gas flowing through the laser cavity, but sealed operation is possible at lower powers.

The major lines lie in windows of good atmospheric transmission, but so far applications have been limited to research.

Visible Xenon Lasers

Triply ionized xenon [Xe^{3+} or Xe(IV)] has several blue-green laser lines. Unlike the rare gas argon and krypton ion lasers described in Chap. 8, these xenon lines are pulsed. Experimental peak powers have reached several tens of kilowatts, but commercial versions deliver peak powers on the order of 1 kW. The main emission lines are at 526.0, 535.3, and 539.5 nm, with a few weaker lines in the blue-green. Experimental xenon lasers have emitted on lines in the blue and ultraviolet as well (Bridges, 1982).

The pulses of 0.1 to 0.4 microsecond (μs) can be repeated at rates to around 200 Hz and have peak powers that are high by ion laser standards. The combination of that high power with the short wavelength has made pulsed xenon lasers desirable for applications requiring the focusing of moderate powers onto small spots. Examples include trimming of thin-film resistors in microelectronic circuits, and making the masks used in semiconductor fabrication. Other potential applications include holography, studies of turbulence, and research requiring such short, moderate-power blue-green pulses. So far most pulsed visible xenon lasers have been incorporated in electronics manufacturing equipment. The lasers normally are made by the company producing the complete system and generally are not marketed as individual devices. Other manufacturers uses different lasers for the same application.

Iodine Lasers

The 1.315-μm atomic iodine laser exists in two basic variations: the chemical oxygen-iodine laser and the optically pumped photodissociation iodine laser. Both operate on the same laser transition; the difference is in the excitation technique.

The chemical oxygen-iodine laser depends on energy transfer from excited molecular oxygen to atomic iodine. The excited O_2 is produced by a chemical reaction, and transfers its energy to iodine atoms, generating the upper laser level. Many different approaches have been investigated (Basov et al., 1988). A driving force behind this work has been military research on high-energy laser weapons.

The photodissociation iodine laser is an older approach which has been used both for high-power pulsed lasers for fusion research (Brederlow et al., 1983), and in smaller commercial iodine lasers used for laboratory research (Bannister and King, 1984). Flashlamp pumping of large iodine lasers in oscillator-amplifier configurations can produce nanosecond pulses with peak powers in the terawatt range, as demonstrated in the Asterix III laser at the Max Planck Institute for Quantum Optics in Garching. Normally operation is pulsed, but continuous-wave operation at powers to 38 mW has been demonstrated in a flowing gas mixture excited by a mercury arc lamp (Schlie and Rathge, 1984).

The photodissociation iodine laser relies on ultraviolet light to break up iodine-containing molecules such as C_3F_7I, producing atomic iodine in an excited state, as shown in Fig. 16.4. The excited iodine atoms can be

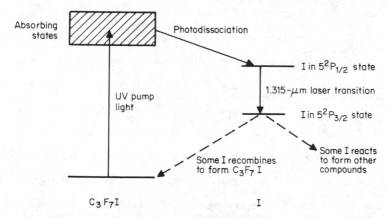

Figure 16.4 Transitions involved in iodine photodissociation laser. Ultraviolet excitation raises the starting molecule, C_3F_7I, to an excited state in which it quickly photodissociates to free excited I, which is the upper laser level. After the I drops to the lower level, it recombines with other molecular fragments to form C_3F_7I or other compounds.

stimulated to emit photons at 1.315 μm. After the laser pulse, much of the iodine recombines with the fragments of the original molecules, so it is often possible to reconstitute the starting compound.

Two approaches have been used to supply C_3F_7I in commercial lasers. One is to place ampoules of the iodine compound in the laser and exhaust the vapor with a vacuum pump; this allows production of up to 1000 shots at repetition rates no more than one pulse per 10 seconds (to allow for removal of the spent gas). The other is to regenerate the iodine compound by recycling it through a sealed laser tube; this allows production of up to a million shots at 1 to 10 Hz.

Commercial iodine lasers have generated pulses of up to a few joules, but so far have found few uses outside of specialized research laboratories. The highest powers from photodissociation iodine lasers have been generated by oscillator-amplifier configurations. Few details on high-power chemical oxygen-iodine lasers have been reported in public, and they have not become important enough to be mentioned even in vague terms in reports to Congress by the Strategic Defense Initiative Organization. Research continues on improving iodine laser performance and developing more practical devices.

Bibliography

J. J. Bannister and T. A. King: "The atomic iodine photodissociation laser," *Laser Focus* 28(8):88–97, August 1984.

N. G. Basov et al.: "A theoretical analysis of oxygen-iodine chemical lasers," in A. N. Orayevskiy (ed.), *Research on Laser Theory*, Nova Science Publishers, Commack, N.Y., 1988.

R. Beck, W. Englisch, and K. Gurs: *Table of Laser Lines in Gases & Vapors*, 3d ed., Springer-Verlag, Berlin and New York, 1980.

Mani L. Bhaumik: "High-efficiency electric-discharge CO lasers," in E. R. Pike (ed.), *High-Power Gas Lasers 1975*, Institute of Physics, Bristol and London, 1976, pp. 243–267.

G. E. Brederlow, E. Fille, and K. J. Witte: *The High Power Iodine Laser*, Springer-Verlag, Berlin and New York, 1983.

W. B. Bridges: "High optical gain at 3.5 μm in pure xenon," *Applied Physics Letters 3*: 45–47, 1963.

W. B. Bridges: "Ionized gas lasers," in Marvin J. Weber (ed.), *CRC Handbook of Laser Science & Technology*, vol. 2, *Gas Lasers*, CRC Press, Boca Raton, Fla. 1982, pp. 171–269.

Christopher C. Davis: "Neutral gas lasers," in Marvin J. Weber (ed.), *CRC Handbook of Laser Science & Technology*, vol. 2, *Gas Lasers*, CRC Press, Boca Raton, Fla., 1982.

C. K. N. Patel: "CW laser action in N_2O (N_2–N_2O) system," *Applied Physics Letters 6*: 12–13, January 1,1965.

C. K. N. Patel and R. J. Kerl: "Laser oscillation on $X'\Sigma^+$ vibrational-rotational transitions of CO," *Applied Physics Letters 5*:81, 1964.

C. K. N. Patel, W. L. Faust, and R. A. McFarlane: "High-gain gaseous (Xe–He) optical maser," *Applied Physics Letters 1*:84–85, 1962.

L. A. Schlie and R. D. Rathge: "Long operating time CW atomic iodine probe laser at 1.315 μm," *IEEE Journal of Quantum Electronics QE-20*(10):1187–1196, 1984.

J. T. Thomas et al.: "Stable CO_2 and N_2O laser design," *Review of Scientific Instruments 51*(2):240–243, 1980.

Marvin J. Weber (ed.): *CRC Handbook of Laser Science & Technology*, vol. 2, *Gas Lasers*, CRC Press, Boca Raton, Fla., 1982.

Dye Lasers

The organic dye laser has found many applications in scientific research because of its unusual flexibility. Its output wavelength can be tuned from the near-ultraviolet into the near-infrared; the use of frequency-doubling crystals can extend emission further into the ultraviolet. Dye lasers can be adjusted to operate over an extremely narrow spectral bandwidth, producing ultrapure light for studies of optical properties of materials in very narrow wavelength regions. They also can produce ultrashort pulses, with durations much shorter than a picosecond.

This versatility comes at a cost in complexity. Individual dyes can emit light at a broader range of wavelengths than other lasers, but a range that is still narrow compared with the entire visible spectrum. Tuning wavelength across the entire visible range requires several changes of dye. The dyes degrade, and because they must be dissolved in a solvent, require a complex liquid-handling system. They require excitation with a separate source of intense light, another laser or a flashlamp, which adds to the cost and complexity, although the choice of pump sources does give some added flexibility. Complex optics are needed to produce either ultranarrow-linewidth output or picosecond pulses, and the two cannot be produced simultaneously. Designers of dye lasers have made great strides to build systems which can be used readily by biologists, chemists, and others whose specialties are far from laser technology, but dye lasers are still more scientific instruments than industrial tools.

Dye lasers have made vital contributions to spectroscopy. It was not long after the first demonstration of a dye laser, by Peter P. Sorokin and J. R. Lankard at the IBM Watson Research Center in Yorktown Heights, N.Y. (Sorokin and Lankard, 1966), that optics were developed to adjust wavelength and produce narrow-bandwidth output.

Powerful techniques were soon developed to probe energy levels and physical processes within atoms and molecules. Arthur L. Schawlow of Stanford University, who made major contributions to this research, was cited for his contributions to laser spectroscopy when he shared in the 1981 Nobel prize in physics.

The recent emergence of the tunable solid-state lasers described in Chap. 24 has led some observers to claim that "dye is dead," but that seems too strong and too broad a statement. Solid-state tunable lasers have an important advantage in ease of operation, and can generate higher output powers. They avoid the hazards of toxic dyes, which can present significant disposal problems. However, they do not cover as much of the spectrum as do tunable dye lasers, and have yet to match the ability of dye lasers to generate extremely short pulses. Nor can they be packaged as readily into comparatively inexpensive systems with modest output powers. Functionally, tunable solid-state laser and tunable dye lasers may become part of a larger family of tunable lasers. Some companies already offer dual-mode lasers, in which the user can interchange a laser dye cell with a titanium-sapphire crystal to obtain tunable output at different wavelengths.

Internal Workings

Active Medium. The active medium in a dye laser is a *fluorescent* organic compound (a dye) dissolved in a liquid solvent. Intense illumination by light from a separate source—another laser or a flashlamp—excites the dye molecules, producing a population inversion. The dye then produces stimulated emission, generating a laser beam. The process is called optical pumping, and inherently involves some energy losses, so the output wavelength is longer than the absorbed wavelength. The dye solution is housed in a transparent cell or flowed in a jet through the pumping region.

Dyes are large molecules containing multiple ring structures, and they have complex spectra. Most important dyes fall into a number of families with chemically similar structures. Members of these families differ in the end groups attached to their outer edges, and these chemically superficial differences lead to important differences in characteristics such as laser emission wavelength, tuning range, absorption wavelengths, and tolerance of operating conditions. Listings of dyes in which laser action has been demonstrated go on for many pages (Maeda, 1984; Steppel, 1982). Many of these dyes are offered commercially for laser use.

Dye degradation is an important practical issue. Complex dye molecules can be decomposed by the pump light, with the degree of decomposition increasing with pump intensity and with shorter pump wavelength. Short ultraviolet wavelengths, present in the light from flashlamps and from ultraviolet pump lasers, trigger some photochemical reactions. Some degradation is thermal; and some, a combination of photochemical and thermal. Operating lifetimes also depend on chemical structures because some molecules are relatively fragile.

One useful index of dye lifetime is how many watt-hours of pumping that a liter of dye solution can withstand before output declines to a particular level (often 50 or 75 percent of the original level). Lifetimes may be as short as a few watt-hours, but the more common blue and green dyes have lifetimes in the 100-watt-hour (Wh) range. Rhodamine 6G, an efficient dye with peak output near 580 nanometers (nm), has a lifetime of about 2000 Wh, and many other longer-wavelength dyes have lifetimes of several hundred watt-hours. New dyes have been developed which promise much longer lifetimes. In practice, intense ultraviolet pumping may degrade some dyes in a matter of hours, while stable dyes pumped with moderate powers at longer wavelengths may last over a year.

Optical properties and degradation of a laser dye depend on the solvent and additives included in the solution, as well as on the dye itself. Many different solvents are in use. Few dyes are water-soluble, so most are dissolved in organic solvents such as methanol and dimethyl sulfoxide. In some cases, alcohols may be mixed with water to form a dye solution. Certain chemicals may be added to the dye solution to improve energy transfer properties of some dyes.

If the dye solution is to be used in a flowing jet (standard when pumping with a continuous-wave laser) the solution must be sufficiently viscous to maintain an optically flat surface. The usual solvent for such purposes is ethylene glycol, but many dyes are not soluble in the pure solvent. In practice, the dye is dissolved in a small quantity of another organic solvent, which is mixed with ethylene glycol to give a dye solution with the desired properties.

Efforts have been made to develop dye lasers in which the dye is carried in a medium other than a liquid. Laser action has been demonstrated with dyes in the vapor phase and embedded in a solid plastic host. Both approaches have some conceptual attractions, but they suffer some serious practical drawbacks and have yet to find practical applications.

Energy Transfer.   The energy that drives a dye laser comes from an external light source. Absorption of a photon raises the dye molecule to a highly excited state, nonradiative processes drop it to the upper level of the laser transition, then after the laser light is emitted nonradiative processes remove molecules from the lower level of the laser transition. The actual energy-level structure is quite complex, and even the diagram in Fig. 17.1 is oversimplified.

Dyes can both absorb and emit light over a range of wavelengths because interactions among electronic, vibrational, and rotational energy levels create a continuum of levels in some energy ranges. The electronic transitions are those between groups of states in Fig. 17.1, and their energies normally correspond to wavelengths in or near the visible region. Superimposed upon the electronic transitions are vibrational transitions of the molecule, represented in the diagram by darker lines within each group of levels. The smallest energy shifts are caused by rotational transitions, represented by light lines. Although the levels are shown separated for clarity, in practice they are spaced closely enough to form a continuum when subjected to normal line-broadening mechanisms.

It is this continuum of energy levels that lets dyes absorb and emit light at what by laser standards is a broad range of wavelengths, as shown in Fig. 17.2. Initial excitation to the first excited electronic level is followed by nonradiative relaxation to the bottom of that

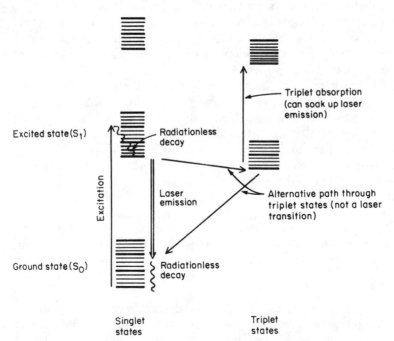

Figure 17.1   Typical energy-level structure in a laser dye. The electronic energy levels (groups of lines) contain vibrational (dark lines) and rotational (light lines) substates.

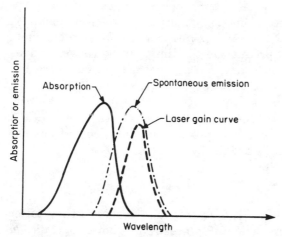

Figure 17.2 Representative shapes of absorption, spontaneous emission (fluorescence), and laser emission curves for a laser dye.

group of energy levels, where a population inversion accumulates. Energy storage in the upper laser level is small because that state's lifetime is only a few nanoseconds. Stimulated emission occurs when a molecule makes a transition from that upper level to a level in the lower band. Spontaneous emission can occur in a wide range of wavelengths in that band, with the highest probability of emission at central wavelengths. With an untuned resonant cavity, amplification is strongest at the peak of the laser gain curve for the dye where the difference between emission and absorption is greatest. Insertion of wavelength-tuning optics allows only a limited range of wavelengths to oscillate within the cavity, so amplification occurs at these wavelengths.

In practice, the picture is complicated by the parallel set of electronic energy bands shown in Fig. 17.1, known as *triplet* states, while the ones from which laser emission occurs are known as *singlets*. These states arise because electrons in dye molecules are bonded together in pairs. In the ground state, the two electrons have different spins and occupy the same energy level. If the electrons maintain opposite spins as one ascends the energy-level ladder, they stay in singlet levels. If the spin of the higher-energy electron reverses to become parallel to that of the lower-energy electron, they enter the triplet states, which have slightly lower energy than the corresponding singlet levels. Because the triplet states have lower energy, the molecules tend to drop into them. That creates problems because those levels have such long lifetimes that they trap molecules, which thus cannot take part in laser emission, but which can cause some absorption. Pump pulses in the 10-ns range or shorter—from excimer, frequency-shifted neodymium-YAG, or nitrogen lasers—are too short to produce triplets. Triplet production can limit tuning ranges of dyes with longer-duration pumping, and can prevent many from operating in the continuous-wave mode, despite the use of rapidly flowing dye jets to control triplet production by moving the dye quickly out of the excitation region, and addition of triplet quenchers to the dye solution. Much has been written on how triplet production limits dye performance (Steppel, 1982), but the details are rarely relevant to laser users.

Internal Structure

Although all dye lasers share the same active medium, they differ widely in internal structure. The differences reflect both the choice of pump source and the intended applications of the laser.

The pump source strongly influences design of a dye laser because of the energy-transfer kinetics of the dye. The upper laser level has a lifetime of a few nanoseconds, making the pulsed dye laser a high-gain high-loss system when it is pumped by pulses with high peak power. Pumping is more difficult with a continuous-wave laser which does not generate high peak powers in pulses. To reach the high power densities needed to exceed laser threshold, the moderate power beam from a continuous-wave pump laser must be focused very tightly, concentrating its power in a small volume. Even so, the continuous-wave pumped dye laser remains a low-gain low-loss system. Modelocking and/or cavity dumping allow a continuously pumped dye laser to generate pulsed output.

Flashlamp-pumped dye lasers come with integral flashlamps, and plug directly into a source of electrical energy. On the other hand, most laser-pumped dye lasers are packaged separately from the pump lasers needed to drive them. This arrangement gives users the flexibility to pick the pump source best suited to their needs, within certain limits. (A pulsed pump laser cannot be used with a dye laser designed for a continuous-wave pump laser, and vice versa.) It also meets the economic need to minimize prices, and allow customers to use dye lasers with existing equipment. The exception to the rule is the packaging of inexpensive nitrogen lasers as dedicated pump sources for some low-end dye lasers.

The choice of cavity optics and cavity structure also depends on the intended application and mode of operation. Different versions of the same continuous-wave pumped dye laser may generate extremely pure continuous single-wavelength output for spectroscopy, or extremely short pulses with broad bandwidth to study ultrafast processes. Both applications and pump sources can vary widely, so while the descriptions that follow cover major types of dye lasers, they cannot cover all possible configurations.

Optics. Wavelength-selective optics are used in most dye lasers. In the common configuration shown in Fig. 17.3, a wavelength-dispersive element (a prism or diffraction grating) is placed inside the laser cavity. The dispersive element is aligned so light at one wavelength is reflected back along the cavity axis, while other wavelengths are dispersed. This ensures that the laser will oscillate only at the selected wavelength (if it is within the dye's gain bandwidth). In the simple examples shown, turning the grating or rear cavity mirror changes the resonant wavelength. The range of oscillating wave-

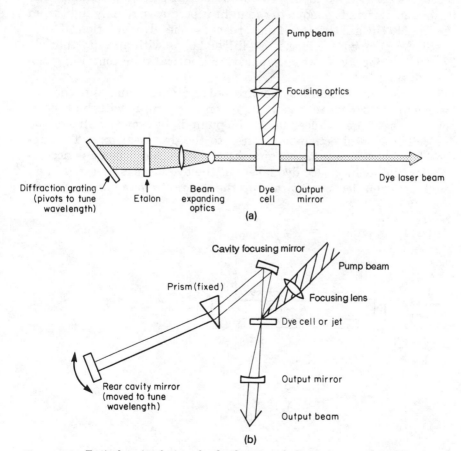

Figure 17.3 Typical cavity designs for dye lasers with dispersive wavelength-tuning elements. (*a*) Grating-tuned dye with pulsed pump laser; (*b*) prism tuning of a simple continuous-wave dye laser.

lengths can be further restricted by inserting etalons or other wavelength-selective elements in the laser cavity.

Another way of tuning is to use a tunable wavelength-selective element. One example is a tuning wedge, which forms a Fabry-Perot cavity and limits oscillation to a narrow wavelength range by interference effects. Moving the wedge vertically changes the spacing. A newer approach is the birefringent tuner, which changes polarization to an extent dependent on wavelength. If the cavity allows oscillation only of linearly polarized light, the laser can oscillate only at the wavelength at which the birefringent tuner transmits linearly polarized light.

Optics also usually focus pump light into the dye. An exception is the coaxial flashlamp, shown in Fig. 17.4, in which dye solution flows through a bore surrounded by an annular cross-section flashlamp. The flashlamp plasma is opaque to the light it produces, so only light from the inner surface of the lamp reaches the dye solution. Other flashlamp pumped dye lasers use linear lamps with a cavity that reflects the pump light onto a separate cylindrical tube containing the dye solution.

The pump laser beam may be directed at different angles to the dye laser axis, as shown in Fig. 17.3. When pumping is with high peak power pulses from a pulsed laser, the pump light typically strikes the dye cell from a direction transverse to the dye laser axis. This approach simplifies alignment, and the sacrifice in efficiency is acceptable for high-gain pulsed dye lasers. If the dye laser is operated in an oscillator-amplifier configuration, the beam from a single pulsed

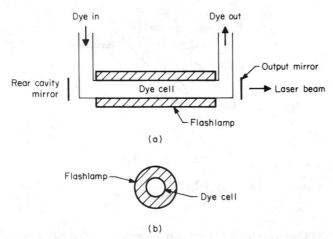

Figure 17.4 Design of a coaxial flashlamp dye laser. (*a*) Side view; (*b*) cross section.

pump laser can be split and directed transversely to the oscillator and one or more amplifiers. Peak powers are sufficiently high that special focusing optics are not required.

Longitudinal pumping, with the pump beam directed along or close to the laser axis, can help improve the efficiency of low-gain continuous-wave pumping. The pump beam may be directed at a small angle to the axis, passing around the cavity mirrors. In some designs, such as shown in Fig. 17.3b, intracavity optics may direct the pump beam along the axis. Special optics are needed to focus the continuous pump beam to a narrow waist a few micrometers across in the dye. The power levels in the dye are high enough to reach laser threshold. They also can damage conventional optical widows, so the dye normally flows as a jet between a nozzle and a collecting receptacle.

An additional component is needed in dye lasers with the ring cavity configuration shown in Fig. 17.5 and described in more detail below. Laser light can oscillate in either of two directions in a ring cavity, but normally the first one to oscillate will suppress the second. However, brief fluctuations can interrupt lasing and allow the laser to switch to oscillating in the opposite direction. This poses problems in extracting the laser beam because light traveling in different directions around the ring will emerge through the output coupler at different angles. That problem can be avoided by inserting into the ring cavity a device which contains two polarization rotators. One is a Faraday rotator, in which the direction of rotation depends on the sign of a magnetic field applied across the device, but not on the direction light travels through it. The other is a standard polarizer, which rotates light in a direction that depends on the direction in which the

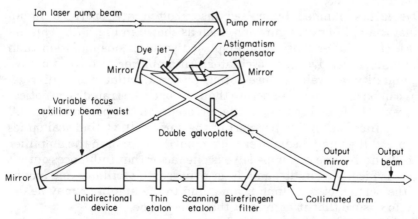

Figure 17.5 Configuration of a ring dye laser with a traveling wave inside the laser cavity. (*Based on drawing from Spectra-Physics Inc.*)

light is traveling. The combination of the two devices makes losses different for light oscillating in different directions around the ring, so the laser oscillates in the direction of lower loss.

Cavity Length and Configuration.   Dye lasers have many different cavity configurations. The choice depends on pump source, intended applications, and the nature of the dye. In most dye lasers, the dye cell is only a small part of the cavity length. (Flashlamp-pumped lasers are an exception, because the dye is contained in a tube along the laser axis.) Except for low-power models, the dye solution flows constantly through the dye cell. The lasers are designed so that dyes can be interchanged to change wavelengths.

The cavities of dye lasers pumped by pulsed lasers typically are no more than tens of centimeters long, so the dye emission can make several round trips of the cavity during a pump pulse lasting on the order of 10 nanoseconds (ns). The pumped region of the dye cell accounts for only a small fraction of the cavity length; it may be a millimeter across. The round trips both increase gain and improve beam quality. Normally 70 to 90 percent of the light is reflected back into the cavity, with the rest coupled out through an output mirror or by other means.

Figure 17.3a shows one standard simple configuration for a dye laser transversely pumped by a pulsed laser. Turning the diffraction grating tunes the wavelength; an optional etalon provides further wavelength control. Because the beam makes only a few round trips through the cavity, the tuning mechanism must strongly select the desired wavelengths. A set of prisms may replace the beam-expanding optics shown to the left of the dye cell. In some designs, the output beam may be light reflected from a surface of one of the beam-expanding prisms.

Dye lasers pumped by pulsed lasers often are operated in an oscillator-amplifier configuration, such as shown in Fig. 17.6. This approach offers lower noise from amplified spontaneous emission than building a larger dye laser oscillator. An amplifier also can improve beam quality, as well as raise output power. Oscillator-amplifier designs can be much more elaborate than the one illustrated. The optical system can be folded so that oscillation occurs in one part of a dye cell and amplification in another part of the same cell. Cylindrical optics can spread the pump beam over a long, narrow region in the amplifier dye cell; the volume of pumped dye is one factor that influences output power. Typically a single pump laser drives both oscillator and amplifier; the portion of the pump beam going to the amplifier may be delayed to ensure that it arrives at the proper time.

Flashlamp-pumped dye lasers have long linear cavities, with dye

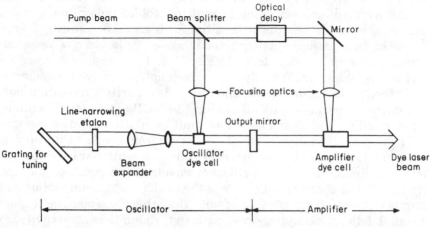

Figure 17.6 Oscillator-amplifier design for a pulsed dye laser.

flowing down the length of a cylinder that runs parallel to a
flashlamp. Laser oscillation is along the length of the cylinder. As
mentioned earlier, the dye may flow through the central bore of an
annular flashlamp, or through a cylinder pumped by separate cylindrical lamps. To produce pulses of several joules, a commercial
flashlamp-pumped dye laser would have a 50-cm-long tube with 1-cm-
bore diameter. The large volume of excited dye solution contributes to
the high average power possible with flashlamp-pumped dye lasers.

Continuously pumped dye lasers have their own characteristic designs and dye-flow systems. Instead of illuminating the dye in a sealed
cell, the continuous pump beam is focused on the narrowest part of a
free-flowing jet of dye solution. This approach avoids optical damage
to solid windows that could be caused by the high power levels from
the tightly focused pump beam. Only a very small volume is illuminated at any instant.

The cavity design shown in Fig. 17.3b is one of many for continuous-
wave dye lasers. Like many designs, the cavity is folded rather than
linear, and includes a wavelength-dispersive element (a prism). Movement of the rear cavity mirror selects the wavelength which oscillates
in the laser cavity.

Dye lasers designed for continuous pumping can be pumped synchronously by a modelocked continuous-wave laser. For synchronous
pumping, the cavity length of the dye laser must match that of or be
an integral multiple of the path length in the pump laser (usually argon or Nd–YAG), so the cavity round-trip times match as required for
modelocked pulse generation. Typical pump lasers have cavities 1 to 2

meters (m) long, so the dye laser cavity must be stretched to that length with a linear extension or multiply folded delay line. Careful matching of cavity lengths is essential for high performance.

The highest performance continuous-wave dye lasers now use variations on the ring cavity shown in Fig. 17.5. Three or four mirrors define a ring, which includes dye jet, wavelength-selective components, and other optical components; one mirror is a partly transparent output mirror. The ring cavity allows light to oscillate in both directions, but normally a device is inserted to limit oscillation to one direction.

The major advantages of ring lasers come from the fact that they do not establish a standing-wave pattern like conventional laser resonators. In a linear cavity, the oscillation wavelength depends on the cavity length; the round-trip distance of the cavity must equal an integral number of wavelengths. This means that the electromagnetic field strength follows a standing-wave pattern through the entire cavity, as shown in Fig. 17.7. At nodal points (where the field strength is zero), the standing wave cannot extract energy from the laser medium. The waves in a ring laser circulate continually around the cavity, never forming standing patterns, so they can extract energy from the entire laser medium. This allows ring lasers to generate single-wavelength output up to about 15 times higher. Those higher powers also are accessible in the laser cavity and permit more efficient harmonic generation inside the laser.

A second subtle but important advantage is the prevention of oscillation at secondary wavelengths. If a standing-wave laser is pumped strongly enough, sufficient gain can build up for the laser to oscillate at a second wavelength, with peaks where the first has nulls. This cannot happen in a ring laser.

Inherent Trade-offs

Three characteristics figure high on the wish lists of many dye laser users: extremely narrow linewidth, high average power, and ultrashort pulses. An individual dye laser can offer one of these features, but not all three. The inability to combine narrow linewidth and short pulses is fundamental because the minimum attainable linewidth and shortest possible pulse duration are related according to the formula:

$$\text{Pulse length} = \frac{0.441}{\text{bandwidth}}$$

(Svelto, 1989). This expresses a Fourier transform relationship between time and frequency domains; it is a fundamental physical limit.

The problem of combining high average power with either of the

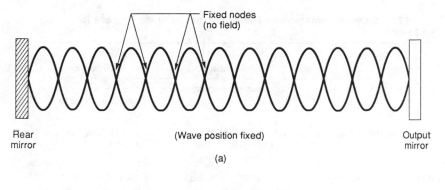

Fixed nodes
(no field)

Rear
mirror

(Wave position fixed)

Output
mirror

(a)

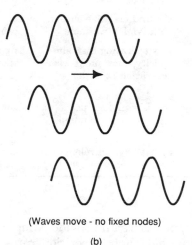

(Waves move - no fixed nodes)

(b)

Figure 17.7 In a standing-wave cavity (a), the waves are fixed in position, forming nodes where the field is at a null point and no gain is extracted from the laser medium. In a traveling-wave cavity (b), the light waves move continually through the cavity, so there are no nulls.

other goals is more practical. In practice, the highest average powers come when the dye is pumped with high average powers from a pulsed laser (generating pulses lasting tens of nanoseconds), or from a flashlamp (generating dye pulses lasting microseconds). Neither mode of operation produces either ultranarrow bandwidth or ultrashort pulses. Continuous-wave operation is needed to produce extremely narrow bandwidth, but it yields limited output power. Although ultrashort pulses in the picosecond or femtosecond regimes have high peak powers, they are so short that they contain little energy, thus limiting average power.

Several trade-offs are inherent in the choice of pump sources, some of which are summarized in Table 17.1. Copper vapor, excimer,

TABLE 17.1   Some Advantages and Disadvantages of Pump Sources for
Dye Lasers

Pump source and wavelength	Advantage	Disadvantage
Copper vapor lasers; 511, 578 nm	High average powers, high efficiency	Long wavelengths, not continuous-wave, capital cost
Excimer lasers; XeCl, 308 nm	High repetition rate, high average power, low operating cost	High initial cost, gas handling
Flashlamps (broadband)	Low capital costs, high average power	Some replacements costly, beam quality limited
Ion lasers; UV, 488, 515 nm; others	Continuous-wave output, ultraviolet available	Limited output power, capital cost
Neodymium lasers; 532 nm doubled, 355 nm tripled	Low operating cost, simplicity, continuous-wave operation possible	High initial cost, pulse rate lower than that for excimers, harmonic generation required
Nitrogen lasers; 337 nm	Small, simple to use, least-expensive pump laser	Low average power, low efficiency
Ultraviolet pumping	Shorter dye wavelengths, more energy to dye	Shorter dye life, lower efficiency

neodymium, and nitrogen lasers all provide pump pulses lasting a few to tens of nanoseconds, but their repetition rates differ. Flashlamp pulses are considerably longer. Rare gas ion and neodymium lasers can provide continuous-wave pumping, although neodymium pump powers are limited. Trade-offs may be even more complicated in practice. For example, a laboratory which already has a pulsed neodymium laser may decide to use it as a pump laser, rather than buy a separate pump laser or a flashlamp-pumped dye laser.

The choice of dyes offers additional options. The primary variables are available wavelengths and output powers. However, dyes also react differently with different pump sources. Their tuning ranges and output powers depend on the choice of solvent as well as the pump source.

As dye lasers have begun to find applications in medical treatment, designers have been faced with a trade-off between ease of operation and system flexibility. Ease of operation usually is more important to physicians, so medical dye lasers typically are designed to operate only at the wavelength most effective for the desired treatment. This avoids the complexity of tuning wavelengths and changing dyes but restricts the laser to use in at most a few applications.

Variations and Types Covered

As indicated above, there are considerable variations among dye lasers, with several distinct types on the market. The rest of this chapter will concentrate on major types that are commercially available.

Beam Characteristics

Wavelength and Output Power. The wavelength and output power of dye lasers depend on the choice of dye and pump source, as well as on the design of the laser. Fundamental-frequency outputs range from about 310 to about 1200 nm. Table 17.2 lists a sampling of laser dyes and their outputs for various pump conditions.

The shortest wavelengths are possible only from pulsed lasers; the shortest wavelengths available continuous-wave (pumped by the ultraviolet lines of argon lasers) are about 370 nm. Considerably shorter tunable wavelengths, to about 200 nm, can be obtained by using dye laser pulses to drive second-harmonic generation or sum-frequency generation in nonlinear crystals. In sum-frequency generation, the dye laser frequency is added to that of a pump laser; in harmonic generation, the dye laser pulses are combined with themselves to double the frequency. The second harmonic of continuous-wave dye output can be produced by placing a nonlinear crystal inside the dye cavity; the shortest wavelengths available are about 215 nm. β-Barium borate must be used to generate harmonic wavelengths shorter than about 250 nm.

The greatest variety of dyes, with the highest output powers, is available in the visible spectrum, but many dyes also are available for the near-infrared. The development of tunable titanium-sapphire lasers has diminished interest in infrared dye lasers, but the dyes continue to be used with existing dye lasers. Some manufacturers offer dye lasers which also can operate with a Ti-sapphire crystal substituted for the standard dye cell.

Each dye covers only a limited range of wavelengths, with peak output roughly in the middle of that range, where the dye can most efficiently convert pump light to laser output. As can be seen in Table 17.2, tuning range and peak wavelength for the same dye can differ considerably. A careful study of the literature would show considerably larger variations. In practice, dye tuning range, peak wavelength, and output power all depend on the pump source, solvent, dye concentration, and laser operating conditions. If the same dye is used with different pump sources, it may be dissolved in a different solvent and/or in a different concentration. The effects of differences in the optical configurations of different dye lasers can be difficult to identify because those data are difficult to summarize in tabular form. The las-

TABLE 17.2 Characteristics of Selected Laser Dyes Plus Titanium-Sapphire for Different Pump Sources*

Dye name	Wavelength, nm		Pump source	Pump, W	Maximum dye output, W (at peak)
	Range	Peak			
Polyphenyl 2		383	Short-UV argon	3.4	0.25
Stilbene 1		415	UV argon	6.0	0.42
Stilbene 3	408–453	425	Nitrogen	—	—
(Stilbene 420)	410–454	424	XeCl	—	—
	412–444	424	Nd–YAG, 335 nm	—	—
	414–465	435	UV argon	7.0	1.0
Coumarin 102		477	Kr, 407–415 nm	4.8	0.58
(Coumarin 480)	454–510	470	Ar, ultraviolet	—	—
	457–520	478	Flashlamp	—	—
	457–517	478	XeCl	—	—
	459–508	475	Nd–YAG, 335 nm	—	—
	453–495	470	Nitrogen	—	—
Coumarin 30		518	Kr, 407–415 nm	4.6	0.38
Rhodamine 110	529–585	540	Ar, 455–514 nm	23	3.6
(Rhodamine 560)	529–570	541	Cu, 511 nm	—	—
	530–580	554	Flashlamp	—	—
	541–583	563	Nd–YAG, 532 nm	—	—
	542–578	555	XeCl	—	—
Rhodamine 6G	546–592	562	Nd–YAG, 532 nm	—	—
(Rhodamine 590)	563–625	586	Flashlamp	—	—
	563–607	585	Cu, 511 nm	—	—
	566–610	583	XeCl	—	—
	568–605	579	Nitrogen	—	—
	573–640	593	Ar, 455–514 nm	24	5.6
Dicyanomethylene	598–677	644	Cu, 511 nm	—	—
	600–677	635	Flashlamp	—	—
	600–695	637	Nitrogen	—	—
	607–676	635	Nd–YAG, 532 nm	—	—
	610–709	661	Ar, 455–514 nm	20	2.9
Ti-sapphire		790	Ar, 455–514 nm	20	3.6
Styryl 9	775–865	818	Nd–YAG, 532 nm	—	—
	784–900	822	Argon	—	—
	810–860	841	Flashlamp	—	—
Infrared dye 140	866–882	875	Nd–YAG, 532 nm	—	—
	875–1015	960	Kr, 753–799 nm	3.0	0.2
	876–912	884	XeCl	—	—
	900–995	936	XeCl	—	—
	906–1018	964	Nitrogen	—	—

*Pumping conditions and solvents differ, and experimental tuning ranges and power levels may differ dramatically. Note that some dyes are sold under different names.
SOURCES: Coherent Inc., Exciton Chemical Corp.

ing characteristics of dyes also can change if two dyes are mixed in the same solution, which is sometimes done to improve performance. Certain chemical additives also can improve dye performance, altering the tuning range while increasing output power.

Wavelength coverage of individual lasers can be limited by the cavity optics. Typically, cavity mirrors have low losses over a broader range of wavelengths than covered by any one dye, but not as broad as those covered by all available dyes. Extending tuning beyond that range requires additional sets of optics. The use of single-frequency optics, to limit linewidth to extremely small values, can restrict oscillation to only part of the wavelength range covered by a single dye.

Output power depends on the choice of pump laser and the design of the dye laser as well as the choice of dye. Typically, manufacturers specify maximum output of Rhodamine 6G, a particularly efficient dye which can be used in all types of dye lasers and has a peak wavelength near 600 nm. Other dyes generally give lower power levels, as does Rhodamine 6G when operated away from its peak wavelength.

Dye laser power is proportional to pumping power above laser threshold. Output power depends on how well the excitation wavelength matches the absorption bands of the dye, and can be influenced by pulse duration. This makes it important to understand the pump-light source when studying reports of dye laser performance. Table 17.1 summarizes the advantages and disadvantages of various sources and lists important wavelengths.

Efficiency.   The efficiency of laser-pumped dye lasers is measured as the percent of input energy converted to output energy. Conversion efficiencies well over 50 percent are possible for inherently efficient processes such as pumping Rhodamine 6G with Nd–YAG's 532-nm second harmonic, but efficiencies are lower in normal operation. Typical specifications give conversion efficiency as a few percent to a few tens of percent. Actual efficiency can be much lower for operation of an inefficient dye near the limits of its tuning range. Some literature also gives quantum efficiency, the fraction of input photons that generate output photons. Quantum efficiency is inevitably higher than energy-conversion efficiency because output photons have less energy than do input photons. Another common scale, particularly for continuous-wave lasers, is slope efficiency, the ratio of extra output to extra input above laser threshold.

For flashlamp-pumped dye lasers, the efficiency is given as the wall-plug efficiency: the laser output divided by the electrical input to the laser. This efficiency can reach about 1 percent for flashlamp pumping Rhodamine 6G but is lower for pumping less efficient dyes. Coaxial flashlamps produce two to three times more optical power for a given electrical input than do linear lamps (Furumoto, 1986). These figures account for both electrical losses in generating light from the flashlamp and optical losses in converting the flashlamp pulse into laser output. Some extra electrical power may be needed to operate liquid pumps for cooling water and dye solution.

The difference in the definitions of efficiency reflects ease of measurement. For flashlamp-pumped lasers, the easiest input energy to measure is the electrical power to the flashlamp. For laser pumping, the pump laser energy is the easiest quantity to measure—and it can differ depending on the choice of pump laser. However, this does complicate comparisons of dye laser efficiencies. A proper comparison requires accounting for the electrical efficiency of the pump laser in laser-pumped version. Most pump lasers have wall-plug efficiency of no more than 1 percent, so the overall efficiency of laser-pumped dye lasers is lower than that for flashlamp pumping. In keeping with that fact, the highest average powers usually are generated by flashlamp pumping.

Temporal Characteristics of Output.   Dye lasers can produce continuous or pulsed output, depending on the pump source. Dye laser emission is possible only while the pump source delivers light to the dye, although emission can be restricted to shorter pulses by modelockers or cavity dumpers. Dye laser pulses may be somewhat shorter than pulses from the pump source because time is needed to reach laser threshold.

Pulses characteristics generated by specific pump sources are as follows:

- *Rare gas ion laser:*   Normally continuous-wave output. Output can be cavity-dumped to generate pulses of 8 to 15 picoseconds (ps) at 4 megahertz (MHz) or modelocked to generate shorter pulses. The ion and dye lasers also can be synchronously modelocked to generate trains of picosecond pulses at repetition rates of 75 to 200 MHz; those repetition rates can be reduced by cavity dumping or external pulse selection techniques. Cavity dumping of modelocked pulses can produce subpicosecond pulses at 4 MHz; Q switching of modelocked pulses yields repetition rates to 100 Hz to 50 kHz.

- *Continous-wave doubled neodymium lasers:*   Normally produce continuous-wave output, as described for rare gas ion lasers.

- *Flashlamp pumping:*   Pulses normally last from 0.25 to 5 microseconds (μs), with energies from 0.1 to 100 joule (J), but pulses can be extended to about 500 μs. Repetition rates range from single shot or one shot every several seconds to 30 Hz for coaxial flashlamp and to 100 Hz for linear flashlamps, which can operate at higher repetition rates. Generally the highest repetition rates are from lasers with lower pulse energies.

- *Pulsed laser pumping:*   Pulses approximate the pulse length and repetition rate of the pump laser. Excimer and Nd–YAG lasers generate pulses of around 10 ns at repetition rates to a few hundred per second; maximum pulse energies are about 100 mJ. Copper vapor lasers generate similar length pulses at a higher repetition rate, to

12 kHz, with maximum energy about 2 mJ per pulse. Nitrogen lasers generate dye pulses of 0.5 to 7 ns at repetition rates to about 300 Hz; pulse energies are tens of microjoules

Ultrashort pulse generation requires some form of modelocking of a dye laser operated in broadband mode. The shorter the pulses, the broader the emission bandwidth must be. Synchronous modelocking of a dye laser to a modelocked pump laser (either rare gas ion or frequency-doubled neodymium) can generate pulses as short as a few hundred femtoseconds. Addition of a saturable absorber can further reduce pulse length. Alternatively, a saturable absorber can by itself passively modelock a dye laser pumped by a continuous-wave laser. In each case, the saturable absorber allows gain only briefly while it is switching between off and on states.

Passive modelocking with a saturable absorber is an essential element of the colliding-pulse modelocked ring laser, shown in Fig. 17.8, the commercial dye laser which generates the shortest pulses. This type of ring laser does not include components to restrict laser oscillation to one direction, and produces pulses going in opposite directions around the ring. One-quarter of the way around the ring from the dye jet, the cavity includes a saturable absorber which has lowest loss when the two opposite-direction pulses pass through it at the same time (causing deeper saturation and hence lower loss). Pulse lengths can be under 100 femtoseconds (fs) in commercial versions and tens of femtoseconds in laboratory systems. Details of operation are complex (Siegman, 1986, pp. 1125–1127).

Pulses from a dye laser can be further compressed by a two-stage optical system which first spreads the pulse out in time (by passing it through a segment of optical fiber) and then compresses it spatially

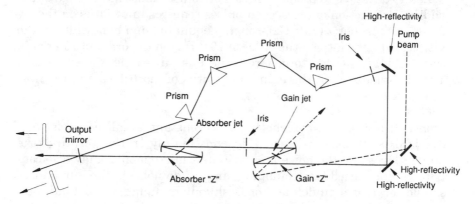

Figure 17.8 Schematic of a colliding-pulse modelocked dye laser. Note that pulses travel in both directions around the ring. (*Courtesy of Clark Instrumentation Inc.*)

(by passing it between a pair of prisms or diffraction gratings). This requires extremely broad bandwidth pulses but can generate pulses as short as 6 fs (Fork et al., 1987).

Spectral Bandwidth. In principle, spectral bandwidth of a dye laser depends on the dye's gain bandwidth and on the cavity optics, but in practice the cavity optics usually are the dominant factor. Most dyes have a gain bandwidth of over 10 nm, but normal gain-narrowing processes limit laser linewidth to a few nanometers in untuned cavities.

Adding wavelength-selecting optics can narrow bandwidth greatly, depending on pulse length. The fundamental transform limit on bandwidth mentioned above is about 1 GHz for a 10-ns pulse, or roughly 0.001 nm at visible wavelengths. Standard wavelength-selecting optics can readily reduce bandwidth to 0.01 nm or less. The addition of an etalon can reduce bandwidth to about 0.001 nm, or 0.025 inverse centimeter (cm^{-1}), for lasers with pulse lengths on the order of 10 ns. The shorter pulses generated by nitrogen pumping cannot have such narrow bandwidths. However, the longer pulses of flashlamp-pumped dye lasers can have spectral bandwidths under 1 GHz with suitable line-narrowing optics. Flashlamp-pumped dye lasers also can generate narrow-line pulses when injection-locked to amplify pulses from narrow-line dye lasers.

The narrowest linewidths are possible with continuous-wave dye lasers. Linear models have linewidths of several gigahertz (roughly 0.01 nm in the visible). Ring cavities allow much more stable operation, with linewidths of commercial versions as low as 0.5 MHz. Custom-built laboratory systems with elaborate stabilization equipment have produced linewidths on the order of a hertz, but those special-purpose systems are not needed for most applications.

Lasers designed to produce ultrashort pulses must have broad bandwidths. Production of picosecond pulses requires linewidths on the order of one terahertz (10^{12} Hz), roughly equal to 1-nm bandwidth in the visible. A bandwidth on the order of 10^{14} Hz (on the order of 100 nm in the visible) is needed to generate the 6-fs pulses mentioned earlier. That is broader than the gain bandwidths of most dyes and requires special accessories.

Amplitude Noise. If wavelength is held constant, amplitude noise in a dye laser well above threshold depends mainly on the nature of the pump source and the noise it contributes; microphonics may also contribute. For pulsed laser pumping, amplitude noise in the pump laser should serve as a guideline. For flashlamp pumping, typical pulse-to-pulse variations are a few percent. Operation at high repetition rates can cause inhomogeneity in pulsed laser media. Intensity stabilization of continuous-wave dye lasers can help restrict output-power variations to less than 1 percent for an hour of operation.

However, changing the output wavelength causes independent changes in output power because dye gain depends strongly on wavelength. Output power can vary by a large factor while tuning across a dye's gain bandwidth. Accessories can help stabilize power near the middle of a dye's operating range, but sharp variations are likely near the edges.

Beam Quality, Polarization, and Modes. Both pump source and optical cavity influence beam quality of dye lasers. In practice, beam quality is often secondary because the primary criteria for dye laser performance usually are tuning range, bandwidth, and pulse duration. The beams rarely need to travel long distances, but they may need to be focused onto small areas or harmonic generators, or coupled into optical fibers.

Grating tuning, common in dye lasers, or the use of any intracavity component at Brewster's angle, produces linearly polarized output, with the angle and degree of polarization dependent on the device. Intra- or extracavity polarizers can be added if stronger polarization is needed.

Continuous-wave dye lasers typically emit a linearly polarized TEM_{00} beam and oscillate in a single longitudinal mode. Dye lasers pumped by pulsed lasers have broader beams, usually two to three times the gaussian diffraction limit, reflecting their short cavities and comparatively high gain. Manufacturers do not specify output modes.

Most flashlamp-pumped dye lasers have large beams because the long tube containing the dye normally is on the order of a centimeter in diameter. With standard plane-parallel mirrors, flashlamp-pumped dye lasers generate a multimode beam; an unstable resonator can be used to produce a single-mode, lower-divergence beam. Tuning optics can polarize the output beam.

Coherence Length. Coherence length of dye lasers is highly variable because it depends on spectral bandwidth. A rough value can be calculated from the formula

$$\text{Coherence length} = \frac{\text{speed of light}}{\text{frequency bandwidth}}$$

Beam Diameter and Divergence. Typical values of beam diameter and divergence for different types of dye lasers are

- *Continuous-wave-laser-pumped:* Beam divergence of 0.5 to 2 milliradians (mrad) and beam diameters of 0.5 to 1 mm.

- *Pulsed-laser-pumped:* Beam divergence of 0.5 to 10 mrad and beam diameters of 0.5 to 5 mm. (Most are 1 to 3 mm.)

- *Flashlamp-pumped:* Typical beam diameters are 5 to 20 mm, although some high-power versions can have beam diameters as large as 50 mm. Divergence typically is 0.5 to 3 mrad, comparable in magnitude to the divergence of other dye lasers, but up to 40 times the diffraction limit because of the large beam diameter. Much smaller divergence, down to 0.1 mrad, is possible if the flashlamp-pumped laser is used as an injection-locked amplifier.

Stability of Beam Direction.   This quantity is not normally specified and is of little importance because most dye laser beams travel only short distances.

Suitability for Use with Laser Accessories.   Dye lasers are routinely used with many accessories. The most important are harmonic generators, sum-frequency generators, modelockers, cavity dumpers, and pulse compression optics.

Second-harmonic generators can shift visible output of pulsed dye lasers into the ultraviolet, obtaining wavelengths in the 200- to 335-nm range which dyes cannot produce directly. This gives a continuously tunable source of ultraviolet light. Harmonic generators are used inside the cavities of continuous-wave dye lasers, where circulating power is high enough for harmonic generation. Pulsed dye lasers generate peak powers adequate for extracavity harmonic generation.

Sum-frequency generators can generate slightly shorter ultraviolet wavelengths than second-harmonic generators by adding the frequency of a tunable dye laser to that of a fixed-wavelength laser. Changing the wavelength of the dye laser tunes the ultraviolet output. For example, a sum-frequency generator might combine a neodymium-YAG harmonic with a dye laser beam to generate ultraviolet output.

Dye lasers are well suited for modelocking because of the broad gain bandwidths of dyes. Modelocked pulses can be generated from continuous-wave dye lasers by synchronous pumping with an actively modelocked ion or neodymium laser; repetition rate of the modelocked pulses is determined by the cavity round-trip time. Dye lasers also can be modelocked passively with a saturable absorber in various configurations. The shortest laser pulses are produced by the colliding-pulse modelocked ring lasers described earlier; commercial versions emit pulses to about 100 fs, while laboratory versions have produced pulses lasting only tens of femtoseconds.

Cavity dumping can be used with either continuously pumped or flashlamp-pumped dye lasers. (Pulse durations of laser-pumped dye lasers are too short.) In a flashlamp-pumped laser with normal pulse duration under 0.5 μs, cavity dumping serves to compress the energy normally contained in the longer pulse into a pulse lasting only about 20 ns. It effectively stores the energy as photons circulating in the cavity and then switches them out in a short pulse lasting roughly one cavity round-trip time. For a continuously pumped dye laser, cavity dumping can produce a similar pulse lasting one round-trip time. Cavity dumping also can be combined with synchronous modelocking to produce pulses with peak power about 30 times higher than ordinary synchronously modelocked pulses, and with a slower repetition rate with is preferable for many applications.

External pulse compression optics can shrink femtosecond pulses further, to a present limit of 6 fs. They first pass the pulse through an optically dispersive medium, with refractive index which varies with wavelength. This spreads the pulse out in time as a function of wavelength (e.g., the shorter wavelengths might arrive first). Then the pulse is passed through a set of prisms or gratings which recombine the different wavelengths to make a shorter-duration pulse.

Dye lasers are not Q-switched to build up high pulse powers because the upper state lifetime of a few nanoseconds is too short for energy storage.

Operating Requirements

Input Power. Dye lasers pumped by an external laser have modest electrical power requirements because the external light generates the beam. However, power is needed for optical and electronic controls, dye solution pumps, and other equipment. Data sheets for some low-power dye lasers do not mention electrical power requirements, and some such lasers may not need electrical power. Others, especially those operating at higher powers or pumped by a continuous-wave laser, need modest electrical power for dye-circulation pumps, computer controls, or other equipment. All pump lasers need electrical power; their requirements are described in other chapters.

Flashlamp-pumped dye lasers *do* need significant electrical power to drive the flashlamp that pumps the dye. The power supply stores electrical energy, which it discharges through the flashlamp to produce the pulse of light which excites the dye. Power requirements range from 15 amperes (A) at 110 volts (V) to 100 A at 208/220 V. Linear flashlamps often are driven with a simmer power supply, which keeps a moderate current flowing through the tube between pulses. This current is not sufficient to generate light, but does preionize the gas, increasing output and decreasing pulse rise time. Coaxial flashlamps cannot be operated with a simmer power supply because the low current does not flow uniformly through the annular lamp.

Cooling. Flashlamp-pumped dye lasers typically remove excess heat by flowing water through the system at rates to 20 liters per minute (L/min). (Slower flows are more typical.) The water cools both the flashlamp and the dye solution.

In laser-pumped dye systems, the dye-flow system usually cools the dye solution as well as keeping it circulating through the dye cell to avoid degradation. Some flow systems include heat exchangers which transfer heat from the dye solution to slowly flowing water or some other medium. A few dye lasers which operate at low average power use sealed dye cuvettes without a liquid flow system (although the dye solution may be stirred).

Consumables. The prime consumable in a dye laser is the dye solution itself. The complex organic dye molecules suffer both thermal and photochemical degradation. The breakdown rate increases both as the pump wavelength becomes shorter (because higher-energy photons can better break bonds) and with the light intensity. Stated lifetimes range from around an hour to well over 1000 hours and vary considerably among dyes and depending on operating conditions. Under favorable conditions, the most durable dyes can last over a year. Adding fresh dye to a solvent often can replenish a dye solution so that it can operate normally, but breakdown products can accumulate to detrimental levels.

Dyes are offered either as premixed solutions or as compounds which can be added to solvents. Dye laser manufacturers typically offer dyes, but the chemicals often are prepared by independent companies which specialize in their production.

Both dyes and solvents can be toxic, as described below, so waste disposal can become an issue with dye lasers. Spent dye solutions may require handling as toxic wastes.

Required Accessories. The wide variation in commercial offerings makes it hard to define "required accessories" for dye lasers. All except flashlamp-pumped dye lasers require a pump lamp. Stabilization loops may be needed to connect some pump lasers with the dye laser. Other equipment usually is offered as options to the laser or depends on the application.

Operating Conditions and Temperature. Dye lasers are designed to function in a normal laboratory environment at room temperature, and some have been designed for field use. However, many models could encounter problems if bounced about in the back of a truck or operated at extreme temperatures. Many solvents have low boiling points; thus solvent evaporation is a potential problem at high temperatures.

Mechanical Considerations. Dimensions and weights vary significantly among the different types of dye lasers, so only general indications of size can be given.

A few small flashlamp-pumped models are packaged as integral units, but typically there are two or more modules. The laser head usually is separate from the power supply and dye circulation, which may be packaged together or separately. The laser head typically is made to mount on an optical bench, while the power supply and dye circulation pump may be rack- or floor-mounted. Controls for the laser also may be in a separate module. Laser heads are 0.9 to 1.3 m long and weigh 13 to 200 kilograms (kg). The power supply and dye circulator are more compact and boxy, typically weighing 20 to 70 kg each, although some dye circulators can be smaller. The largest flashlamp-pumped dye lasers require separate power supply and dye circulator modules, each 1.75-m-tall floor-mounted units.

By avoiding the need for internal flashlamps and power supplies, pulsed or continuous laser-pumped dye lasers can be rendered much more compact; the pump lasers may be much larger. The smallest model measures 7.4 by 11.7 by 6.4 cm and weighs 0.9 kg; it has a sealed dye cell and is designed for pumping with a nitrogen lasers. One moderate-sized dye laser, also designed for pulsed laser pumping, measures 25 by 20 by 13 cm and weighs 5 kg. Some other laser-pumped lasers are 0.4 to 0.5 m long, but many require cavities 1 to 2 m long for synchronous modelocking and other optical components. Oscillator-amplifier configurations also are long. Some of these lasers have separate dye circulators and/or control modules.

Safety. Dye lasers present four potential types of hazard: the laser beam itself, electric shock, toxic dyes and solvents, and explosion and fire hazards.

The optical hazards from dye laser beams are similar to those from other lasers operating at the same wavelengths (Sliney and Wolbarsht, 1980). The principal danger is eye damage, because the visible, near-infrared, and ultraviolet light from dye lasers generally can penetrate the eye. High-power pulses are particularly dangerous because a single pulse can do serious damage, particularly in the ultraviolet. One special hazard of the dye laser is a direct result of its tunability: because the emitted wavelength can change, there are no safety goggles that can block just the narrow range of wavelengths emitted by a dye laser. It is also desirable to block any stray pump light, or any of the fundamental frequency light left after frequency doubling. To meet these requirements, goggles must block a comparatively broad chunk of the visible spectrum. Unfortunately, this is like wearing sunglasses inside, and it can obstruct vision, presenting other hazards as well as frustrating some wearers enough that they abandon the goggles. Also, if the entire visible spectrum is being covered, users will have to change goggles at least once during the experiment for proper eye protection.

Electrical hazards are presented by the high voltages needed to fire flashlamps and to operate pump lasers. In commercial systems, the points carrying high voltages are enclosed, but the high voltages may be exposed in some laboratory systems. Special care must be taken to avoid contacting these points if safety goggles significantly impair vision.

Both laser dyes and solvents can be toxic (Kues and Lutty, 1975), although the hazards are not well quantified and dye solutions generally lack detailed warning labels. They do not have to be ingested to present dangers; spills can be absorbed through the skin, and some organic solvents [notably dimethyl sulfoxide (DMSO)] enhance the skin penetration by dangerous dyes. Solvent vapors pose some hazards, and small amounts of the dye may be carried into the vapor phase. Care should be taken to ensure adequate ventilation while handling dye solutions and when dye solutions are exposed to air. People handling dye solutions should take precautions such as wearing butyl rubber gloves to ensure that the solutions do not contact skin. Spent dye solutions should not be poured down drains both because of their potential toxicity and because organic solvents could react with other chemicals in the drainage system. Some dye solutions may require handling as toxic wastes.

Most organic solvents are highly flammable, both in liquid and vapor forms, and present potential fire hazards (see, e.g., Winburn, 1990, p. 201). The main dangers are sparks, contacting objects at high temperatures, and flashlamp explosions. Flashlamps can explode if driven at excessive current levels or if the driver is not properly matched to the lamp; such explosions can ignite dye solutions (Grant and Hawley, 1975).

Reliability and Maintenance

Lifetime. As mentioned above, lifetimes of laser dye solutions are limited and depend on the dye and operating conditions. Progress has been made in extending dye lifetime, but high pump energy and short pump wavelength do cause dyes to degenerate faster.

Flashlamps also have limited lifetimes, which differ considerably among different types. Lifetime decreases very rapidly as pulse energy increases. For example, a 500-J coaxial flashlamp will last only 10,000 to 20,000 shots if operated at 500 J, but if energy loading is decreased to 64 percent of that value, lifetime will exceed 750,000 shots (Furumoto, 1986). Linear flashlamps can produce millions of shots. Sometimes lifetimes are given in terms of total energy that a lamp can deliver before failure, typically in the range of 10^7 to 10^8 J. Lamp failures occur when output energy decreases below a required level, but lamps driven too hard can explode.

Except for dyes and flashlamps, lifetime figures generally are not given for dye lasers.

Maintenance and Adjustments Needed. Several types of maintenance may be needed for dye lasers. These include

- Periodic replacement of dye solutions or replenishment of dye.

- Replenishment of volatile components of dye solution lost by evaporation.

- Cleaning the dye system when switching dyes to avoid optical losses from impurities.

- Cleaning of dye-flow system and filters.

- Cleaning and adjustment of optics. (The more complex the optical system, the more often adjustment is usually needed.)

- Replacement of flashlamps in flashlamp-pumped systems.

- Checking for optical damage, particularly to dye cells. Damaged components must be realigned to avoid damaged zones or replaced.

- Maintaining alignment of the pump laser with the dye cell.

- Maintaining adjustment of tuning optics.

- Periodic maintenance of liquid flow pumps, filters, and electronic components.

Mechanical Durability. Some dye lasers are fragile instruments intended only for laboratory environments, but others are rugged enough for use in the field, with reasonable care and handling. (Shocks could knock optics out of alignment or damage the liquid flow system.) A significant fraction of dye lasers are built explicitly for medical applications and are designed to meet hospital requirements. Some dye lasers are designed for use in instrumentation and measurement.

Failure Modes and Causes. Dye lasers can fail to meet a variety of specifications. If the pump source or the laser dye is not at fault, the problem is probably in the optics. The cavity could have drifted out of alignment, key components could be out of adjustment, or optical damage could have occurred at key points such as dye-cell walls. Note that continuous-wave dye lasers generally have flowing dye jets rather than dye cells to avoid the possibility of optical damage there.

Users should be aware that flashlamps and pump lasers can degrade in ways that reduce output power to lower levels than specified, and possibly to levels below the threshold for laser action in the dye. Proper diagnosis of the problem may require measuring power of the pump source.

Possible Repairs. Other than fixing the light source or replacing the dye solution, the main repairs are optical realignment and replacement or repair of damaged optics. If damage has occurred at a small area, such as on the surface of a dye cell, it may be possible to move the component so the damaged area is not in the optical path. Some high-power pulsed dye lasers have dye cells designed so damaged areas can be moved out of the pump laser focal spot.

Commercial Devices

Standard Configurations. There are many dye laser configurations on the market, each with its own advantages and advocates. Many models are designed for use with a specific type of pump source, but some can accept different types of pulsed lasers. The major configurations offered commercially are

- Coaxial-flashlamp-pumped, in which the dye cell is in the central bore of a coaxial flashtube, as shown in Fig. 17.4

- Linear-flashlamp-pumped, in which the dye cell is in the same reflective cavity as a linear flash tube

- Nitrogen-laser-pumped, sometimes in a compact, integrated package

- Excimer-laser-pumped, often with the 308-nm output of xenon chloride, which offers the best combination of long excimer lifetime and limited dye degradation, but sometimes with other lines

- Pulsed-neodymium-YAG-pumped, with the 532-nm second harmonic, the 355-nm third harmonic, or for some infrared dye YAG's 1.06-μm fundamental wavelength in the near-infrared

- Pulsed-laser-pumped, in a version designed to accept YAG, excimer, or nitrogen pumping

- Copper-vapor-pumped, similar to other types pumped with pulsed lasers, but designed to operate at higher repetition rates and average powers, with pump lines in the green and yellow

- Continuous-wave, pumped with an ion laser (usually argon but sometimes krypton) or a neodymium laser, using a linear, folded, or ring cavity

- Synchronously pumped with a modelocked ion or doubled neodymium laser

- Colliding-pulse modelocked lasers, with ring cavities, which generate ultrafast pulses in the 100-fs range

- Hybrid modelocked lasers, in which the more intense parts of a modelocked pulse are preferentially amplified, shortening the pulse

Pulsed dye lasers are available in oscillator-amplifier configurations that give higher output power than is readily available from an oscillator alone. These are supplied in several variations, depending on the pump source and pulse duration. In flashlamp-pumped versions, separate lamps are used for oscillator and amplifier, with the pulse-generating electronics controlled synchronously. In laser-pumped versions operating in the nanosecond-pulse regime, a single pulse from a powerful pump laser may be split to pump both oscillator and amplifier. In femtosecond-pulse lasers, separate pump lasers are used. For example, a modelocked ion laser may drive the oscillator, which produces a picosecond pulse that is compressed to the femtosecond regime by external optics and then amplified separately in an amplifier pumped by a copper vapor laser. This approach makes it possible to obtain reasonable energies in ultrashort pulses.

Some dye lasers are packaged together with pump lasers, but unless both are inexpensive, the two usually are sold separately. Many companies manufacture both dye and pump lasers, and these companies typically design their dye lasers to mate directly with their pump own lasers. However, pump lasers are similar enough that this does not prevent the use of competing models.

Although dye lasers are versatile, users should remember that commercial products are designed to meet specific needs. Some dye lasers are designed as general-purpose instruments. Others are optimized to specific design goals, such as extremely narrow linewidth, extremely short pulse length, high power, or production of high power at a single stable wavelength. No single laser can meet all these goals, and some are mutually exclusive.

Options. To meet the wide range of user requirements for dye lasers, manufacturers offer many options beyond the simple choice of dye and pump source. Major options are listed below, some of which may be standard features on certain models:

- Active stabilization of output frequency or power
- Amplifiers for pulsed lasers

- Automatic frequency scanning across a range of wavelengths
- Beam delivery systems for special requirements (e.g., medical treatment)
- Cavity dumpers
- Cavity-extending optics to match length of a dye laser cavity to that of a synchronously modelocked pump laser
- Computer interfaces and controls
- Difference-frequency mixing to generate tunable infrared light by mixing tunable dye output with output of a fixed-frequency laser
- Dual-dye operation to allow simultaneous operation at two wavelengths for certain applications
- Dye circulators
- Dye coolers (increase lifetimes of certain dyes by reducing operating temperature)
- Dye filters to reduce scattering by removing particles larger than the wavelength of light
- Etalons, to narrow spectral bandwidth
- Frequency-summing accessories to generate short wavelengths by adding the frequency of a tunable dye laser to that of a fixed-frequency laser
- Harmonic generators to produce ultraviolet output from visible dyes
- Modelockers
- Optical output monitors
- Optics with broad spectral range to reduce the need for interchange of optics while tuning wavelength
- Polarizing optics
- Pulse compression optics for generating femtosecond pulses
- Raman shifters
- Titanium-sapphire operation of the laser, by substituting Ti-sapphire crystal for the dye solution to generate tunable near-infrared light (for pumping with ion or doubled neodymium lasers)
- Unstable resonator optics (for pulsed lasers)
- Wavelength-tuning optics (standard on many models), including tunable birefringent filter, tuning wedges, prisms, and gratings

Special Notes. The relative merits of different pump sources for pulsed dye lasers have been argued at great length, generally by companies which manufacture dye lasers and/or pump sources. However, each type appears to have earned its own niche in the complex market. Some users may find that particular requirements, such as repetition rate, pulse energy, pricing limits, or wavelength range, may dictate selection of one type. Others may find that two or more types are viable options.

Pricing. Pricing of dye lasers can be confusing because of the many configurations available. Flashlamp-pumped dye lasers are sold with internal pump lamps, and a *few* laser-pumped dye lasers are sold as systems incorporating pump lasers. However, the attractively low prices of some sophisticated dye lasers may only indicate that the expensive pump lasers are sold separately. Note also that standard optical configurations vary widely; some models come complete with highly sophisticated optics, while others come equipped with mirrors suitable only for broadband operation.

Some manufacturers have become reluctant to publish prices, particularly for continuous-wave pumped dye lasers. The following prices are estimates for the typical range of lasers offered in 1990, based on offerings for which prices have been quoted:

- Flashlamp-pumped dye: $6000 to $190,000

- Pulsed-laser pumped (without pump laser): circa $2000 to $70,000

- Continuous-wave pumped (without pump laser): $10,000 and up, depending on options

Suppliers. The dye laser market is quite competitive, although no company offers all types, and most specialize in one or two types. Virtually all types are available from two or more manufacturers. Competition is most intense in lasers pumped by pulsed lasers (except for lasers designed specifically for copper vapor pumping, a limited market). Some companies entered the dye market after developing a pump laser often used with dye lasers. Fewer companies supply continuous-wave-pumped and flashlamp-pumped dye lasers. A few specialized dye laser systems are at this writing available from only one company.

Many companies have been manufacturing dye lasers for many years, but new suppliers do appear, and existing ones sometimes change. Users would be wise to consult a current industry directory to check the list of suppliers.

There are enough suppliers to make pricing competitive for most dye lasers for scientific applications. However, some specialized markets have little competition, notably dye lasers for medical treatment.

Applications

The vast majority of dye lasers traditionally have been sold for research and development, but applications in medical treatment are increasing rapidly. A few dye lasers are used for other applications, including medical diagnostic measurements, other types of inspection, and entertainment displays.

The major scientific applications of dye lasers are in spectroscopy and other types of measurement, both in the time and frequency (or wavelength) domains. The ability to tune dye laser wavelength and limit emission to a narrow spectral bandwidth makes dye lasers extremely useful for studying the absorption and emission of light by various materials. Although other light sources can be used in spectroscopy, dye lasers offer higher light intensities in narrower spectral bands than do other tunable, nonlaser sources. They have been particularly useful in the study of atomic and molecular physics, and in stimulating fluorescence. Dye lasers can be precisely tuned to match specific absorption and emission bands, allowing atoms to be slowed down to speeds equivalent to temperatures below 1 kelvin (K).

Ultrashort picosecond and femtosecond pulses from dye lasers have been invaluable in studies of time response, such as measurements of excited-state lifetimes and decay rates. Ultrashort pulses have been used to study semiconductor properties, chemical reaction kinetics, photosynthesis, time-resolved spectroscopy, and biomolecular processes. The scope of these time and frequency domain studies is far beyond the realm of this book.

In medicine, dye lasers are used to treat skin disorders and to shatter stones in the urinary system and gall bladder. Dye lasers can be built to emit at lines which are strongly absorbed in the target tissue, such as the dark birthmarks called "portwine stains," which are caused by surface networks of abnormal blood vessels. Selection of the proper wavelength causes the laser energy to concentrated in the target tissue, while doing minimal damage to surrounding tissue. Dye laser light can be directed through an optical fiber and delivered inside the body to shatter solid stones in the kidneys and gall bladder; a series of laser pulses generally can reduce the stone to fragments small enough to be passed naturally from the body. Such treatment has gained rapid acceptance because it avoids the need for surgery. Some dye lasers are used for diagnostics, such as cell sorting. Dye lasers have also been studied for treatment of cancer and eye disease.

Dye lasers have been considered for some measurement and inspection applications requiring specific wavelengths, but at this writing no such applications have gained wide acceptance. The cost and complexity of dye lasers have helped to limit such uses.

Dye lasers have sometimes been used for displays because they can generate varicolored beams. Although dye lasers offer a unique way to scan the entire visible spectrum, their practical display applications are limited by their operational complexity and high cost.

The tunability of dye lasers makes them attractive for photochemical processing techniques in which the product depends critically on wavelength. The most important such process in development is the enrichment of isotopes, based on tuning a dye laser so that its output excites one isotope but not the slightly shifted lines of a second. Throughout the 1980s, the Lawrence Livermore National Laboratory conducted two parallel programs using high-power dye lasers pumped by copper vapor lasers. One uses the dye lasers to ionize uranium-235 (^{235}U), but not the more common uranium-238 (^{238}U). Collection of the ionized material yields uranium enriched in the fissionable ^{235}U, as required for nuclear reactor fuel. Livermore also developed a process to purify plutonium produced in nuclear reactors for use in nuclear bombs by removing certain isotopes. At this writing, neither process is operating on a commercial scale.

Bibliography

F. J. Duarte and L. W. Hillman (eds.): *Dye Laser Principles*, Academic Press, Boston, 1990 (compilation of review articles).

Michael J. Feld, John E. Thomas, and Aram Mooradian (eds.): *Laser Spectroscopy IX*, Academic Press, Boston, 1989 (conference proceedings; includes many reports on dye laser applications in spectroscopy).

R. L. Fork et al.: "Compression of optical pulses to six femtoseconds by using cubic phase compensation," *Optics Letters* 12:483–485, 1987 (research report).

Horace Furumoto: "State of the art high energy and high average power flashlamp excited dye lasers," in SPIE (Society of Photo-Optical Instrumentation Engineers), *Flashlamp Pumped Laser Technology: Proceedings of the SPIE*, vol. 609, SPIE, Bellingham, Wash., 1986, pp. 111–128.

W. B. Grant and J. G. Hawley: "Prevention of fire damage due to exploding dye laser flashlamps," *Applied Optics* 14(6):1257–1258, 1975 (technical note).

Ken Ibbs: "Ultrafast laser systems—comparing technologies," *Lasers & Optronics* 9(2): 39–50, February 1990 (review article).

T. J. Johnston, Jr.: "Tunable dye lasers," in Robert A. Meyers (ed.), *Encyclopedia of Physical Science and Technology*, vol. 14, Academic Press, Boston, 1987, pp. 96–141 (review of dye laser technology).

Henry A. Kues and Gerard A. Lutty: "Dyes can be deadly," *Laser Focus* 11(5):59–61, May 1975 (listing of health hazards).

Mitsuo Maeda: *Laser Dyes: Properties of Organic Compounds for Dye Lasers* Academic Press, Orlando, Fla., 1984 (description of laser dyes and dye lasers).

F. P. Schafer (ed.): *Dye Lasers*, 2d ed., Springer-Verlag, Berlin and New York, 1977 (collection of review articles).

Anthony E. Siegman: *Lasers*, University Science Book, Mill Valley, Calif., 1986 (textbook covering laser fundamentals in detail).

David Sliney and Myron Wolbarsht: *Safety with Lasers and Other Optical Sources*, Plenum, New York, 1980 (comprehensive handbook).

Peter P. Sorokin and J. R. Lankard: "Stimulated emission from an organic dye, chloroaluminum phlalocyonine," *IBM Journal of Research & Development 10*:162, 1966 (research report of first dye laser).

Richard Steppel: "Organic dye lasers," in Marvin Weber (ed.), *CRC Handbook of Laser Science & Technology*, vol. 1, *Lasers & Masers*, CRC Press, Boca Raton, Fla., 1982 (review article; update in process for a supplement to this handbook).

Orazio Svelto: *Principles of Lasers*, 2d ed., Plenum, New York, 1982, p. 191.

B. Wilhelmi and F. Weidner: "Liquid lasers," in Kai Chang (ed.), *Handbook of Microwave and Optical Components*, vol. 3, *Optical Components*, Wiley-Interscience, New York, 1990, pp. 399–450 (review of dye lasers and other liquid lasers).

D. C. Winburn: *Practical Laser Safety*, Marcel Dekker, New York, 1990 (short guide to laser safety in general).

Semiconductor Diode Lasers—Structures

The 1980s saw an explosion in semiconductor lasers. The sheer numbers sold overwhelm all other types—worldwide unit sales passed 20 million a year in 1988, compared to less than half a million of all other lasers combined (*Lasers & Optronics*, 1990). The technology has advanced dramatically, with major advances in laser power, efficiency, speed, bandwidth, and wavelength range.

The first semiconductor lasers were developed in 1962, when three groups independently and nearly simultaneously demonstrated gallium arsenide diode lasers (Hall, 1976; Hecht, 1991). However, it was nearly a decade before researchers at Bell Laboratories and in the Soviet Union made the first semiconductor lasers able to produce a continuous beam at room temperature. The pace of development has accelerated since then, and semiconductor lasers have found many applications, including fiber-optic communications, compact-disk audio players, and laser printers.

Semiconductor lasers represent a large family of devices, with many different structures, made of a variety of materials. The variety is too large to cover in a single chapter, so it is grouped in Chaps. 18 to 21. This chapter is devoted to internal structures in commercial use and in development. These structures are common to most major semiconductor material systems, although certain structures are used only in certain materials. They are fundamental to understanding the discussions of specific semiconductor materials in Chaps. 19 to 21.

Chapter 19 covers short-wavelength semiconductor lasers, loosely defined as types emitting wavelengths shorter than about 1000 nanometers (nm). The principal members of this family have long been GaAs and GaAlAs devices emitting at about 750 to 900 nm. Recent years have seen an important addition to this family, the first visible diode lasers, InGaAlP and InGaP types emitting in the red. The emerging technologies of quantum wells and strained-layer superlattices are adding new lasers to this family at longer wavelengths as well. These lasers serve a range of applications, including compact-disk players, laser printers, laser pointers, and pumping solid-state lasers.

Chapter 20 covers semiconductor lasers made of indium, gallium, arsenic, and phosphorous, which emit at roughly 1.1 to 1.65 micrometers (μm), and related types being developed for long wavelengths. These lasers are used primarily in fiber-optic communications, leading to differences from GaAs lasers, although both are made from elements in groups III and V of the periodic table.

Chapter 21 covers semiconductor lasers made from different compounds (often called "lead salts") from groups II, IV, and VI of the periodic table. They produce longer infrared wavelengths, roughly 3 to 30 μm. Their primary use is to generate specific tunable wavelengths for spectroscopy.

Description

As the name implies, all semiconductor diode lasers are made of semiconductors, and have the electrical characteristics of diodes. Passage of an electric current through the diode causes current carriers to recombine at the junction between regions with different dopings. This recombination process releases energy as light in some semiconductors. Silicon and germanium lack the energy level structures required, so diode lasers are made from binary semiconductors such as gallium arsenide or more complex semiconductors which are compounds of three elements (ternary) or four elements (quaternary).

Diode lasers are similar in structure to incoherent light-emitting diodes (LEDs), and generally can function as LEDs—albeit inefficient ones—at low drive currents. Modest drive currents suffice to cause some recombination, which generates spontaneous emission at an LED junction. Higher current levels are needed to sustain the population inversion needed for laser action. A diode laser also requires a resonant cavity, not present in an LED, and an internal structure able to withstand high laser drive currents. Only certain compound semiconductors with so-called "direct bandgaps," such as GaAlAs and InGaAsP, are suitable for diode lasers, but LEDs can be made from other compound semiconductors which have indirect bandgaps. (The terms *direct* and *indirect* bandgap refer to energy levels of the semiconductor; the details are not relevant here.)

Individual diode lasers are tiny blocks of semiconductor material less than a millimeter on each side. Most emit light from the edge of the wafer, but some newer types emit light from the surface. Diode lasers also are made in monolithic arrays or "bars." The most powerful diode laser sources today are linear arrays of parallel laser stripes on a wafer which emit light from the side, and two-dimensional stacks of such arrays. Arrays which emit from the wafer surface are in development.

Because of their small size, most commercial diode lasers are packaged in housings which simplify handling and control beam direction. Some housings are designed for general-purpose use, but many are made for specific applications. For example, diode lasers for fiber-optic communications may be coupled directly to a fiber-optic "pigtail" or built into a connector. Lasers are packaged singly, in monolithic arrays, in arrays of single-element lasers, or as multiple monolithic arrays.

Internal Workings

As in other semiconductor devices, current in a diode laser is carried by electrons free to move within the crystal when an electric field is applied, and by holes, vacancies for valence electrons within the crystal lattice. The semiconductor is doped with impurities to control the type and density of current carriers. Material doped with elements that produce holes (or, equivalently, accept electrons) is called a *p-type* semiconductor (for positive carrier). Material doped with donors of excess electrons is called *n-type* (for negative carrier).

A semiconductor diode is a two-terminal device with a junction between *p* and *n* materials. (It is sometimes called a *junction* laser because of the junction or an *injection* laser to describe the process of current injection.) Application of a negative voltage to the *n* material and a positive voltage to the *p* material creates a local excess of minority carriers on each side of the junction. This is called forward-biasing the *pn* junction; it lowers internal potential barriers and causes carrier injection from one side of the junction to the other. At the junction, excess electrons fall to lower energy states, i.e., holes in the valence band. Figure 18.1 shows the process, called *electron-hole recombination*.

The electrical result of forward-biasing a semiconductor diode is current flow, with the recombination of electrons and holes releasing energy at the junction. In silicon diodes, most energy is released as heat, and no light is produced. In LEDs and diode lasers, much of the energy is released as photons with energy roughly equivalent to the bandgap—the energy difference between the conduction and valence bands—as shown in Fig. 18.2.

The two key elements needed to produce stimulated emission from a diode laser are a population inversion and optical feedback. Passing a

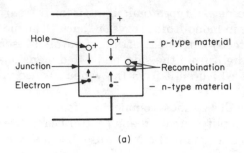

Hole

Junction

Electron

− p-type material

Recombination

− n-type material

(a)

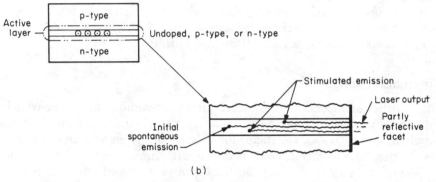

Active layer

p-type

Undoped, p-type, or n-type

n-type

Stimulated emission

Laser output

Partly reflective facet

Initial spontaneous emission

(b)

Figure 18.1 Operation of a diode laser (simplified). (a) Forward bias produces current flow and recombination; (b) stimulated emission in active layer.

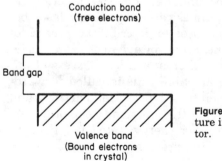

Conduction band (free electrons)

Band gap

Valence band (Bound electrons in crystal)

Figure 18.2 Energy-level structure in an unbiased semiconductor.

high drive current through a diode laser produces a population inversion at the junction. (Lower drive currents produce incoherent spontaneous emission from LEDs and from diode lasers operated below the threshold for laser action.) The optical feedback comes from reflective structures which form the diode laser resonator. In traditional edge-emitting diode lasers, facets cut on each end of the active junction layer reflect light back into the cavity. Other arrangements are described in more detail later in this chapter.

The bandgap depends on the crystalline structure and chemical composition of the semiconductor material, so diode laser wavelengths depend on their composition. Table 18.1 lists the important semiconductor laser compounds and their normal wavelength ranges. Binary compounds, such as GaAs or InP, have a fixed bandgap and wavelength. Adding a third element, such as substituting aluminum for some gallium in GaAs, changes the lattice constant and bandgap, thus producing a different wavelength. Such mixtures are often labeled in the manner $Al_xGa_{1-x}As$, where the x denotes the fraction of one element, in this case aluminum. The number of gallium and aluminum atoms together must equal the number of arsenic atoms, so the gallium fraction is $1-x$. Many quaternary compounds contain two group III elements and two group V elements and are written in the form $In_{1-x}Ga_xAs_yP_{1-y}$. Others may contain different mixtures, such as $In_xAl_yGa_{1-x-y}P$, in which indium, gallium, and aluminum are all group III elements, and must combine to equal the number of phosphorous atoms. In practice, only some compositions can be used in diode lasers; others cannot grow on standard substrates, are difficult to produce, or suffer other problems. Specific materials systems are described in Chaps. 19 to 21.

Only part of the electrical energy used by a diode laser emerges in the laser beam; the rest is dissipated as heat. The need to minimize

TABLE 18.1   Some Semiconductor Laser Materials

Compound	Wavelength, nm	Notes
AlGaInP	630–680	Shortest wavelengths are recent developments; higher power at longer wavelengths.
$Ga_{0.5}In_{0.5}P$	670	Active layer between AlGaInP layers; long room-temperature lifetime
$Ga_{1-x}Al_xAs$	620–895	$x = 0$–0.45; lifetimes very short for wavelengths < 720 nm
GaAs	904	
$In_{0.2}Ga_{0.8}As$	980	Strained-layer superlattice, on GaAs substrate
$In_{1-x}Ga_xAs_yP_{1-y}$	1100–1650	InP substrate
$In_{0.73}Ga_{0.27}As_{0.58}P_{0.42}$	1310	Major fiber communication wavelength
$In_{0.58}Ga_{0.42}As_{0.9}P_{0.1}$	1550	Major fiber communication wavelength
InGaAsSb	1700–4400	Possible range, developmental, on GaSb substrate
PbEuSeTe	3300–5800	Cryogenic
PbSSe	4200–8000	Cryogenic
PbSnTe	6300–29,000	Cryogenic
PbSnSe	8000–29,000	Cryogenic

and remove this waste heat constrains diode laser operation, particularly because operating characteristics degrade as temperature increases. Heat dissipation once was so severe a problem that early diode lasers were limited to pulsed and/or low-temperature operation. Great improvements have made semiconductor lasers the most efficient types available, leading to much higher output powers and shorter wavelengths, but heat dissipation remains an important consideration.

Internal Structure

In its simplest form, a diode laser consists of a block of semiconductor in which part is doped with electron donors to form *n*-type material and the other part is doped with electron accepters (hole generators) to produce *p*-type material. In practice, the structure contains additional layers. A separate undoped active layer is usually formed at the junction, which usually is sandwiched between other layers of different composition to improve its operation. Diode lasers also have lateral as well as vertical structures.

Diode laser fabrication starts with a substrate of a binary semiconductor (usually GaAs or InP) with a controlled impurity level. Additional layers may be formed by diffusing different impurities into selected areas of the substrate surface. For example, if the substrate is *n*-type material, electron accepters may be added in sufficient numbers to overwhelm the electron donors and create *p*-type material, producing a junction at the boundary of the diffusion layer and the substrate. More typically, layers are formed by epitaxial growth of new material on top of the substrate (which itself may be etched in selected regions to control the geometries of the added layers). Various epitaxial growth techniques allow the deposition of extremely thin layers of precise geometry, with each layer having a distinct composition. The combination of multiple layers and surface patterns controls current flow, guides laser light, and produces laser oscillation. The most important variations are described under the section on types of diode laser structure.

Laser diodes incorporate structures which make their output much more directional than the spontaneous emission from an LED. Figure 18.1 is a side view of a simplified edge-emitting laser diode. The two end facets of the laser chip are polished or cleaved to form a smooth surface that reflects some light back into the semiconductor. (Some lasers have one facet coated to reflect all incident light; others emit a weak beam from one facet and a strong beam from the other.) As in other lasers, initial spontaneous emission is reflected back and forth between the cavity mirrors, stimulating the emission of more light along the axis of the cavity—i.e., the junction plane in most diode lasers. If the net optical gain in a round trip of the cavity is higher than the losses, the result is laser oscillation.

Stimulated emission increases the optical power level in a diode laser. Spontaneous emission in LEDs is inefficient because the energy from many recombination sites is dissipated as heat rather than emitted as light. Stimulated emission can extract this energy as light before it can be dissipated. The stimulated emission process also amplifies most strongly the wavelengths where emission is strongest, effectively suppressing other wavelengths and causing a laser diode to have much narrower spectral bandwidth than an LED.

A diode laser will produce stimulated emission only if enough current is passing through the junction to produce a population inversion. This is called *threshold current*. It is a figure of merit for diode lasers; the lower the threshold current, the less energy is dissipated in the device and the higher its efficiency. The critical physical parameter is the current density, measured in amperes per square centimeter of junction area. If the current density is below the threshold, a diode laser functions like an LED, producing spontaneous emission.

Although a single value usually is given for threshold current, the transition from LED to laser action is not instantaneous. As shown in Fig. 18.3, a plot of output power vs. drive current changes slope rapidly at the nominal threshold current. This change reflects the much higher slope efficiency of laser emission. Low-resolution plots may make light output seem to be zero below laser threshold, but careful measurements show that output is not exactly zero. Practically speaking, threshold current usually is defined by extrapolating the linear part of the laser output-vs.-current curve back to the zero-output line.

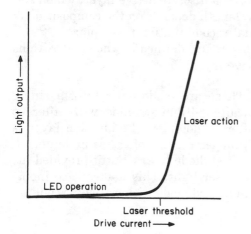

Figure 18.3   Output of a laser diode below threshold (when it operates as an LED) and above threshold (when it operates as a laser). The much steeper slope for laser operation indicates higher-efficiency emission.

Gain in semiconductor lasers is very high, so only a small length of active medium is needed. Typical edge-emitting laser chips have active layers about 500 μm long, and virtually all are within a factor of 3 of that length. (High-speed edge emitters may be as small as 50 to 100 μm.) The active layers normally are under 1 μm thick, but considerably wider, so emitting areas of single-stripe lasers typically are about 1 μm tall and several micrometers wide.

Junction Structures. The variety of diode laser structures is far too extensive to cover completely here. The list of structures seems to grow longer with every volume of such journals as *Applied Physics Letters*, *Electronics Letters*, and *Photonics Technology Letters*, and nothing published in book form could avoid instant obsolescence. To keep this discussion manageable, the focus here will be on broad categories and the factors which have the most important effects on performance.

There are several ways to view diode laser structures. Rather than cataloging them all, successively finer levels will be discussed, as shown in Fig. 18.4, stopping short of making fine distinctions among the state-of-the-art designs described in the current literature. Inevitably, some distinctions do not fit neatly into this two-dimensional pattern, but it can serve as a conceptual framework. The descriptions that follow are in a loosely evolutionary sequence, starting with single-element edge-emitting lasers, working through to arrays and cavity designs, and then covering surface-emitting lasers and arrays.

Figure 18.5 shows important edge-emitting diode laser structures. The consistent trend has been to more complex structures which better confine light to the active layer where current carriers recombine to emit light. In the simplest lasers, this layer is simply the junction between *p* and *n* materials. However, developers soon found that performance improved when they used a discrete active layer, which seldom is doped. The emission wavelength depends on the composition of the active layer, where the actual recombination takes place.

Three major classes of diode lasers are defined by the compositions of the layers adjacent to the active layer:

- *Homojunction diode lasers:* These are made entirely of a single semiconductor compound, typically gallium arsenide, with different parts of the device having different dopings. The junction layer is the interface between *n*- and *p*-doped regions of the same material. This structure was used in the first diode lasers, but it provided little vertical confinement of the laser light. This makes it so inefficient that it can operate only in pulsed mode with cryogenic cooling, so it is no longer used.

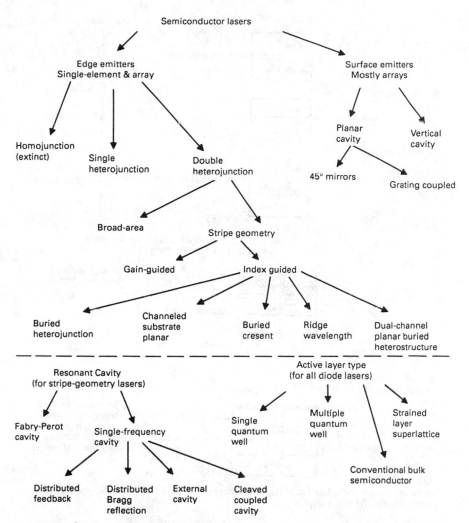

Figure 18.4 Categorization of diode laser structures, showing major relationships. The two classifications at the bottom are not part of the family tree; the breakdown of active layers applies to all diode lasers, while that for resonant cavities is for stripe-geometry lasers. The list of laser types is far from complete.

- *Single-heterojunction (or single-heterostructure) lasers:* In these lasers a material with a different bandgap is adjacent to one side of the active layer. This means that the active layer is sandwiched between two layers of different composition, typically GaAs and GaAlAs, which have different bandgaps. The refractive index difference between GaAlAs and GaAs helps confine the light to the active

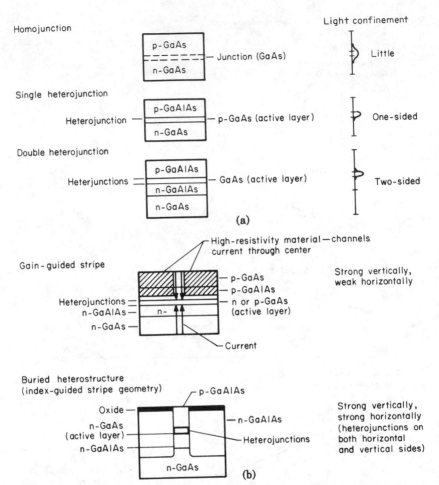

Figure 18.5 Major categories of edge-emitting diode lasers. (*a*) Three types shown in side views; (*b*) end views.

layer on the GaAlAs side, giving single-heterojunction lasers much better performance then homojunction lasers. Single-heterojunction lasers can produce pulses with high peak power at room temperature, both singly and in arrays of discrete devices packaged together. However, they cannot operate continuously or at high duty cycles. They remain in production for some military applications which specify them, but they rarely are used for other applications.

■ *Double-heterojunction (or double-heterostructure) lasers:* In these lasers the active layers are sandwiched between two layers of different material—for example, GaAs between two GaAlAs layers. This confines the laser light on both top and bottom of the active layer, making double-heterojunction lasers much more efficient than single-heterojunction types. The double-heterojunction diode laser was the first type capable of continuous operation at room temperature, and it has become the prevalent type for most applications.

A number of categories and subcategories of double-heterojunction lasers have been developed, and several are in use. The broadest division is by the width of the emitting region of the active layer. Early double-heterojunction lasers had active regions 50 μm or more wide. Today, most widely used diode lasers are *stripe-geometry* types, in which the active region is a stripe only 1 to 10 μm wide.

The stripe-geometry laser has important advantages for most applications. Because it concentrates the drive current in a small area, the threshold current is lower than that of wider-stripe lasers. The narrow stripe also improves beam quality by limiting oscillation to a single transverse mode and often to a single longitudinal mode. Such higher-quality beams can be readily focused to a small spot, important for applications such as compact-disk playback and laser printers. The narrow stripe has a small emitting area, which makes it easier to couple the output to a the small core of a single-mode optical fiber.

Waveguiding in the Active Layer.   Laser action can be limited to a narrow stripe in two fundamentally different ways, which involve structures both in the active-layer plane and elsewhere in the device. Restricting the flow of current to a small portion of the active layer is called *gain guiding*. Confining the light by refractive-index changes similar to those produced by a heterojunction is called *index guiding*.

Figure 18.5 gives one example of a gain-guided stripe laser. The formation of high-resistivity materials on the surface directs the drive current through a narrow region in the center of the laser, so it reaches only a narrow stripe in the active layer. The population inversion is produced only in this narrow stripe, limiting the region of laser action. A more subtle effect also helps confine light to the active stripe. The existence of gain creates an effective waveguide, constraining laser light to the region where the current flows, and there is gain. The waveguiding effect is weaker than if the active layer contained structures with refractive-index boundaries like those of heterojunctions.

Gain-guided lasers are simpler to make, but their weak light confinement limits beam quality and makes it difficult or impossible to obtain stable output in a single longitudinal mode. They can also suffer from instabilities and kinks or nonlinearities in their output-vs.-current characteristics. However, their inability to concentrate their output beam tightly is an advantage in generating high powers, because it disperses the light over a large enough area of the facet to avoid surface damage. Index-guided lasers have emission areas only a few micrometers across, and concentrating powers above about 200 milliwatts (mW) on such a small area can degrade the surface. Overall, gain-guided lasers generally can provide higher powers.

At lower powers, many problems of gain guiding can be avoided with an index-guided structure, such as the buried-heterostructure design shown in Fig. 18.5. In such lasers, the laser stripe is bounded on both sides by material of a different refractive index. Often the active

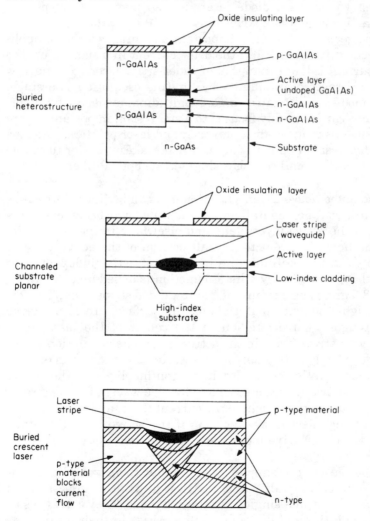

Figure 18.6   Major types of index-guided diode lasers described in text.

region index is higher than in the surrounding material, producing a waveguide analogous to an optical fiber, but the active stripe also may have an index lower than that of the surrounding material. Commercial index-guided lasers have low threshold currents, around 20 to 30 mA, and are more efficient than gain-guided lasers; they also tend to have a narrower spectral bandwidth and often lase in a single longitudinal mode.

Index-guided lasers are the types used for most diode laser applications, including compact-disk players, laser printers, and fiber-optic communications. They are available in many different forms which are sufficiently similar functionally that most manufacturers do not identify the type other than as index-guided; a sampling is shown in Fig. 18.6. The most important types include

- *Buried-heterostructure lasers*, in which the active region is a narrow stripe completely surrounded by material of lower refractive index. Such lasers can be formed from ordinary double-heterostructure material by first etching away all but a narrow mesa, often only 1 to 2 µm across, containing the region that will become the active stripe.

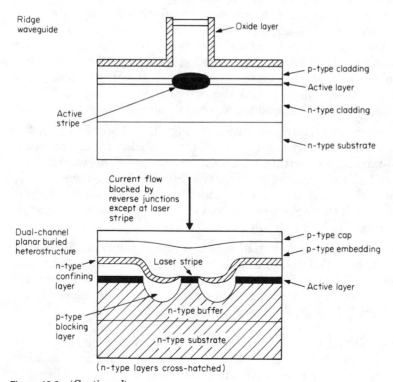

Figure 18.6 (*Continued*)

Then appropriate material is deposited around the heterostructure stripe, burying it. Current is prevented from flowing through the surrounding material either by depositing an insulating layer, or by forming a reverse *pn* junction. This resulting current concentration combines with the waveguide effect of the refractive index differential to give buried-heterostructure lasers high efficiency and low thresholds.

- *Channeled substrate planar lasers*, in which the active layer is grown on a substrate with a channel etched into it. The substrate has a high refractive index, but the cladding layer deposited between it and the active layer has a low index. The thick cladding which fills in the etched channel (creating a planar surface for the active layer) isolates the active layer from the substrate, so the active layer in that region has low loss. However, on the sides, where the cladding layer is thin, the active layer experiences higher loss, and the interaction causes an effective decrease in the refractive index which is large enough to confine light laterally in the active layer.

- *Buried-crescent lasers*, which, like channeled substrate planar types, have a groove or channel etched into the substrate. However, the groove is not filled completely to form a planar surface for the active layer. Instead, the active layer is grown on the grooved surface using liquid-phase epitaxy, forming a stripe in the active layer with a crescent-shaped cross section. This structure serves as a waveguide to confine light in the active stripe. Additional layers cover the crescent and produce an overall planar structure.

- *Ridge waveguide lasers*, in which a ridge is left above the active stripe, while the surrounding areas are etched close to the active layer—typically to within about 0.2 or 0.3 μm. The edges of the ridge reflect light guided in the active layer, forming a waveguide. Insulating coatings on the surrounding areas help confine current flow to a path through the ridge and active stripe. Stable output requires ridges only 2 to 3 μm across.

- *Double-channel planar buried-heterostructure lasers*, in which the active stripe is isolated in a mesa by etching two parallel channels around it. Material is then grown around it to bury the heterostructure. This structure is attractive for high-power lasers made from InGaAsP, described in Chap. 20.

A survey of the research literature from the mid-1980s on will reveal many other types. Many of these are minor variations on these basic types, in some cases with the most obvious difference being in the name chosen by the developers. Others are intended for specific purposes, such as high-power single-mode output.

Quantum Wells.   One of the more powerful concepts in semiconductor physics to emerge during the 1980s was the *quantum well*. In a sense, it is a logical continuation of the trend to shrink active structures to smaller and smaller dimensions, made possible by sophisticated growth techniques such as metal-organic chemical vapor deposition and molecular beam epitaxy. When layer thicknesses are reduced to about 20 nm or less, the quantum mechanical properties of electrons become significant, changing the energy-level structure of the semiconductor in ways that improve laser operation.

A single quantum well is a thin layer of semiconductor sandwiched between two layers with larger bandgaps, as shown at the top of Fig. 18.7. This confines electrons to the quantum-well layer, which is so thin that energy states in the valence and conduction bands are quan-

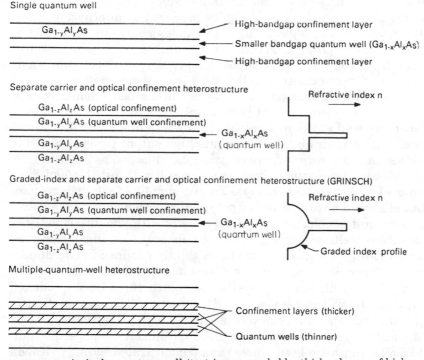

Figure 18.7   A single quantum well (top) is surrounded by thicker layers of higher-bandgap material. Forming layers with lower refractive index on top and bottom produces a separate carrier and optical confinement heterostructure (SCH). Grading the refractive index of the confinement layer produces the graded-index and separate carrier and optical confinement heterostructure (GRINSCH) laser. Multiple quantum wells also can be stacked in the active layer, as shown at bottom.

tized, and no longer form the continuum present in bulk semiconductors. This quantization combined with high carrier concentration helps reduce threshold current to only about one-half to one-third the level in comparable double-heterostructure lasers and allows higher output powers. Predictions of higher modulation speeds have yet to be realized; as of 1990, the highest 3-decibel (dB) bandwidth was 24 gigahertz (GHz) for a bulk 1.3-μm InGaAsP laser, compared to 14 GHz for the best quantum-well lasers.

Layers adjacent to the quantum-well structure confine the laser light vertically. In the separate carrier and optical confinement heterostructure (SCH) design shown in Fig. 18.7, the layers adjacent to the quantum well have a refractive index intermediate between that of the quantum well and the surrounding material, forming a waveguide. Alternatively, the refractive indexes of the surrounding layers may be graded, also shown in Fig. 18.7, forming a graded-index and separate carrier and optical confinement heterostructure—given the acronym "GRINSCH," which bears a suspicious resemblance to the stealer of Christmas in the Dr. Suess story.

Multiple quantum wells can also be produced by alternating layers with low and high bandgaps. The high-bandgap barrier layers must be thick enough to prevent quantum tunneling through them. Adjustment of layer thickness and composition gives some control over wavelength. (Multiple-quantum-well structures sometimes are called "superlattices," but that term also is used in a broader sense to cover multilayer structures without true quantum wells.)

Quantum-well structures are used in both gain-guided and index-guided lasers. Their ability to generate high output powers has led to their use in high-power arrays of gain-guided lasers, as described below. Index-guided versions have extremely low thresholds and have been used in development arrays. From a simplified viewpoint, quantum wells could be considered as potential replacements for the active layers in most double-heterostructure lasers with planar geometry.

Quantum wells are tightly confined in only one dimension. Researchers also are studying structures tightly confined in two dimensions (quantum lines or quantum wires) and three dimensions (quantum dots). As with quantum wells, these structures have quantized energy levels which promise useful properties. The combination of high gain and small material volume should lead to low threshold currents, narrow spectral linewidth, and high modulation speeds. Individual quantum-wire lasers have been demonstrated recently (Kapon et al., 1989), with threshold currents as low as 3.5 mA. Somewhat lower threshold currents have been reported for arrays of three quantum-wire lasers, and developers believe they can reduce thresholds further (Simhony et al., 1990).

Strained-Layer Structures.   All the diode lasers described thus far obey a major constraint on semiconductor devices—the interatomic spacing of all layers must be within about 0.1 percent that in the substrate. This means that device designers must select layer compositions not only to have a specific bandgap and refractive index but also to match the lattice constant of the substrate. The standard substrates are GaAs and InP, with lattice constants of 0.566 and 0.587 nm, respectively. Fortuitously, GaAlAs compounds have lattice constants close to that of GaAs for reasonable aluminum concentrations. InGaAsP, however, requires careful matching to the InP lattice constant by adjusting fractions of all four elements simultaneously. The lack of stable lattice-matched compounds for the available substrates long made it difficult to produce semiconductor lasers at certain wavelengths, notably between about 0.9 and 1.1 μm.

Recent work has shown that semiconductor layers below a critical thickness—a few tens of nanometers—can accommodate the strain of a lattice mismatch of more than 1 percent without forming the misfit dislocations which degrade the performance of thicker layers. This gives developers much greater flexibility in selecting materials, enabling them to manufacture devices not possible if strict lattice-matching requirements had to be met. Figure 18.8 shows such a strained layer in somewhat exaggerated form. In semiconductor lasers, the strained layer forms a quantum well in the active layer, which need not be precisely lattice-matched to the adjacent layers. Typically, the only strained layers in a semiconductor laser are quantum wells; the laser may contain one or more strained layers, depending on the number of quantum wells it contains. Deposition of many strained layers produces a strained-layer superlattice, which can accommodate lattice mismatches in the percent range (Osbourn et al., 1987) and is being investigated for other electronic applications.

The first practical application of strained-layer structures to diode lasers has been production of InGaAs lasers emitting at 980 nm with GaAlAs confinement layers and GaAs substrates. That wavelength is needed to pump erbium-doped optical amplifiers (described in Chap. 26), but it had not been available from lattice-matched diode lasers.

The strain alters properties of the semiconductor, in some cases

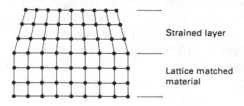

Strained layer

Lattice matched material

Figure 18.8   Strained-layer superlattice accommodates a modest difference in lattice constant between the substrate and active layer; the degree of distortion in the lattice is exaggerated.

making it better suited for laser operation. Dark-line defects do not appear to propagate as fast in InGaAs strained layers as in GaAs, so strained-layer lasers may have longer lifetimes than conventional types. So far researchers report documented lifetimes of 20,000 hours, and output powers have approached 100 mW per uncoated facet before device failure (Coleman, 1990).

Even for wavelengths where diode lasers already are available, strained-layer lasers may be attractive because of their high powers, high efficiencies, and low threshold currents. For example, 200-mW output at 1550 nm has been reported from strained-layer lasers with cavities 1500 μm long (Thijs et al., 1990). Early life test results showed that threshold current increased less than 1 percent after 5400 hours of low-power operation at 60°C.

Strained layers have been tested in a number of different semiconductor laser structures. At this writing, they are only beginning to emerge as developmental 980-nm commercial devices. However, they may find applications at other wavelengths if they live up to early predictions of potential for high bandwidth and output power, with excellent reliability.

Edge-Emitting Arrays. The output available from single-diode lasers is inherently limited by the volume of the devices, the need to remove excess heat, and optical damage to the laser facets. Single index-guided lasers are limited to about 200 mW because they tightly focus their output, although commercial versions are not rated for such high powers. Gain-guided lasers can generate somewhat more power, but they, too, are limited. Production of much higher powers requires multiple diode laser elements. High-power commercial arrays are electrically coupled in order to emit in unison; some are optically phase-coupled to form a partially coherent beam. In some small arrays and research devices, each stripe may have its own electrodes so that it can be addressed and controlled individually. Commercial arrays are made with gain-guided laser stripes and typically have single- or multiple-quantum-well active layers.

The idea of making arrays of diode lasers to obtain higher powers dates back many years. Pulsed single-heterostructure lasers have long been packaged in arrays, but commercial versions were discrete devices packaged together. The successful extension of semiconductor technology to the fabrication of monolithic arrays of diode lasers came in the 1980s. Monolithic arrays were so successful in generating high output powers that they were soon introduced commercially, and now account for the highest power diode lasers.

An edge-emitting diode laser array is made up of many parallel laser stripes, as shown in Fig. 18.9. Each stripe is a few micrometers

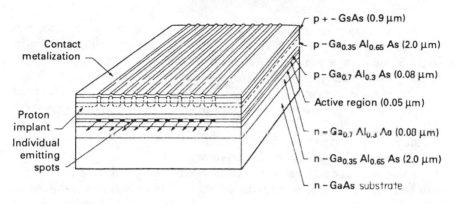

Figure 18.9 Monolithic array of diode laser stripes. (*Courtesy of Spectra Diode Laboratories.*)

wide, and with spacing between stripes a few times the stripe width (10 μm center-to-center spacing is typical). In practice, up to about 200 laser stripes can be fabricated on a 1-cm-wide bar. Bars with so many laser stripes typically are made with the stripes in clusters, with each cluster separated by a distance of a few times the stripe width. For example, a 10-W continuous-wave bar array (the highest power offered commercially at this writing) consists of 10 clusters of 20 laser stripes on 10-μm centers; each cluster is 200 μm wide, with intercluster spacing 800 μm on the 1-cm-wide bar. Continuous output powers above 50 W have been produced in the laboratory, but arrays do not last long at that power level.

Although semiconductor lasers are exceptionally efficient by laser standards, heat removal becomes a problem as the number of laser stripes and the output power increase. This limits continuous-wave output powers from commercial lasers to about 10 W from a monolithic bar at this writing. (That bar is about 25 percent efficient, so it must dissipate 30 W of heat.) Higher peak powers, to about 60 W, are possible from single bars designed for pulsed operation at duty cycles of about 2.5 percent. Manufacturers also can stack bars together to generate higher powers; at this writing, the highest power from a pulsed stack is 300 W, from five bars, specified for 1 percent duty factor. Lifetime is critically dependent on heat removal and packaging; continuous-wave powers over 50 W have been obtained in the laboratory at room temperature, but the devices did not last long.

The degree of optical coupling among stripes differs among arrays. Some arrays have enough coupling to emit a partly coherent beam; others generate incoherent beams. Like other semiconductor lasers, they have large divergence by laser standards. The long, narrow emitting stripe produces a beam of unusual shape in the near field; the highest-power arrays emit from an area 1 cm wide but only 1 μm high. Some arrays are packaged with fiber-optic beam collection systems, which effectively generate a beam of different shape. Beam quality can be improved by phase locking an array to an external single-mode diode laser. Researchers also are investigating prospects for controlling direction and quality of array output by modulating individual laser stripes so that the array would produce an electronically steerable beam like a phased-array radar.

Dual-beam diode lasers are an interesting example of a simple array made for a specific application. Two individually addressable laser stripes are made on a single substrate, spaced and aligned so closely that one optical system can focus both beams. They are used in laser printers which scan two lines simultaneously with the separate beams, writing at twice the speed possible with a single-beam laser.

Edge-Emitting Resonant Cavity Structures. The simplest resonant cavity for an edge-emitting diode laser is a pair of reflective end facets on the chip itself. These reflective surfaces are perpendicular to the active stripe and reflect light back and forth through it. They may be coated to reflect all incident light or to suppress reflection, or may be left uncoated to reflect only some of the incident light. Diode lasers with such Fabry-Perot resonators can emit from one or both cavity mirrors, depending on the design. (Diode-laser chips which emit from both facets usually are packaged in cases with only one output; the other output beam may be either absorbed in the container or directed to a photodiode which monitors laser output.)

Cavity design, like the difference between index guiding and gain guiding, influences output power. Simple cleaved facets reflect about 30 percent of incident light, giving adequate feedback for modest-power diode lasers. The output power can be roughly doubled by applying an antireflection coating to the front facet, reducing reflectivity to 5 to 10 percent and increasing the fraction of energy transferred out of the laser. (It is the intracavity power that causes optical damage to the facet surface, so this can reduce the likelihood of damage for high output powers. Damage is a problem in GaAs materials, but not in long-wavelength lasers.)

Another way to increase output is by reducing absorption at the mirror ends of the laser cavity. This can be done by keeping the drive current away from those regions and by creating regions with high bandgap near the mirror. The high-bandgap material is transparent at the laser wavelength because absorption of a photon would not raise a valence electron to the conduction band. This can increase damage threshold in short-wavelength lasers by a factor of 5.

Simple Fabry-Perot diode laser cavities have some important limitations. The gain curve in semiconductor lasers is broad enough to span several longitudinal modes of the laser cavity. Gain-guided lasers often oscillate in multiple longitudinal modes. Narrow-stripe index-guided lasers may limit oscillation to one longitudinal mode at a time when they are driven with a steady current, but two or more modes can oscillate simultaneously, and multimode output is common when the laser is modulated at high speeds, as is typical in fiber-optic communication systems. In addition, changes in temperature or drive current can cause a laser to "hop" between modes. This leads to rated spectral bandwidths of 2 to 4 nm. Shortening the laser cavity can reduce the tendency to multimode operation by increasing mode spacing but does not eliminate it.

Some applications benefit from broad bandwidth and short coherence length, such as playing compact disks, where the lack of coherence reduces the coupling of reflected light back into the laser. Indeed, some companies manufacture diode lasers with broadband, multimode output specifically for compact-disk players. However, broad bandwidth can prove troublesome for other applications, notably high-speed fiber-optic communications. To avoid them, researchers have devised several types of resonant cavities which produce stable single-frequency output. Four major types are shown in Fig. 18.10.

- *Distributed-feedback lasers* include a grating adjacent to the active layer in the region through which current flows. The periodic grating scatters light in the active layer, producing optical feedback along the length of the laser cavity, without cleaved end facets. The feedback process selects the particular wavelength which oscillates in the laser cavity. Commercial distributed-feedback lasers can provide a stable single wavelength output with spectral bandwidth in the 20-MHz range; linewidths as narrow as 170 kHz have been achieved in the laboratory (Okai et al., 1990). They are the type of single-frequency laser most readily available commercially.

- *Distributed-Bragg-reflection lasers* are similar to distributed-feedback lasers in their use of a periodic grating. However, the grating is placed only in part of the optical cavity through which no drive current passes (i.e., a "passive" part of the cavity). This simplifies fabrication by avoiding the need for epitaxial growth of layers on top of the grating. However, it requires formation of a guiding layer transparent at the laser wavelength on the grating outside the active region. Laboratory linewidths have been measured as low as 3.2 MHz (Takahashi et al., 1989).

- *External-cavity lasers* include external optics which tune the diode laser and stabilize its output wavelength, analogous to the

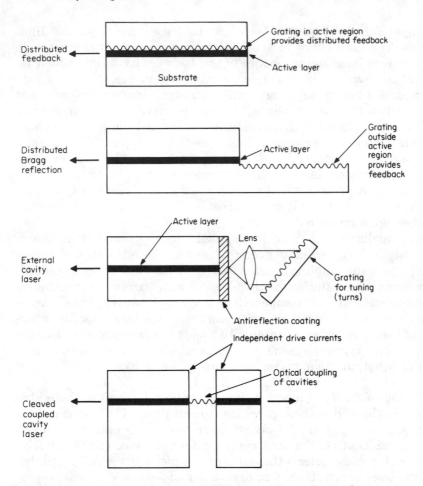

Figure 18.10 Cavities used for narrow-line semiconductor lasers; from top to bottom: distributed-feedback laser, distributed-Bragg-reflection laser, external-cavity laser, and cleaved-coupled-cavity lasers.

wavelength-selective optics used for dye lasers. Like dye lasers, they can produce both narrow-line and broadly tunable output, unlike distributed-feedback diode lasers, which are tunable only across a limited range by changing their temperature. In the example shown, one facet is antireflection-coated, so light passing through it can reach a diffraction grating, which can be turned to adjust wavelength. While standard diode lasers are less than a millimeter long, external-cavity diode lasers can have cavities up to a meter long. This structure can produce very narrow linewidths; a developmental commercial version has "typical" spectral bandwidth under 100 kHz. Tuning has been demonstrated across wavelength ranges to 200 nm (Lidgard et al., 1990).

- *Cleaved-coupled-cavity lasers* (C^3 lasers) consist of two lasers which are coupled optically but pumped by independent drive currents. Like external-cavity lasers, they can be tuned in wavelength (distributed-feedback and distributed-Bragg-reflection lasers are fixed in output). However, they have proved complex to build.

So far, distributed-feedback lasers are the most common commercial type, but external-cavity lasers are also offered. The main interest in such narrow-line lasers is for 1.3- and 1.55-μm lasers for high-performance fiber-optic communications. Available products are described in more detail in Chap. 20.

Surface-Emitting Lasers. The past few years have seen the emergence of novel semiconductor lasers which emit beams from their surface. These devices are attractive because they could be packed densely onto a semiconductor wafer, would not have to be cleaved to expose an end facet, and are compatible with requirements of integrated optics. Because they emit light from the chip surface rather than the side, they can form two-dimensional as well as one-dimensional arrays. Potential applications include optical interconnection of integrated circuits, optical computing, generation of laser beams steerable without moving parts, fiber-optic communication, and some applications now served by conventional end-emitting diode lasers.

There are two principal types: those with laser cavities in the plane of the active layer which include optics to deflect the beam through the surface of the wafer and those with vertical cavities. Both can be made in arrays as well as singly. The differences between them are significant.

Planar-cavity surface-emitting lasers. As in edge emitters, the cavities of these lasers are in the plane of the active layer. A structure in that plane deflects the laser light upward through the surface of the device. This approach means that each laser in an array requires at least as much area as is occupied by a conventional edge-emitting semiconductor laser.

Figure 18.11 shows one example in which a mirror is etched into the wafer at an angle of 45° to the active layer to deflect laser light vertically. This approach has the attraction of conceptual simplicity, and requires little room on the wafer other than that needed for the active region of the laser. Advocates say it is compatible with the processing needed for conventional optoelectronic devices. Recent experiments show that threshold currents are higher and output powers lower for such surface-emitting lasers than for similar edge-emitting devices with cleaved ends, but developers have demonstrated maximum pulsed powers of 380 mW (Kim et al., 1990).

Alternatively, distributed Bragg reflection in the plane of the active

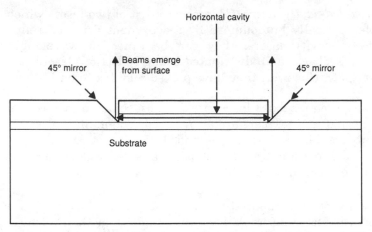

Figure 18.11 Surface-emitting laser with a planar laser cavity, and 45° mirrors reflecting out vertically from both ends of the cavity.

layer can scatter laser light through the surface of the wafer. The area occupied by the Bragg reflector in experimental versions is relatively large. However, if elements in an array of lasers are coupled through the distributed-Bragg-reflection region, the entire array can be made to generate coherent output or the direction of the output beam can be steered (Carlson et al., 1990; Evans et al., 1988). Although only a modest degree of steering has been demonstrated, the ability to electronically steer the beam from a stationary array of diode lasers could prove useful for a variety of applications that require moving beams.

Vertical-cavity surface-emitting lasers. These lasers have reflective surfaces above and below the active layer, forming a very short laser cavity perpendicular to the active-layer plane, as shown in Fig. 18.12. Each laser occupies a very small area on the chip, as small as a few micrometers across, so the lasers can be tightly packed in an array. One group has packed over a million such lasers on a single GaAs chip at a density above two million lasers per square centimeter (Scherer et al., 1989). Threshold currents of individual lasers are extremely low—values below 1 mA have been reported (Geels and Coldren, 1990; Yang et al, 1990), and researchers have projected "ultimate" threshold currents as low as 1 μA (Iga and Koyama, 1990).

The unusual operating characteristics of vertical-cavity diode lasers come from their unusual structure. The axis of the resonant cavity is perpendicular to the active layer, so only a small thickness of active medium is contained in the cavity. Most use quantum-well structures, so the beam travels through only tens of nanometers of active medium on each pass. This contributes to the low threshold currents observed. Although the gain in the quantum-well material is high, the cavity contains only a small amount of active medium, so cavity optics must

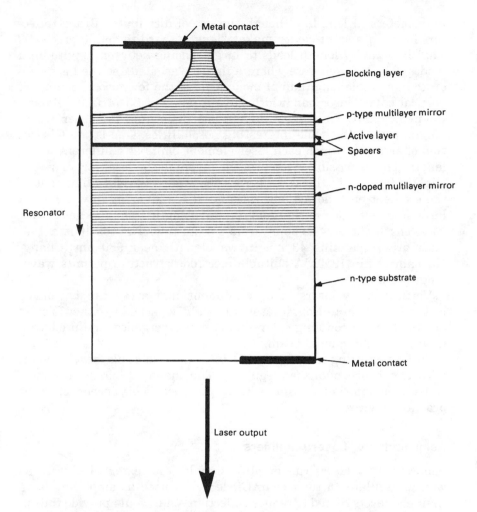

Figure 18.12 Vertical-cavity surface-emitting laser with mirrors above and below the active layer. In this version, the mirrors are multilayer structures fabricated as part of the diode structure; in other versions, one mirror may be a reflective metal or oxide film on the semiconductor surface.

be highly reflective to sustain laser oscillation. Such mirrors can be made by stacking multiple layers of alternating composition in the laser structure to form a distributed Bragg reflector. At this writing, stacks of 10 to 20 layer pairs are typical, but recent work suggests that fewer layers may suffice. Alternatively, the laser surface may be coated with a reflective layer, which can also serve as electrode if it is metallic. The beam is coupled out through surface or substrate, sometimes through a hole etched in the material.

The vertical cavity is exceptionally short by laser standards, several micrometers in length, although the use of distributed Bragg reflectors for such a short cavity makes "length" hard to define precisely. The short cavity length leads to large spacing between longitudinal modes (20 nm in one case where a figure was cited), so the lasers operate in a single longitudinal mode. One of the few measurements of vertical cavity laser bandwidth gave a value of 85 MHz (Geels and Coldren, 1990). The cavity dimensions determine the laser wavelength, suggesting some interesting possibilities. One is the fabrication of arrays in which each laser element emits at a different wavelength. This is possible if a few of the layers in the laser structure are slightly tapered in thickness across the wafer. Careful alignment of the gradient has allowed fabrication of a 7 × 20 array in which each laser is about 0.3 nm different in wavelength from the next, with wavelength changes stepping from row to row. The result is a 140-laser array spanning 43 nm in wavelength near 980 nm (Chang-Hasnain et al., 1990). A suitable electrode structure permits wavelength tuning.

Vertical-cavity lasers in a two-dimensional array can be phase-locked to emit in unison, if lasers are tightly packed together (Yoo et al., 1990). Such coupling can reduce beam divergence, and produce a high-power, directional beam.

Surface-emitting lasers are still in a stage of rapid development, with new reports published regularly. They have yet to reach commercial production, but at this writing prospects look encouraging for practical devices.

Semiconductor Laser Amplifiers

Semiconductor lasers can amplify signals from external sources as well as oscillate on their own. Diode lasers used as amplifiers have their end facets coated to reduce reflections that would provide the optical feedback needed to oscillate. There are two major potential uses for semiconductor laser amplifiers: signal amplification in fiber-optic communication systems and injection locking of powerful laser arrays.

Fiber-optic signal amplification. Light signals experience attenuation as they pass through optical fibers, and eventually require amplification to travel any farther. Through the 1980s, the standard approach was to build an electro-optic regenerator, which first converted the optical signal to electronic format and then amplified it to drive a diode laser transmitter. In the last several years, extensive work has been devoted to developing purely optical amplifiers, which could amplify the optical signals without first turning them into electrical format.

One approach is to use a semiconductor laser, with its facets coated to suppress reflections that could induce noise or internal oscillation.

The internal structures of semiconductor amplifiers are similar to those of diode lasers. Gains to about 30 dB have been recorded, but that does not count losses in fiber coupling or other factors. In practice, a 500-μm-long semiconductor amplifier typically can produce about 10 dB of gain from input fiber to output fiber. Specified gains for developmental commercial products are even lower, about 6 dB.

Semiconductor amplifiers have bandwidths in the 40-nm range, so they can amplify signals over a broad range. However, the degree of amplification depends on polarization, a problem which can introduce noise in a communication system. Crosstalk also can introduce noise. Despite an early head start, semiconductor amplifiers seem to have fallen behind the doped-fiber amplifiers described in Chap. 26.

Externally injection-locked arrays. The beam quality of a diode laser array can be improved by having it amplify the signal from an external single-mode laser rather than oscillating on its own. This approach is attractive for applications such as intersatellite communication in space, which require high-power modulated signals with narrow beam divergence. It remains primarily in the laboratory.

Bibliography

Dan Botez: "Laser diodes are power-packed," *IEEE Spectrum*, June 1985, pp. 43–53 (review article).

J. E. Bowers and M. A. Pollack: "Semiconductor lasers for telecommunications," in Stewart E. Miller and Ivan P. Kaminow (eds.), *Optical Fiber Telecommunications II*, Academic Press, Boston, 1988, pp. 509–568 (review paper).

N. W. Carlson, et al.: "Demonstration of a grating-surface-emitting diode laser with low threshold current density," *Applied Physics Letters* 56(1):16–18 (1990) (research report).

C. J. Chang-Hasnain et al.: "Multiple wavelength tunable surface emitting laser arrays," *IEEE Journal of Quantum Electronics*, in press (research report).

J. J. Coleman: "Strained laser InGaAs–GaAs quantum well heterostructure lasers," paper SDL6.1 in *LEOS '90 Conference Digest*, IEEE Lasers and Electro-Optics Society, Boston, November 1990 (research report).

Gary A. Evans et al.: "Coherent, monolithic two-dimensional (10 × 10) laser arrays using grating surface emission," *Applied Physics Letters* 53(22):2123–2125, November 28, 1988 (research report).

Luis Figueroa: "High-power semiconductor lasers," in Peter K. Cheo (ed.), *Handbook of Solid-State Lasers*, Marcel Dekker, New York, 1989, pp. 113–226 (review of high-power semiconductor lasers).

Luis Figueroa: "Semiconductor lasers," in Kai Chang (ed.), *Handbook of Microwave and Optical Components*, vol. 3, *Optical Components*, Wiley-Interscience, New York, 1990 (review of semiconductor lasers, concentrating on high-power types).

Randall S. Geels and Larry A. Coldren: "Submilliamp threshold vertical cavity laser diodes," *Applied Physics Letters* 57(16):1605–1607, October 15, 1990 (research report).

Klaus Gillessen and Werner Schairer: *Light Emitting Diodes: An Introduction*, Prentice-Hall, Englewood Cliffs, N.J., 1987 (introductory book).

Robert N. Hall: "Injection lasers," *IEEE Transactions on Electron Devices ED-23*:700–704, July 1976 (historical recollections).

Jeff Hecht: "*Laser Pioneers*, Academic Press, Boston, 1991 (historical recollections and overview).

K. Iga and F. Koyama: "Vertical cavity surface emitting lasers, device technology and future prospects," pp. 517–520 in *Extended Abstracts of the 22nd Conference on Solid State Devices and Materials*, Sendai, Japan, August 22–24, 1990, conducted by Japan Society of Applied Physics (research report).

Eli Kapon: "Semiconductor diode lasers: Theory and Techniques," in Peter K. Cheo (ed.), *Handbook of Solid-State Lasers*, Marcel Dekker, New York, 1989, pp. 1–112 (review).

Eli Kapon et al.: "Single quantum wire semiconductor lasers," *Applied Physics Letters* 55(26):2715–2717 (1989) (research report).

Jae-Hoon Kim et al.: "High-power AlGaAs/GaAs single quantum well surface emitting lasers with integrated 45 degree beam deflectors," *Applied Physics Letters* 57(20): 2048–2050, November 12, 1990 (research report)

Lasers & Optronics staff report: "The laser marketplace—forecast 1990," *Lasers & Optronics* 9(1):39–57, January 1990.

A. Lidgard et al.: "External-cavity InGaAs/InP graded index multiquantum well laser with a 200 nm tuning range," *Applied Physics Letters* 56(9):816–817, February 26, 1990.

M. Okai et al.: "Corrugation-pitch modulated MQW-DFB laser with narrow spectral linewidth (170 kHz)," *IEEE Photonics Technology Letters* 2(8):529–530 (1990).

G. C. Osbourn, et al.: "Principles and applications of semiconductor strained-layer superlattices," in Raymond Dingle (ed.), *Semiconductors and Semimetals*, vol. 24, *Applications of Multiquantum Wells, Selective Doping, and Superlattices*, Academic Press, San Diego, 1987, pp. 459–504 (review of strained-layer superlattices, not specifically lasers).

A. Scherer et al.: "Fabrication of microlasers and microresonator optical switches," *Applied Physics Letters* 55(26):2724–2726, 1989 (research report).

S. Simhony et al.: "Threshold current reduction in single and multiple quantum wire lasers," paper SDL7.2 in *LEOS '90 Conference Digest, Boston, November 1990*, IEEE Lasers and Electro-Optics Society (research report).

Yasuharu Suematsu et al: "Dynamic single-mode semiconductor lasers with a distributed reflector," in W. T. Tsang (ed.), *Semiconductors and Semimetals*, vol. 22, *Lightwave Communications Technology, Part B, Semiconductor Injection Lasers I*, Academic Press, Orlando, Fla., 1985, pp. 205–255 (review paper).

Mitsuo Takahashi et al.: "Narrow spectral linewidth 1.5 μm GaAsP/InP distributed Bragg reflector (DBR) lasers," *IEEE Journal of Quantum Electronics* 25(6):1280–1287, 1989.

P. J. A. Thijs, T. van Dongen, and W. Nijman: Paper WJ2 in *Technical Digest Conference on Optical Fiber Communications*, San Francisco, January 22–26, 1990, Optical Society of America (research report).

R. G. Waters: "Enhanced longevity strained-layer lasers," paper SDL6.2 in *LEOS '90 Conference Digest, IEEE Lasers and Electro-Optics Society*, Boston, November 1990 (research report).

Ying Jay Yang et al.: "Submilliampere continuous-wave room-temperature lasing operation of a GaAs mushroom structure surface-emitting laser," *Applied Physics Letters* 56(19):1839–1840, May 7, 1990 (research report).

Hoi-Jun Yoo et al.: "Fabrication of a two-dimensional phased array of vertical-cavity surface emitting lasers," *Applied Physics Letters* 56(13):1198–1200, March 26, 1990 (research report).

Jay Zeman: "Visible diode lasers show performance advantages," *Laser Focus World*, August 1989, pp. 69–80 (review article).

19

Short-Wavelength
Semiconductor
Diode Lasers

Convenience dictates that the broad field of semiconductor lasers be broken up into manageable chunks, and it seems logical to group together
those with wavelengths shorter than about 1.1 micrometers (μm). Most
are based on gallium arsenide, the oldest type of semiconductor laser.
Unlike longer-wavelength diode lasers, they have a wide range of applications, including compact-disk players, optical data storage, laser printers, measurement and inspection, laser pointers, and power sources for
pumping other lasers. Short-wavelength lasers themselves break into
distinct families, as do as some general-purpose lasers.

Short-wavelength laser diode technology has become increasingly
versatile as it has matured. Mass-produced milliwatt lasers sell for a
few dollars each in large quantities. Monolithic arrays can generate
watts of power, and command much higher prices. Some lasers have
very high modulation bandwidths for high-speed communications.
Others are packaged with focusing optics to meet stringent beam-
quality requirements. They employ the family of semiconductor laser
structures described in Chap. 18.

New types of lasers have broadened the range of wavelengths beyond those of traditional gallium aluminum arsenide lasers. In late
1990, the shortest wavelength offered was 635 nanometers (nm) in the
red, from InGaP/InGaAlP multiple-quantum-well lasers on GaAs substrates. Only one company offered that wavelength, but more were expected to follow.

At longer wavelengths, developers have made strained-layer
InGaAs lasers (see Chap. 18) emitting at 980 nm. That wavelength
had been sought for pumping erbium-doped fiber amplifiers (see Chap.
26) but had not been available because it was in the lattice-

unmatchable void between lasers made on GaAs and on InP substrates. Similar lasers can emit at 900 to 1100 nm, and early life tests indicate that strained-layer lasers may be more reliable than bulk semiconductor lasers.

This chapter concentrates on specific semiconductor laser devices with wavelengths shorter than about 1.1 μm, particularly those available commercially. The internal structures used in these devices are detailed in Chap. 18. More than 23 million short-wavelength diode lasers were sold in 1989, the vast majority of all lasers. In the same year, sales of long-wavelength diode lasers were only about 100,000, and those of all other nonsemiconductor lasers combined were under 500,000 (*Lasers & Optronics*, 1990).

Active Media

The wavelength of a semiconductor laser is a function of the bandgap in the active layer, where current carriers recombine, producing photons with energy slightly less than that of the bandgap. In double-heterostructure lasers, the active layer is sandwiched between two layers with larger bandgaps (for electrical confinement) and lower refractive indexes (for optical confinement). The entire structure is grown on a substrate of compatible material, which must have lattice constant (interatomic spacing) close to that of the other layers. Success in making diode lasers depends on a host of other factors, including impurity dopants, electrical properties of the semiconductor, and growth processes, but for our purposes here the key parameters are bandgap, refractive index, and lattice constant, which depend on the semiconductor composition.

Virtually all semiconductor lasers are constructed on substrates of binary compounds, such as gallium arsenide. Binary compounds are fairly easy to make, but their composition fixes their lattice constants, refractive indexes, and bandgaps. Other elements must be substituted for one or both of the elements in the binary compound to produce layers with the different bandgaps and refractive indexes needed for diode lasers. These compounds are written in the form $Ga_{1-x}Al_xAs$, where the x denotes the fraction of gallium replaced by aluminum and the total number of gallium and aluminum atoms equals the number of arsenic atoms.

Constraints on Active Media. Some factors constrain the choice of semiconductor compounds for use within a laser structure.

- *Lattice-matching constraints:* In bulk lasers, atomic spacing in all layers must be within about 0.1 percent that in the substrate. This constraint is relaxed to about 1 percent in strained-layer lasers.

- *Direct-bandgap requirement:* The semiconductor must have a direct-bandgap energy-level structure to allow easy transitions between the peak of the valence band and the bottom of the conduction band. Some III–V semiconductors meet this requirement, but others do not. For example, GaAs is a direct-bandgap material, but AlAs is not and cannot sustain laser action. $Ga_{1-x}Al_xAs$ has a direct bandgap for low aluminum levels, but not at high levels.

- *Availability of good materials for both optical and electrical confinement.*

- *Material miscibility:* Some materials do not dissolve readily in each other, making device fabrication difficult and limiting production yields.

- *Dopant availability to produce good p- and n-type material.*

Gallium Arsenide. Gallium arsenide was the first material used for diode lasers but now is used mainly in single-heterostructure lasers produced for military applications.

Gallium Aluminum Arsenide. Adding aluminum to the active layer increases bandgap energy, producing lasers with shorter wavelengths. $Ga_{1-x}Al_xAs$ is a direct-bandgap material for aluminum concentrations of $x = 0–0.45$. This corresponds to a wavelength range of about 870 to 620 nm, but good lasers cannot be produced at the shorter wavelengths because changes in the energy-level structure make laser action less efficient. The shortest wavelength offered commercially is 750 nm; GaAlAs lasers can operate at shorter wavelengths, but their lifetimes are too short for practical use. Most commercial lasers have wavelengths of 750 to 850 nm, with the most common wavelength 780 nm, the wavelength used in compact-disk players. Pulsed double-heterojunction lasers which are at least nominally GaAlAs types are available with wavelengths as long as 895 nm.

Increasing the aluminum content decreases the refractive index of GaAlAs as well as increasing the bandgap. This means that higher-aluminum layers provide both optical and electronic confinement, as required for good laser operation. Lasers with $Ga_{1-x}Al_xAs$ active layers have $Ga_{1-y}Al_yAs$ confinement layers, which contain more aluminum, so y is larger than x. The entire structure is grown on a GaAs substrate. GaAlAs lasers have been grown on silicon (Egawa et al., 1990) and glass (Yablonivich et al., 1989) substrates in the laboratory, but commercial versions are not available.

InGaAsP/GaAs Lasers. The quaternary InGaAsP compound used in long-wavelength lasers can emit much shorter wavelengths when lattice-matched to a GaAs substrate. Lasers emitting near 800 nm have been demonstrated, but are still in an early stage of development.

GaInP Visible (670-nm) Lasers. Researchers had to develop new materials for lasers with reasonable lifetimes at wavelengths shorter than 750 nm. Their first major success was the GaInP/GaInAlP laser (Kobayashi et al., 1987) shown in Fig. 19.1. Lasers with an active layer of $Ga_{0.5}In_{0.5}P$ emit at 670 nm. The confinement layers are formed by substituting aluminum for some gallium in the compound $(Ga_{1-x}Al_x)_{0.5}In_{0.5}P$ to raise the bandgap and lower the refractive index. The entire structure is lattice-matched to a GaAs substrate.

$Ga_{0.5}In_{0.5}P$ lasers were the first visible diode lasers with reasonable lifetimes. First reported in the mid-1980s, they were soon introduced commercially and have rapidly gained acceptance. The beam is easily visible, although the eye is less sensitive to 670 nm than to the 633-nm helium-neon wavelength. However, 670 nm is the short-wavelength limit for the ternary compound.

AlGaInP Lasers. As in GaAs, substituting aluminum for gallium in $Ga_{0.5}In_{0.5}P$ increases the bandgap, so the quaternary compound $(Ga_{1-x}Al_x)_{0.5}In_{0.5}P$ can produce wavelengths shorter than 670 nm. This technology lags slightly behind that of simple $Ga_{0.5}In_{0.5}P$ lasers, partly because it is difficult to make the bandgap difference large between the active and cladding layers. The shortest wavelength available commercially in 1990 from bulk lasers was 660 nm, from $(Ga_{1-x}Al_x)_{0.5}In_{0.5}P$ lasers containing small amounts of aluminum. However, quantum-well lasers introduced in 1990 with eight $Ga_{0.5}In_{0.5}P$ quantum wells separated by $Al_{0.25}Ga_{0.25}In_{0.5}P$ barriers op-

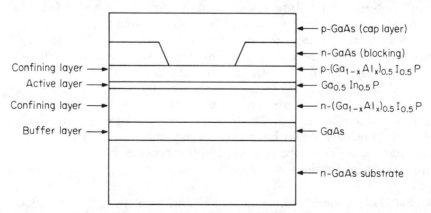

Figure 19.1 Basic structure of a 670-nm $Ga_{0.5}In_{0.5}P$ laser, showing active layer and confinement layers fabricated on a GaAs substrate.

erate continuously at 635 nm, essentially the same as the He-Ne wavelength (Valster et al., 1990a).

Bulk lasers have operated near 633 nm in the laboratory. Continuous output of 2 milliwatts (mW) at 637.8 nm was achieved in a laser with an 80-nm active layer of $(Al_{0.15}Ga_{0.85})_{0.5}In_{0.5}P$ surrounded by cladding layers of $(Al_{0.7}Ga_{0.3})_{0.5}In_{0.5}P$ (Ishikawa et al., 1990). A modified version achieved continuous output to 3 mW at temperatures to 48°C (Itaya et al., 1990). A laser with a 96-nm active layer of $(Al_{0.19}Ga_{0.81})_{0.5}In_{0.5}P$ and cladding layers of $(Al_{0.6}Ga_{0.4})_{0.5}In_{0.5}P$ has emitted 2 mW at 632.7 nm continuously at 20°C (Kobayashi et al., 1990).

Shorter wavelengths can be produced by multiple quantum wells of either $Ga_{0.5}In_{0.5}P$ or $(Ga_{1-x}Al_x)_{0.5}In_{0.5}P$. Researchers have reported continuous operation at temperatures to 5°C of 626-nm lasers with eight 3-nm quantum wells of $Ga_{0.5}In_{0.5}P$ separated by 4-nm confinement barriers of $Al_{0.25}Ga_{0.25}In_{0.5}P$ (Valster et al., 1990b). A multiple-quantum-well laser has operated continuously at 555 nm, but only when cooled to 77 K (Valster et al., 1990c).

InGaAs Strained-Layer Lasers. We saw in Chap. 18 that compounds which do not meet strict lattice-matching criteria can be produced using thin strained layers in the semiconductor structure. Thus far, this concept has been applied mostly to fabricating $In_xGa_{1-x}As$ lasers on GaAs substrates. The system is capable of laser action between 900 and 1100 nm, a gap between the conventional GaAlAs and InGaAsP systems, but the major effort has been to make 980-nm lasers to pump erbium-doped fiber amplifiers.

The $In_xGa_{1-x}As$ layer is a quantum well sufficiently thin to accommodate strain in a strained-layer structure. It is confined by GaAlAs layers with aluminum fraction ranging from 0.20 to 0.85. The wavelength depends on the number and arrangement of quantum wells, the indium and gallium content of the quantum well, the aluminum fraction in the confining layer, and the cavity length (Chen et al., 1990a). As mentioned in Chap. 18, the strained-layer structure appears to improve lifetime. Single buried-heterostructure lasers with single-quantum-well structures have produced 330 mW at 980 nm (Chen et al., 1990b), and five-stripe arrays have generated total powers to 730 mW at 1010 nm (Beernink et al., 1990).

Zinc Selenide Lasers. The zinc selenide family of materials are semiconductors with direct bandgaps much larger than GaAs, making them attractive for short-wavelength visible lasers using quantum-well structures. However, formidable material problems remain to be overcome to make practical lasers.

Laser action has been obtained by optically pumping in ZnSe (Zonudzinski et al., 1990), but output powers, duty cycles, and operating temperatures all are limited. Multiple $Zn_{0.86}Cd_{0.14}Se$ quantum well lasers have produced 490-nm pulses at room temperature when pumped optically (Jeon et al., 1990). Outputs were sufficiently high that a bright spot of blue light could be seen when the laser operated at 100 K. The lasers have ZnSe barrier layers and $In_{0.04}Ga_{0.96}As$ substrates.

Internal Structure

Semiconductor laser structure is described in detail in Chap. 18. All short-wavelength diode lasers have double-heterostructure designs except for a few pulsed single-heterostructure lasers used in military applications.

Index guiding is used in most low- and moderate-power single-element short-wavelength diode lasers to produce the good beam quality needed for applications such as laser printers and compact-disk players. Several different index-guided structures are in common use. Gain-guiding is used in most high-power diode lasers, including both single-element and monolithic arrays, because it reduces power density at the facet, thus reducing the likelihood of optical damage.

Virtually all commercial short-wavelength lasers have edge-emitting Fabry-Perot cavities. Most emit a strong beam from one facet and a weaker beam from the rear facet, which is directed to a monitoring photodiode inside the laser package. Only the strong beam emerges from the packaged laser. Surface emitters remain in development. Distributed-feedback designs remain in the laboratory because no important applications of short-wavelength lasers require single-frequency emission.

Variations and Types Covered

Short-wavelength semiconductor lasers are used for diverse applications. The same basic technology is used for most, but some require special devices, and others require special packaging of fairly standard devices.

The economics of large-scale semiconductor production influence the choice of devices for many applications. The huge volume of low-power 780-nm GaAlAs lasers produced for compact-disk (CD) playback drove their prices down to a few dollars each in large quantities. That has made CD lasers (sometimes in special packages) the most economical choice for applications where that wavelength and power will suffice. However, other applications may require higher powers, different wavelengths, or other capabilities available only from other devices.

Compact-Disk Lasers. Nearly 20 million CD audio players are sold each year, creating a tremendous market for inexpensive diode lasers emitting a few milliwatts at 780 nm. The laser is not modulated in a CD player, but it should emit a single transverse mode and have low noise. Figure 19.2 shows how the beam is focused onto the disk by the same optics that collect the scattered light and return it to a detector, which reads the disk and collects the data needed to track the spots on the disk. The same digital recording format is used for other read-only optical disks, including CD-ROM (compact-disk read-only memory, for computer data), CD-I (compact-disk interactive), and CD-V (compact-disk video). Videodisk players operate much like compact-disk players but require lower-noise lasers because their recording format is analog rather than digital.

Lasers for Write-Once and Erasable Optical Storage. Other optical data storage systems write new data as well as read prerecorded data. The storage medium may be a disk or card and may store data optically or magneto-optically. Details differ, but laser requirements are similar. A continuous beam of a few milliwatts reads data, while recording requires higher powers—tens of milliwatts—in a beam that is modulated in intensity at a speed which depends on the hardware design.

Laser Printers. Millions of diode lasers are used in laser printers, in which a moving mirror scans the beam across a photoconductive

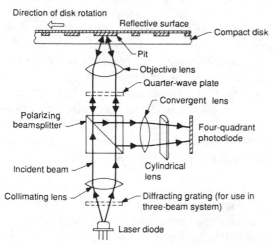

Figure 19.2 Read head focuses light from a 780-nm diode laser onto a compact disk, then directs reflected light back to a quadrant photodetector, which reads data and tracks the beam. The quarter-wave polarizer rotates reflected light 90° after it makes two passes, so the polarizing beamsplitter directs it to the photodetector. (*Courtesy of Sharp Corp.*)

drum, writing a pattern which is transferred to paper in the same way as a photocopier image. Photoconductive materials are more sensitive to short wavelengths, so lower powers are required at 750 nm than at longer wavelengths. The faster the laser printer, the higher the power required. A high-speed laser printer may need 5 mW at 750 nm or 10 to 15 mW at 780 nm, while 5 mW at 780 nm may suffice for a slower model. Diode lasers with two, three, or four independently modulatable channels are made for laser printers. Lasers made for printers also can be used in facsimile machines, bar-code readers, or laser light pens.

Fiber-Optic Lasers. Although optical fibers have much higher loss at wavelengths shorter than 900 nm than at 1300 or 1550 nm, GaAlAs lasers are used for short-distance communications because they cost less than InGaAsP lasers. These lasers generally are mounted in cases with fiber pigtails for direct connection to optical fibers, or with connectors which mate directly to fiber-optic cables. Losses decrease at longer wavelengths, so 810- or 840-nm lasers have some advantages over 780-nm CD lasers. Two or more different laser wavelengths can be used for wavelength-division multiplexing, so one fiber can carry multiple channels.

High-Power Laser Diodes. The most important brute-force application requiring high-power semiconductor lasers is pumping neodymium lasers (described in Chap. 22), which have strong absorption bands between 790 and 820 nm. GaAlAs diode lasers can pump neodymium lasers at these wavelengths with much higher overall electrical efficiency than can broadband sources such as flashlamps. Single-stripe lasers may be used, but monolithic arrays of many laser stripes offer higher power; pumping may be pulsed or continuous.

High-power diodes and arrays also can be used for other applications in medicine, space communications, and research. Most commercial types operate in the 790- to 820-nm range used for neodymium pumping, but some have wavelengths as long as 850 nm. High powers are hard to achieve at wavelengths shorter than about 780 nm without sacrificing laser lifetime.

High-Speed Lasers. Some short-wavelength lasers are designed specifically to have very broad modulation bandwidths for applications such as analog microwave communications. In most cases, they are packaged to transmit through optical fibers, but they also may transmit through air. Many applications are military.

Visible Lasers. InGaP lasers emitting at 670 nm and 635- to 660-nm InGaAlAs lasers produce red beams visible to the human eye for applications such as laser pointers and bar-code scanners. Those wavelengths represent a compromise between ease of making longer wavelength lasers and the rapid increase in sensitivity of the human eye at shorter wavelengths. The short wavelengths also are advantageous for other applications because many optical recording materials become more sensitive as wavelength decreases, allowing a lower-power beam to do the same job as higher powers at a longer wavelength.

Pulsed Single-Heterostructure Lasers.   For many years, the only diode lasers able to deliver high-power pulses were single-heterostructure types emitting pulses near 900 nm with a low duty cycle. These lasers were designed into military systems for such applications as measuring the range between an air-to-air missile and its target. Such lasers still are made for such applications but are rarely used elsewhere.

Research Thrusts

Researchers are making rapid strides in several areas of short-wavelength semiconductor technology. Some of their work may yield important results in the 1990s. This includes

- Shorter-wavelength, long-lived diode lasers from conventional III–V materials, such as InGaAlP.

- Development of new visible-wavelength laser materials, such as ZnSe and related compounds.

- Direct harmonic generation from high-power GaAlAs diode lasers, to produce blue, green, and ultraviolet light.

- Upconversion of diode laser energy by pumping other lasers, in essence expanding on present neodymium laser pumping.

- Reducing laser thresholds and improving laser efficiency.

- Packaging diode lasers to efficiently remove excess heat, so that a small package can generate high powers.

- Avoiding facet damage, which is a key limit on power from GaAlAs lasers (which are much more sensitive to optical damage than are longer wavelength InGaAsP lasers). Approaches include development of nonabsorbing mirrors, division of laser output among many stripes, and chemical passivation of facets.

- Increasing power levels available from diode laser arrays.

- Improving the quantum-well and strained-layer structures described in Chap. 18, which should offer new wavelengths, such as the recently developed 980-nm diode laser.

- Developing arrays which can be controlled in wavelength and/or phase to produce a high-quality high-power beam. The beam also might be steerable, depending on array structure.

- Developing arrays with multiple beams that can be modulated independently to increase effective speeds of laser printers and optical data storage systems.

- Developing vertical-cavity laser arrays (an approach promising for multiwavelength and/or steerable beams).

- Integrated optical circuits including lasers along with other optical devices, for optical signal processing and computing.

- Making lasers to provide optical interconnections between electronic chips.

The rest of this chapter concentrates on lasers with wavelengths below 1.1 μm available commercially at the end of 1990. Longer-wavelength III–V semiconductor lasers are covered in Chap. 20; tunable lead salt diode lasers are covered in Chap. 21.

Beam Characteristics

Wavelengths and Output Power. The wavelength of a diode laser is determined by its structure and the bandgap of its active layer. Figure 19.3 plots bandgaps of important III–V materials and the corresponding wavelengths against their lattice constants. Ternary compounds have characteristics that lie on the line connecting binary compounds, such as GaAs and AlAs. Quaternary compounds lie within the areas defined by the four possible binary compounds. The need for lattice-matching of bulk lasers constrains wavelengths, but strained-layer structures and quantum wells give developers more freedom.

Table 19.1 lists the nominal wavelengths of commercial lasers and maximum continuous-wave power levels. Some companies specify different wavelengths, but typical specifications have a range of ±10 to 20 nm. Manufacturers do not state precise compositions of GaAlAs active layers.

The specified wavelength range is a margin of error inherent in manufacture. A precise wavelength can be obtained by testing fabricated lasers and selecting one that meets the desired specifications. The operating wavelength changes with temperature and drive current, so the laser can be tuned over a limited range by controlling operating conditions.

Powers of short-wavelength diode lasers have climbed steadily in recent years, but manufacturers remain cautious in the specifications listed in Table 19.1, which are recommendations for long-term operation. The highest rated powers for single-element GaAlAs lasers are

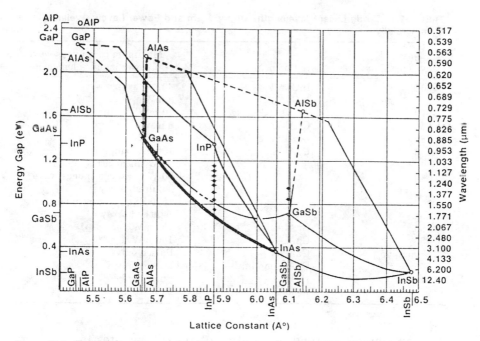

Figure 19.3 Energy bandgap and lattice constants for III to V bulk semiconductors. Dashed lines indicate indirect-bandgap materials not suitable for laser operation; solid lines are direct-bandgap materials. Lines that connect points for binary compounds represent values of intermediate ternary compounds; for example, GaAlAs characteristics fall along the line connecting GaAs with AlAs. Possible characteristics of quaternary compounds fall inside the area defined by the four possible binary compounds (e.g., GaAs, GaP, InAs, and InP for InGaAsP). Bulk lasers must be lattice-matched to a substrate. |*Courtesy of P. K. Tien. From Kapon (1989), with permission*|

150 mW, although life tests indicate that average life for 100-mW operation at 50°C should be 20,000 hours (Welch et al., 1990). Researchers have demonstrated much higher powers in the laboratory, but reliability remains a concern. A buried-heterostructure laser with nonabsorbing facets and a $Ga_{0.96}Al_{0.04}As$ active layer 1 μm wide and 1000 μm long has emitted continuous powers to 500 mW in a single spatial mode (Ungar et al., 1990).

Monolithic GaAlAs bar arrays containing many parallel laser stripes are rated for continuous powers of up to 10 W. That high power emerges from 200 stripes, each delivering 50 mW. The combined emitting area of the stripes is 1 cm wide by 1 μm high, and the output is incoherent. Smaller arrays, containing fewer elements, may generate partially coherent beams at lower powers.

Monolithic laser arrays can produce pulses with higher peak powers than are available in a continuous beam. Linear arrays of 200 laser stripes can produce up to 60 W in "quasi-continuous" operation—gen-

TABLE 19.1   Diode Laser Wavelengths under 1 μm and Power Levels Available Commercially

Nominal wavelength, nm	Compound	Maximum continuous-wave power (single-element)
635	InGaAlP	3 mW
660	InGaAlP	3 mW
670	$Ga_{0.5}In_{0.5}P$	10 mW
750	GaAlAs	8 mW
780	GaAlAs	35 mW
810	GaAlAs	100 mW single
		10 W linear array
		60 W quasi-continuous-wave (pulsed) array
		1500 W quasi-continuous-wave stacked array
830	GaAlAs	150 mW
850	GaAlAs	100 mW
880 or 895	GaAlAs	Pulsed only
905	GaAs (nominal)	Pulsed only
910	InGaAs	Pulsed only
980	InGaAs	50 mW

erating 200-microsecond (μs) pulses at repetition rates to 100 hertz (Hz). (That corresponds to an average power of 1.2 W, lower than the 10 W available in a continuous beam from similar arrays.) Peak powers to 1.5 kW are available from a stack of 25 multistripe laser "bars."

Efficiency. The overall efficiency of a laser diode (the fraction of the electrical drive power converted to optical output) depends on two factors, the threshold current and the slope efficiency—the percentage of each added milliwatt of electrical input converted into light. Because of the threshold effect, overall efficiency typically increases with output power.

Typical threshold currents are 30 to 60 milliamperes (mA) for commercial index-guided GaAlAs lasers emitting up to about 20 mW. Gain-guided lasers have thresholds roughly 50 percent higher. Thresholds are comparable for GaInP and AlGaInP lasers emitting at 660 and 670 nm. Multistripe arrays have considerably higher thresholds because the drive current must push each stripe past its threshold. The 200-stripe 10-W array mentioned above has a rated threshold current of 4 A, 20 mA per stripe. Threshold currents below one milliampere have been reported in some exotic laboratory lasers (e.g., vertical-cavity and quantum-wire lasers) which are not ready for commercial use.

GaAlAs lasers efficiently convert drive power above laser threshold into light. Typical differential efficiencies range from 0.2 to over 1 mW/mA, or roughly 10 percent to above 50 percent. Total power-conversion efficiency ranges from a few percent for low-power lasers to 30 percent in commercial arrays. Researchers have demonstrated 60 percent overall power conversion in a 500-mW single-element laser (Welch et al., 1990).

InGaP lasers have somewhat higher threshold currents than do GaAlAs lasers, about 50 mA for index-guided and 70 to 80 mA for gain-guided lasers. Differential efficiencies are almost as large as for GaAlAs lasers, but the high threshold currents and low operating powers mean that overall efficiency is typically a few percent.

InGaAs diode lasers emitting at 980 nm have low threshold currents (15 mA) and high differential efficiency (about 60 percent), giving an overall efficiency of 30 percent at 50 mW output.

Increasing temperature reduces differential efficiency and increases threshold current. Because both factors work in the same way, and the lost energy is converted to heat in the semiconductor laser, thermal runaway is possible at excessive operating temperatures.

Temporal Characteristics of Output. Most double-heterojunction lasers are nominally continuous at room temperature, except for a few high-power arrays. However, single-heterojunction diode lasers can produce pulses no longer than 0.2 μs at repetition rates to a few kilohertz at room temperature. Stacked arrays of single-heterojunction lasers are limited to lower repetition rates. Some high-power monolithic arrays operate in quasi-continuous mode, generating pulses up to 200 μs long at repetition rates to about 100 Hz with peak power several times above that possible in continuous operation, useful for pumping neodymium lasers. The need to dissipate waste heat prevents continuous operation at such high powers.

Double-heterojunction lasers can be modulated at high speeds by varying the drive current. Rise times typically are under 1 ns, but often are not specified. Typical modulation bandwidths are 1 to 2 gigahertz (GHz), but some high-speed lasers have bandwidths above 10 GHz.

Diode lasers can be modelocked in external cavities, and experimental diode lasers have been made with internal structures that generate modelocked pulses. For example, a passively modelocked two-section GaAlAs multiple-quantum-well laser has generated 2.4-ps pulses at 108 GHz. However, modelocked diode lasers are not available commercially.

Transient ringing or relaxation oscillations can cause output power to fluctuate briefly after a diode laser is turned on. Typically these oscillations last less than a nanosecond and are considered in specifications for frequency response and rise time.

Spectral Bandwidth. The instantaneous spectral bandwidth of short-wavelength diode lasers typically runs from 0.1 nm for a single-transverse-mode double-heterojunction laser well above threshold to around 4 nm for one just above threshold or designed to emit multiple longitudinal modes. Figure 19.4 shows evolution of the spectral profile of a typical laser as power increases. At low power, it oscillates in several longitudinal modes that fall within its gain bandwidth. As power increases, one longitudinal mode becomes dominant, until well above threshold emission is in a single longitudinal mode. (Increasing drive current also heats the laser, so emission shifts to longer wavelengths.) Spectral bandwidth is about 4 nm at low power but decreases dramatically as power rises and emission is limited to one longitudinal mode. Typical instantaneous linewidths are on the order of 10 MHz for 100-mW commercial single-stripe lasers. Lasers with less then 10 MHz spectral width and no longitudinal mode hops to 100 mW are available for a premium.

Lasers for applications such as videodisk playback and analog signal processing have cavities which oscillate in multiple longitudinal

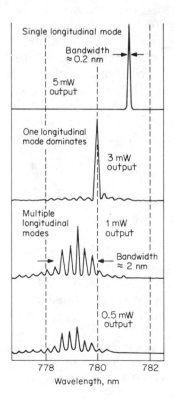

Figure 19.4 At low powers, diode lasers oscillate simultaneously in several longitudinal modes within the gain curve, but as power increases, a single longitudinal mode dominates. Output wavelength increases with power level because the semiconductor is being heated.

modes even at high currents, to avoid coherence noise. They maintain a 4-nm spectral bandwidth even as power increases. Multistripe laser arrays also have spectral bandwidths in the 2- to 4-nm range.

Pulsing or modulation increases the bandwidth of diode lasers, integrated over time, because of shifts in wavelength caused by changes in junction temperature. However, instantaneous bandwidths remain narrow. Pulsed spectral bandwidths seldom are specified for double-heterojunction lasers; for pulsed single-heterostructure lasers they are in the 15-nm range.

GaAlAs diode lasers have been demonstrated with distributed feedback and external cavities, which can produce much narrower emission bandwidths. However, most serious work on narrow-line lasers has concentrated on longer-wavelength InGaAsP lasers for high-performance fiber-optic communications; in some cases, GaAlAs was a technological testbed for researchers not familiar with InGaAsP.

Frequency and Wavelength Stability. The center wavelength of a GaAlAs diode laser typically increases about 0.25 nm/°C for each longitudinal mode. Raising the drive current increases the junction temperature, so wavelength also increases with increasing drive current and output power, as shown in Fig. 19.4.

The picture is somewhat more complex for lasers which oscillate in a single longitudinal mode. Several longitudinal modes actually fall within the laser's gain curve, but one dominates in a single-mode laser. Changes in drive current or temperature can make the laser "mode hop" to another longitudinal mode, abruptly shifting its output wavelength. This combines with the gradual wavelength shift caused by temperature to produce a noncontinuous change in wavelength like that shown in Fig. 19.5. Frequency instabilities and excess noise can occur at mode-hopping points. Active temperature compensation can stabilize wavelength, but it is rarely needed in most GaAlAs laser applications.

Amplitude Noise. Diode lasers suffer from two major types of noise: internal noise generated by mode competition and optical feedback noise, caused by external sources reflecting part of the beam back into the laser. Many manufacturers do not specify noise figures because they do not know what external reflections to expect. Some manufacturers quote signal-to-noise ratios of 60 to 80 decibels (dB) for lasers designed to suppress external reflection noise.

Data sheets typically indicate linearity by plotting light output against drive current. In practice, diode lasers can be modulated digitally with no noise problems, but analog modulation may require compensation for nonlinearities.

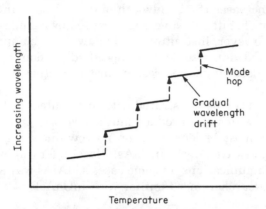

Figure 19.5 Peak wavelength of a diode laser varies slowly with temperature or drive current until emission hops to another longitudinal mode.

Beam Quality, Polariziation, and Modes. With a cavity that is very short by laser standards, and a very small emitting area that is wider than it is high, a diode laser produces a rapidly diverging beam as shown in Fig. 19.6. Divergence is much larger in the plane perpendicular to the thin active layer than in the plane of the active layer because of diffraction effects at the tiny emitting area. Cylindrical optics can make the beam circular, and collimating optics can focus it into a tight beam similar to that from a helium-neon laser.

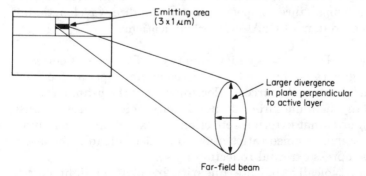

Figure 19.6 Individual narrow-stripe diode lasers emit from a region a few micrometers wide and about a micrometer high. This produces a beam that diverges more rapidly in the direction perpendicular to the active layer than in the plane of the active layer.

Beam quality of narrow-stripe lasers depends on the guiding mechanism. Index guiding concentrates the output on a narrow spot on the facet, which is desirable for tight focusing but not suitable for high-power lasers because the high power density might damage the surface. Gain-guided lasers emit over larger areas, limiting power concentration and optical damage, at the cost of larger spot size.

Although manufacturers normally do not specify polarization, the output of single-stripe lasers becomes more linearly polarized as power increases. The dominant polarization is parallel to the junction plane.

The terminology used for diode laser modes can be confusing because some data sheets may say only "single mode" without specifying whether it is transverse (spatial) mode or longitudinal mode. Single narrow-stripe double-heterostructure lasers generally emit a single transverse mode, and may emit a single longitudinal mode when operated well above threshold. Some lasers are designed to oscillate in multiple longitudinal modes at high powers to reduce coherence noise. Multistripe lasers emit multiple longitudinal modes.

Virtually all commercial monolithic arrays are edge emitters which emit a single beam, which is similar to that from a single-element laser, but comes from a broader area and generally is less coherent. The only exceptions are lasers in which two, three, or four independently modulated laser stripes are produced on the same substrate for use in laser printers and similar applications. The beams emerge from spots on the chip tens of micrometers apart, close enough to pass through the same optics and beam deflection systems.

Coherence Length.   Coherence lengths of GaAlAs lasers range from a fraction of a millimeter for multistripe arrays to several meters for single narrow-stripe lasers operating in a single longitudinal mode.

Coherence can cause undesirable effects in many diode laser applications, so coherence length can be an important specification. Problems occur when laser light is reflected from external objects (e.g., the surface of a optical disk) back to the laser. If the reflected light is coherent with the laser beam (i.e., less than a coherence length away), it can stimulate the emission of more light, generating noise. Operation in multiple longitudinal modes reduces the possibility of such noise.

Beam Diameter and Divergence.   The beam from a diode laser starts out as a small oval and diverges very rapidly by laser standards. A single-element narrow-stripe GaAlAs laser emits from an area along the junction a few micrometers long and about a micrometer high. (The active layer is thinner than one wavelength, so light is emitted from a larger area on the facet.) That is small enough to couple the output to the core of a single-mode fiber.

Diode arrays emit from areas of about the same height, but much longer, depending on the number of stripes. Emitting areas range from 100 μm wide for a 10-stripe array to 10 cm wide for a 200-stripe array. Monolithic arrays may have their output coupled into multimode fibers with core diameters comparable to the width of the emitting area.

The high divergence makes beam diameter largely irrelevant beyond a millimeter or two away from the laser. Typical figures for beam divergence from a GaAlAs laser without special collimating optics are 40° in the plane perpendicular to the active layer and 10° in the plane parallel to the active layer. Divergence varies some with operating conditions and tends to decrease with output power; it is about the same for arrays and single lasers.

Stability of Beam Direction.   The stability of beam direction is rarely mentioned in diode laser specifications, presumably because beam divergence is so large. In practice, a more relevant measurement may be misalignment error between the beam and the chip. Typical values are ±2 to 3 percent.

Suitability for Use with Laser Accessories.   The advent of high-power diode lasers and new nonlinear materials has stimulated interest in directly frequency-doubling diode lasers. Pumping a resonant monolithic cavity of potassium niobate with 167 mW at 842 nm from a diode array has generated 24 mW at 421 nm, an efficiency of 14.4 percent (Goldberg and Chun, 1989). Electronic servo locking of the diode laser frequency to that of a monolithic $KNbO_3$ ring cavity allowed generation of 41 mW at 428 nm from a single-mode GaAlAs laser emitting 105 mW, corresponding to 39 percent optical efficiency and about 10 percent conversion of input electrical power into blue light (Kozlovsky et al., 1990).

At this writing, frequency-doubled diode lasers are not available commercially. However, direct doubling of diode laser emission seems an attractive way to produce blue powers in the 10-mW range sought for some applications.

Diode laser energy also can be converted to shorter wavelengths in other ways. The best developed at this writing is by using them to pump 1.06-μm neodymium lasers, which are then frequency-doubled to 532 nm, as described in Chap. 22. Developers also are studying the use of diode lasers to pump other solid-state lasers and the nonlinear sum-frequency mixing of diode-laser output with that of other near-infrared lasers to generate blue light.

Modelocking of diode lasers has been demonstrated with both external cavities and monolithic structures. A monolithic, passively modelocked GaAlAs laser has produced 2.4-ps pulses at a 108-GHz repetition rate, but few applications require that capability and the technology is not available commercially.

Operating Requirements

Input Power. Diode lasers require very modest input powers by laser standards. Voltage drops are about 2 volts (V) across GaAlAs lasers and about 2.5 V for larger-bandgap 660- and 670-nm InGaAlP lasers. Typical thresholds for single-element GaAlAs lasers are 30 to 60 mA, with the lower values for index-guided and the higher values for gain-guided lasers. Typical operating currents are not much larger: 10 to 15 mA above threshold for 3-mW output, or 30 mA above threshold for 10-mW output. Thus a typical 10-mW GaAlAs laser may consume only 150 mW of electrical power. This low power consumption, together with voltages compatible with other semiconductor electronics, are principal attractions of diode lasers.

Higher-power diode lasers and multistripe arrays require more power but generally are more efficient than lower-power lasers. For example, one single-element laser delivers 100 mW at an operating current of 170 mA, corresponding to about 340 mW of electrical power. A 3-W multistripe array has a rated operating current of 5.2 A, for power consumption of only 10 W.

Stacked arrays of laser bars may be assembled in series electrically, multiplying the voltage requirement by the number of bars. Arrays of single-heterostructure lasers also typically are in series.

Their short rise times and low operating voltages render continuous-wave diode lasers very susceptible to damage from voltage and current surges, so input power must be controlled carefully. Special-purpose diode laser drivers are offered for this purpose.

Cooling Requirements. Most short-wavelength diode lasers are packaged to operate at room temperature with heat-sinking to remove excess heat. Passive air cooling is adequate for most applications, but thermoelectric cooling may be needed at high powers or for precise wavelength control. The laser may be packaged with a cooler, or with a heat sink designed to be mated to a cooler.

Consumables. Diode lasers do not require consumables.

Required Accessories. A diode laser or array is a tiny piece of semiconductor that must be packaged into a useful form for practical applications. Semiconductor lasers generally are not sold as bare chips, although some arrays may be sold to companies that package them for specific applications.

Two typical packages for short-wavelength diode lasers are shown in Fig. 19.7. The key elements of a package are

- A housing to permit easy handling and to help remove excess heat. Most housings resemble standard transistor or integrated-circuit cases.

- An output window or optical fiber to direct the laser beam.

- Electrical connections between the chip and the outside environment.

- A photodetector, which usually is mounted within the housing to monitor output from the rear facet of the laser. The detector output can be used with a feedback circuit to keep laser power at a constant level during its operating life.

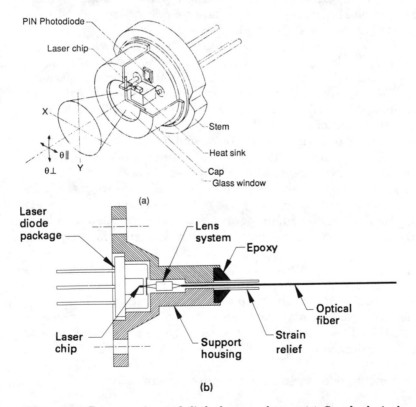

Figure 19.7 Cutaway views of diode laser packages. (*a*) Standard single-element laser packaged for use in compact-disk player, with beam emerging from window at top. (*Courtesy of Toshiba America.*) (*b*) Sketch of a pigtailed laser diode. (*Courtesy of Seastar Optics.*)

Auxiliary optics are needed to generate a narrow collimated beam or a tightly focused spot and may be included with lasers packaged for end-user applications such as research, measurement, and laser pointers.

Most diode lasers are sold without the drive electronics needed to operate the laser, but some packaged for end users do include drive circuitry.

Operating Conditions and Temperature. Specified operating temperatures for GaAlAs lasers range from 0 to 30°C to −30 to +85°C. Normally operating ranges are broader for lower-power lasers and more limited for higher-power lasers and arrays.

Laser power and wavelength vary significantly with temperature, so temperature stabilization may be needed to control output characteristics. Higher temperatures speed diode aging, so limiting operating temperature can pay dividends in increased reliability and operating life. Laser performance generally improves at lower temperatures, but this can be a mixed blessing. A temperature drop can produce unexpectedly high output power if drive current is not reduced, and in some cases the increase might be sufficient to damage the output facets.

A diode laser requires no warm-up, responding in a nanosecond or less to direct electrical modulation. Duty-cycle limitations are intended to avoid excessive heating, which can cause rapid degradation and failure of high-power and single-heterostructure lasers.

Mechanical Considerations. Individual GaAlAs lasers are chips less than 1 mm on each side; monolithic arrays are thin bars up to about 1 cm wide. The mechanical considerations of commercial diode lasers depend on packaging. Typical packages are shown in Fig. 19.7; many are similar to standard transistor housings and mount directly on circuit boards or heat sinks. Except for packaging limitations, diode lasers generally are resistant to normal mechanical shocks.

Safety. Diode lasers pose no significant electrical hazards because of their low operating voltages. The eye hazards of lasers depend on the wavelength, operating power, and beam-focusing optics.

The rapid divergence of diode laser beams gives some protection against eye damage by rapidly dispersing the power over a large area. A meter or two away, power density may be a factor of 10,000 or more lower than at the laser facet. However, collimation or focusing of the beam, required for most applications, removes this safety factor.

635-nm diode lasers deserve the same safety precautions as He-Ne lasers of the same power level.

The 660- or 670-nm output of InGaAlP or InGaP lasers is visible as red light, although the human eye is less than half as sensitive to it as to the 633-nm output of helium-neon lasers. Power levels are modest, 10 mW or less in commercial versions. As with helium-neon lasers, one should avoid staring into a collimated beam because intensities are comparable to that of sunlight. A brief accidental glance at a few-milliwatt beam should cause no permanent damage; the eye's natural aversion response will turn away from the bright light, as it does from the sun. Safety goggles are advisable, but one should be sure to check that the goggles block the desired wavelengths. Goggles designed to protect from GaAlAs lasers may not block 660 or 670 nm effectively.

The 750- to 900-nm emission of GaAlAs lasers can be more hazardous than visible diode lasers. One reason is that much higher powers are available; multistripe arrays can emit several watts, sufficient for eye damage after only brief exposure. Another reason is that the eye is insensitive to those wavelengths, so inadvertent exposure to them does not trigger the eye aversion response. The eye can sense 750 nm as red light, but the response is only about 0.001 as strong as to the 633-nm He–Ne line (Sliney and Wolbarsht, 1980, p. 88), so do not be fooled into thinking that a weak-looking beam must be very low in power. Safety goggles are recommended for GaAlAs lasers, especially at high powers and/or with collimated beams. Most types should block all GaAlAs wavelengths, but it is prudent to verify the wavelength range and degree of blocking because 750-nm types and high-power arrays became available only in the last few years.

The 980-nm emission of InGaAs lasers is invisible to the human eye. Safety goggles are recommended, but take care to select models that block 980 nm adequately. Some goggles made for GaAlAs lasers may not.

Fluorescent screens or infrared viewers shift near-infrared light to visible wavelengths, so users can align diode lasers in optical systems without removing infrared-blocking safety goggles.

Special Considerations.   Advances in diode laser technology have been exceptionally rapid, and new types will continue moving from the laboratory to the marketplace. The most advanced models typically appear as "developmental" products even before life tests have been completed. Continuing developments are likely to include broader wavelength ranges and higher output powers.

Laser diodes are very vulnerable to damage from surge currents, so care must be taken in handling them to avoid electrostatic discharges. Manufacturers stress the importance of static-free environments, including grounded wrist straps for users, grounded work surfaces, and antistatic floors.

Reliability and Maintenance

Lifetime. Lifetime was a serious limitation for early diode lasers but is no longer a problem for standard commercial devices. Few laser manufacturers specify lifetimes, but that reflects an inherent caution, the difficulty of life testing, and the sensitivity of laser lifetime to operating conditions. Accelerated aging tests are performed at high temperatures, and room-temperature lifetimes are extrapolated using the known decline in laser lifetime at elevated temperatures.

It is reasonable to project room-temperature lifetimes of hundreds of thousands of hours to over a million hours for double-heterojunction lasers emitting several milliwatts at 780 to 850 nm at room temperature. Expected lifetimes are shorter for shorter-wavelength lasers but still should be well over 10,000 hours. Lifetimes of high-power single-stripe lasers and multistripe arrays have been extrapolated to be over 10,000 hours (Sakamoto et al., 1990; Welch et al., 1990). Thus far, strained-layer lasers appear to have relatively long lifetimes; stable 20-mW operation for 5000 hours at 50°C has been reported (Okayasu et al., 1990). Continuing improvements seem likely, although projected lifetimes will remain shortest for lasers with particularly high power or short wavelength.

Maintenance and Adjustments Needed. Diode lasers need no special maintenance other than operation under proper conditions and careful cleaning of the output window if needed. However, many applications require stabilization of laser power and/or temperature.

Failure Modes and Causes. According to laser manufacturers, the most common cause of premature diode laser failure is electrostatic discharge. Laser diodes can fail in several other ways:

- Catastrophic optical damage to facets, caused by power densities above 10^6 W/cm^2
- Gradual facet degradation, often caused by oxidation
- Formation of "dark line defects," regions in the active layer which do not emit light
- Failure of metallization and/or bonding solder which makes electrical contacts to the chip
- Degradation of internal structure

Care in handling and operation can avoid most premature failures that are not caused by manufacturing defects. Many manufacturers perform burn-in tests to screen out defective devices.

Possible Repairs.   Once a laser diode has failed, the user can only re-place it and correct any problems which may have caused the failure.

Commercial Devices

Standard Configurations.   Short-wavelength diode lasers are sold in several forms, including

- Low-power models packaged in standard transistor-type cases with output windows, for applications such as compact-disk playback
- Low-power models packaged with optical fiber pigtails or fiber-optic connectors, primarily for optical communications
- High-power lasers and arrays packaged in standard electronic cases with output windows
- High-power lasers and arrays packaged with optical fiber pigtails for beam delivery
- Stacks of laser arrays, pulsed or continuous-wave, to deliver high powers for applications such as pumping neodymium lasers
- Pulsed single-element lasers and arrays, generally with output windows
- Multibeam lasers, with two, three, or four stripes on a single substrate, for laser printers and optical storage systems
- Laboratory lasers, with collimated beams for general research and development purposes
- Laser pointers with collimated visible output beams
- Threaded-stud packages with output windows, mostly for military applications

Options.   The major options are in packaging, as described above.

Pricing.   Prices of short-wavelength lasers cover a broad range. Low-power 780-nm lasers mass-produced for compact-disk players can be bought for a few dollars each in large quantities. InGaP and InGaAlP lasers with red output sell for tens of dollars to over \$100. High-power arrays can carry prices of thousands of dollars. Prices are subject to the economies of scale of semiconductor production, so a large demand for a particular type of laser could reduce prices dramatically. However, high-power arrays are likely to remain significantly more expensive than low-power diode lasers for CD players.

Suppliers.   The market for short-wavelength diode lasers is divided into a number of segments. Some are highly competitive, but others are not.

- *Visible red diode lasers:* Most are made by a few companies in Japan and Europe, but many other companies package them into products such as laboratory lasers and laser pointers. This market is likely to become more competitive as more companies develop the required technology.

- *Low-power 750-nm lasers:* A few companies produce these GaAlAs types, but unless new applications appear, this market may be stunted by the emergence of shorter-wavelength red diode lasers.

- *Low-power 780-nm lasers:* Most of these lasers are used in CD players. The rapid growth of the CD market attracted several Japanese companies, which built up large production capacities. These lasers are readily available at low cost because supplies exceed current demand.

- *Other low- and moderate-power GaAlAs lasers:* American, European, and Japanese companies all offer lasers optimized for applications such as laser printers, optical disk storage, fiber-optic communications, and measurement. The market is reasonably competitive.

- *High-power arrays:* Initially dominated by one American company, this market has attracted other manufacturers, but demand remains limited and prices are high.

- *Special-purpose packaged GaAlAs lasers:* Several small and medium-sized companies buy mass-produced diode lasers and package them for special applications, such as short-pulse generation, interferometry, and general laboratory use. This market is reasonably competitive.

- *Single-heterostructure pulsed GaAlAs lasers:* Only a couple of companies offer these lasers, used almost exclusively in existing military systems. This market is likely to shrink further because better technology is available for most applications.

- *980-nm InGaAs lasers:* Developmental versions are available, but a market has yet to develop.

Applications

The range of applications for diode lasers has increased tremendously as their variety as grown, their power levels have increased, and their prices have dropped. Many fall under the general heading of information handling, including compact-disk players, fiber-optic communications, and laser printers. Other uses include measurement and inspection, displays, medical treatment, general instrumentation, powering other lasers, military systems, and general research and development.

Compact-disk players are designed around unmodulated 780-nm lasers which emit a beam of a few milliwatts which reads data from the disk. The first and so far biggest use of this technology is digital audio equipment. However, the same data storage format is used for computer data storage (CD-ROM), video playback (CD-V), and interactive multimedia systems (CD-I).

Optical data storage systems require higher laser powers and can write data on a disk as well as read it back. There are two major variants: write-once read-many (WORM) storage and magneto-optic storage. In WORM systems, a high-power beam forms indelible marks or pits on disks or cards, and a lower-power beam reads it back. Magneto-optic systems use a laser beam together with a magnetic field to create magnetic data fields on a disk; data is read by a low-power laser beam, which is affected by the magnetic field. Unlike WORM media, magneto-optic media can be erased and reused.

Laser printers scan a modulated laser diode beam across a photoconductive drum, writing a pattern on the light-sensitive surface. The drum is stepped between scans to build up a laser-written image. Like optical storage, this requires higher laser powers than does CD playback. As prices have come down, laser printers have found growing use with personal computers. However, not all "laser printers" contain lasers—some contain linear arrays of LEDs which write tiny spots onto the photoconductive drum without the need for a beam scanner and give about the same print quality.

GaAlAs lasers have been used for several years to read bar codes printed on certain types of packages, such as shipping containers. Usually the lasers are mounted in hand-held wands, which are passed close to the surface. The advent of visible diode lasers promises to extend semiconductor lasers to retail bar-code readers. (The Universal Product Code used in supermarkets is specified for the 633-nm He–Ne wavelength; some materials used in consumer packaging may have different reflectivity at GaAlAs laser wavelengths, which cannot be seen by the human eye at a printing plant.)

The wavelengths of GaAlAs lasers are not well suited for long-distance fiber-optic communications, but they are fine for local area networks or links of a few kilometers or less. LEDs are cheaper, but they cannot generate as much power or carry signals at as high data rates as diode lasers.

Diode lasers are used in a variety of inspection and measurement applications. Some involve simply measuring the distance between points, or watching for interruption of a beam to detect presence or absence of a part or variations of component size on a production line. Diode lasers with long coherence lengths can be used for holographic interferometry or interferometric measurements. Visible-light diode lasers can be used to define lines or planes for construction applications and in other situations where alignment by the human eye is important.

The advent of visible diode lasers has created a market for diode laser pointers for presentations. Units about the size of a fat ballpoint pen, powered by AAA cells, are available for about $200.

Diode lasers have been used for some nontraditional medical applications, such as the illumination of acupuncture points and "biostimulation" with red or infrared beams. Orthodox physicians are testing high-power laser diodes for the treatment of eye conditions such as diabetes-related blindness that had been treated with larger gas lasers.

High-power diode lasers and arrays can be used to pump neodymium lasers (see Chap. 22) for industrial and research applications. GaAlAs lasers emitting at an 810-nm absorption line of neodymium are more efficient pumps than flashlamps, which emit light at a broad range of wavelengths. Recently 680-nm diode lasers have pumped alexandrite lasers (see Chap. 24).

High-power diode lasers and arrays are being studied for possible uses in space communications, such as satellite-to-satellite links (Evans and Ettenberg, 1987).

Pulsed diode lasers are used in military ranging systems to measure distances to potential targets by timing pulse round-trip time.

InGaAs strained-layer lasers at 980 nm are being developed to pump erbium-doped fiber amplifiers for use in long-distance fiber-optic communication systems (see Chap. 26). They also could be used to pump other erbium lasers emitting near 1.55 μm, a wavelength desirable because it poses little threat to the human eye.

GaAlAs lasers could find a variety of computing and signal-processing applications. Surface-emitting lasers could send signals between electronic chips, avoiding interconnection bottlenecks of present high-performance circuits. Lasers in integrated optical devices could directly process optical signals, or perform high-speed, highly parallel computing operations.

All types of diode lasers are used in a variety of research and development as well as for laboratory measurements. Some of this research may lead to new applications for diode lasers as well as to new types of semiconductor lasers.

Bibliography

K. J. Beernink, J. J. Alwan, and J. J. Coleman: "InGaAs–GaAs–AlGaAs gain-guided arrays operating in the in-phase fundamental array mode," *Applied Physics Letters* 57(26):2764–2767, December 24, 1990 (research report).

Dan Botez: "Laser diodes are power-packed," *IEEE Spectrum*, June 1985, pp. 43–53 (review paper).

T. R. Chen et al.: "Cavity length dependence of the wavelength of strained-layer InGaAs/GaAs lasers," *Applied Physics Letters* 57(23):2402–2403, December 3, 1990 (1990a) (research report)

T. R. Chen et al.: "High-power operation of buried-heterostructure strained-layer InGaAs/GaAs single quantum well lasers," *Applied Physics Letters* 57(26):2762–2763, December 24, 1990 (1990b) (research report).

T. Egawa et al.: "Low-threshold continuous-wave room-temperature operation of $Al_xGa_{1-x}As$/GaAs single quantum well lasers grown by metalorganic chemical vapor deposition on Si substrates with SiO_2 back coating," *Applied Physics Letters* 57(12): 1179–1181, September 17, 1990 (research report).

Gary Evans and Michael Ettenberg: "Semiconductor laser sources for satellite communication," in Morris Katzman (ed.), *Laser Satellite Communications*, Prentice-Hall, Englewood Cliffs, N.J., 1987 (review paper).

Luis Figueroa: "High-Power semiconductor lasers," in Peter K. Cheo (ed.), *Handbook of Solid-State Lasers*, Marcel Dekker, New York, 1989, pp. 113–226 (review of high-power semiconductor lasers).

Luis Figueroa: "Semiconductor lasers," in Kai Chang (ed.), *Handbook of Microwave and Optical Components*, vol. 3, *Optical Components*, Wiley-Interscience, New York, 1990 (review of semiconductor lasers, concentrating on high-power types).

Lew Goldberg and Myung K. Chun: "Efficient generation at 421 nm by resonantly enhanced doubling of GaAlAs laser diode array emission," *Applied Physics Letters* 55(3):218–220, July 17, 1989 (research report).

M. Ishikawa et al.: "Short-wavelength (638 nm) room-temperature CW operation of InGaAlP laser diodes with quaternary active layer," *Electronics Letters* 26(3):211–212, February 1, 1990 (research report).

K. Itaya, M. Ishikawa, and Y. Uematsu: "636 nm room-temperature CW operation by heterobarrier blocking structure InGaAlP laser diodes," *Electronics Letters* 26(13): 839–840, June 21, 1990 (research report).

H. Jeon et al.: "Room temperature blue lasing action in (Zn, Cd)Se/ZnSe optically pumped multiple quantum well structures in on lattice-matched (Ga,In)As substrates," *Applied Physics Letters* 57(23):2413–2415, December 3, 1990 (research report).

Eli Kapon: "Semiconductor diode lasers: Theory and Techniques," in Peter K. Cheo (ed.), *Handbook of Solid-State Lasers*, Marcel Dekker, New York, 1989, pp. 1–112 (review of semiconductor lasers).

Kenichi Kobayashi et al.: "AlGaInP double heterostructure visible-light laser diodes with a GaInP active layer grown by metalorganic vapor phase epitaxy," *IEEE Journal of Quantum Electronics* QE-23(6):704–711, 1987 (research report).

Kenichi Kobayashi et al.: "632.7-nm CW operation (20°C) of AlGaInP visible laser diodes fabricated on (001) 6° off toward [110] GaAs substrate," *Japanese Journal of Applied Physics* 29(9):L1669–L1671, 1990 (research report).

W. J. Kozlovsky et al.: "Generation of 41 mW of blue radiation by frequency doubling of a GaAlAs diode laser," *Applied Physics Letters* 56(23):2291–2292, June 4, 1990 (research report).

Lasers & Optronics staff report: "The laser marketplace—forecast 1990," *Lasers & Optronics* 9(1):39–57, January 1990 (market analysis).

Paul Mortensen: "Diode lasers make gains in Japanese markets," *Laser Focus World* 26(7):67–80, July 1990 (status report on visible diode lasers).

M. Okayasu et al.: "Stable operation (over 5000 h) of high-power 0.98-μm InGaAs–GaAs strained quantum well ridge waveguide lasers for pumping Er^{3+}-doped fiber amplifiers," *IEEE Photonics Technology Letters* 2(10):689–691, 1990 (research report).

M. Sakamoto et al.: "High power, high-brightness 2-W (200 μm) and 3 W (500 μm) CW AlGaAs laser diode arrays with long lifetimes," *Electronics Letters* 26(11):729–730, May 24, 1990 (research report).

David Sliney and Myron Wolbarsht: *Safety with Lasers and Other Optical Sources*, Plenum, New York, 1980 (handbook on laser safety).

P. J. Thijs, A., T. van Dongen, and W. Nijman: Paper WJ2 in *Technical Digest, Conference on Optical Fiber Communications*, San Francisco, January 22–26, 1990, Optical Society of America, Washington, D.C. (research report).

J. Ungar et al.: "GaAlAs window lasers emitting 500 mW CW in fundamental mode," *Electronics Letters* 26(18):1441–1442, August 30, 1990 (research report).

A. Valster et al.: "633 nm CW operation of GaInP/AlGaInP laser diodes," paper Cl in *Conference Digest, 12th IEEE International Semiconductor Laser Conference*, Davos, Switzerland, Sept. 9–14, 1990a (research report).

A. Valster et al.: "Short wavelength 626 nm GaInP/AlGaInP laser diode with a multiple quantum well active layer," paper CMC3 in *Technical Digest, Conference on Lasers and Electro-Optics*, Baltimore, Optical Society of America, Washington, D.C., 1990b (research report).

A. Valster et al.: "Green (555 nm) continuous-wave lasing emission at 77 K of GaInP/AlGaInP multiple quantum well laser diode," postdeadline paper 12 at *12th IEEE International Semiconductor Laser Conference*, Davos, Switzerland, Sept. 9–14, 1990c (research report).

D. Welch et al.: "High reliability, high power, single mode laser diodes," *Electronics Letters* 26(18):1481–1483, August 30, 1990 (research report).

Eli Yablonovich et al.: "Double heterostructure GaAs/AlGaAs thin film diode lasers on glass substrates," *IEEE Photonics Technology Letters* 1(2):41–42, 1989 (research report).

Jay Zeman: "Visible diode lasers show performance advantages," *Laser Focus World*, August 1989, pp. 69–80 (review paper).

C. A. Zmudzinski, Y. Guan, and P. Zory: "Room-temperature photopumped ZnSe lasers," *IEEE Photonics Technology Letters* 2(2):94–96, 1990 (research report).

Long-Wavelength III–V Semiconductor Diode Lasers

Semiconductor diode lasers with wavelengths longer than 1.1 micrometers (μm) are used primarily for fiber-optic communications. The only III–V material system used for commercial lasers is InGaAsP, a quaternary compound which can be lattice-matched to InP substrates for laser wavelengths between about 1.1 and 1.65 μm. Researchers are studying other III–V systems, but none are offered commercially. Lead salt semiconductors are used in long-wavelength lasers which have different applications and are treated separately in Chap. 21.

Development of InGaAsP lasers has followed a pattern somewhat different from that for shorter-wavelength diode lasers. Although InGaAsP lasers can be made at a broad range of wavelengths, most work has concentrated on two "windows" of silica optical fibers: 1.31 μm, where step-index single-mode fibers have zero wavelength dispersion and loss of about 0.5 decibel per kilometer (dB/km), and 1.55 μm, where silica fibers have their lowest loss, about 0.15 dB/km. InGaAsP lasers have been designed with properties such as high modulation bandwidth and narrow spectral linewidth needed for high-performance long-distance fiber-optic systems.

Unit sales of InGaAsP lasers are modest compared to those of shorter-wavelength diode lasers—only about 100,000 in 1990. However, the total has been increasing by some 20 percent a year, and at an average price near $500, InGaAsP lasers cost much more than their shorter-wavelength counterparts. That price reflects modest production, difficult technology, exacting performance requirements, and their vital role in expensive high-capacity communication systems.

Active Medium

Semiconductor laser developers began exploring InGaAsP in the 1970s, when they realized that optical fibers had their lowest losses at wavelengths longer than 1 μm, beyond the reach of GaAlAs lasers. They initially were wary of a quaternary compound because it is inherently more complex than ternary materials like GaAlAs. However, they had little choice; no ternary compound has the right combination of band-gap energy and lattice constant for making lasers between 1.1 and 1.65 μm.

Figure 20.1 shows the nature of the problem by plotting band-gap energy (in electron volts) against lattice constant (in angstroms) for selected III–V compounds. Binary compounds, such as GaAs, InP, and InAs, occupy single points because their composition uniquely determines both band-gap energy and lattice constant. Lines connecting points represent the range of ternary compounds, most of which span a range of lattice constants and band gaps. GaAlAs is a special case because gallium and aluminum ions are about the same size, so mov-

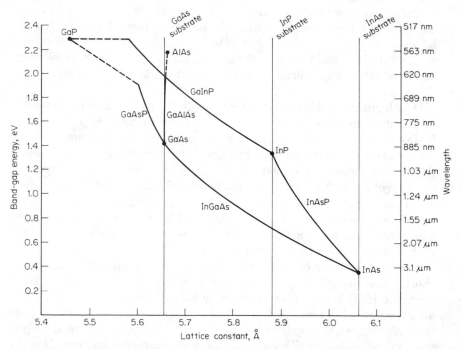

Figure 20.1 Energy band gaps (in electron volts) and lattice constants (in angstroms) of InGaAsP and GaAlAs materials, with lines showing substrates for lattice matching. Wavelengths equivalent to band gaps are shown at the right. See Fig. 19.3 for other materials.

ing along the $Ga_{1-x}Al_xAs$ line from GaAs ($x = 0$) to AlAs ($x = 1$) changes the lattice constant only a little but alters the bandgap considerably.

There are no such equal-sized pairs in the InGaAsP system. Changing the indium composition of $In_xGa_{1-x}As$ from 0 to 1 alters both bandgap energy and lattice constant. The same happens for other ternary compounds. The only way to match the lattice constant of the binary substrate InP is to use the quaternary compound InGaAsP, which occupies the area defined by the points GaP, InP, InAs, and GaAs. The designer must vary both In/Ga and As/P ratios in compound $In_{1-x}Ga_xAs_yP_{1-y}$ to make an active layer lattice-matched to InP with the desired band-gap energy (or, equivalently, emission wavelength).

Most commercial InGaAsP lasers are made with one of two active-layer compositions:

- $In_{0.73}Ga_{0.27}As_{0.58}P_{0.42}$, for operation at 1.31 μm

- $In_{0.58}Ga_{0.42}As_{0.9}P_{0.1}$, for operation at 1.55 μm

However, lasers have been made at other wavelengths for wavelength-division multiplexing and for pumping erbium-doped fibers. Some are available commercially; others would be offered if a market emerged.

InGaAsP lasers have smaller band gaps and lower photon energies than GaAlAs lasers, leading to some important differences in operating characteristics. InGaAsP lasers generally have lower threshold currents, are less vulnerable to optical facet damage, and tend to have longer lifetimes despite the greater complexity of the material system. On the other hand, InGaAsP lasers are more sensitive to temperature increases.

Development of strained-layer structures has stimulated fabrication of InGaAs lasers, with InGaAsP confinement layers on InP substrates. The strained-layer structures appear to offer attractive improvements in performance. Recent results include lasers with continuous-wave output to 206 mW between 1.48 and 1.5 μm (Tanbun-Ek et al., 1990), and with threshold currents as low as 1.1 mA at 1.56 μm (Temkin et al., 1990).

Researchers also are investigating the possibilities of the GaInAsSb system with confinement layers of GaAlAsSb grown on GaSb substrates for 1.7 to 4.4 μm; this can be seen in the plot of III–V materials in Fig. 19.3. Pulsed operation at room temperature has been demonstrated at 1.8 to 2.4 μm (Drakin et al., 1987). Room-temperature continuous-wave operation has been reported at 2.0 to 2.3 μm, and pulsed outputs have reached 290 milliwatts (mW) per facet at 2.29 μm (Eglash and Choi, 1990). InGaAs/InAsPSb diode lasers on InP substrates emit at 2.52 μm at 190 kelvins (K) (Martinelli and Zamerowski, 1990), and other systems also are being investigated (Horikoshi, 1985). Such long-wavelength lasers would be used with nonoxide optical fibers expected to have extremely low loss in the midinfrared (Katsuyama and Matsumura, 1989). However, developers have yet to demonstrate such low loss in practical fibers.

Internal Structures

InGaAsP laser development has concentrated on the semiconductor laser structures described in Chap. 18 which offer characteristics sought for fiber-optic communications. Double-heterostructure index-guided lasers with Fabry-Perot cavities are used most often, with powers of about 10 mW sufficient for most fiber systems. (Higher powers in a the roughly 10-μm core of a single-mode fiber can produce nonlinear effects which degrade transmission.) The most powerful commercial lasers deliver about 100 mW continuous-wave and are used in fiber-optic test and measurement equipment.

The lack of need for higher powers has limited interest in monolithic arrays modulated in unison to generate one beam. However, developers are studying arrays which can be modulated independently to generate different signals at different wavelengths.

Many advanced fiber-optic systems require lasers with extremely narrow spectral linewidths. A main reason is to minimize pulse dispersion arising from the dependence of refractive index with wavelength. For standard single-mode fibers, this wavelength dispersion is zero at 1.3 μm but at 1.55 μm is large enough to severely limit data rates. Narrow linewidth also is needed for coherent communication systems (Koch and Koren, 1990), in which the receiver shifts signals to lower frequencies by heterodyne or homodyne mixing, so they can be amplified electronically. Coherent systems remain in development but offer attractive improvements in receiver sensitivity and wavelength sensitivity. Because of these applications, developers of narrow-line lasers have concentrated on the InGaAsP system. Distributed-feedback InGaAsP lasers are available commercially at both 1.3 and 1.55 μm, where their narrow linewidths can reduce dispersion. Developmental narrow-line external-cavity lasers are available, and researchers are working on more broadly tunable versions for coherent communications. Distributed-Bragg-reflection lasers also are in development.

Quantum-well and strained-layer structures also are being studied. Multiple-quantum-well structures have improved InGaAsP lasers, but not as much as do GaAlAs types (Tsang et al., 1990). Strained-layer structures can improve performance of buried-heterostructure 1.5-μm lasers. Continuous-wave outputs have reached 200 mW in a strained-layer $In_{0.8}Ga_{0.2}As/InGaAsP/InP$ multiple-quantum-well double-channel planar buried-heterostructure laser (Thijs et al., 1990). Threshold current increased only 1 percent after 5400 hours of operation at 5 mW and 60°C. Other strained-layer InGaAs/InGaAsP/InP lasers have threshold currents as low as 1.1 milliamperes (mA) at 20°C (Temkin et al., 1990).

Semiconductor Laser Amplifiers. Virtually all work on semiconductor laser amplifiers (see Chap. 18) has concentrated on the InGaAsP system at 1.3 and (particularly) 1.55 μm. The goal is to directly amplify optical signals without first converting them to electronic form, as must be done in conventional fiber-optic repeaters.

Semiconductor laser amplifiers suffer from important limitations, including sensitivity to the polarization of input light, limited saturation power, and losses in coupling light between the amplifier and the fiber. Developmental versions are specified at only 6 dB of fiber-to-fiber gain. Researchers hope that multiple-quantum-well structures will offer somewhat higher gain, but it is not clear whether such improvements will be sufficient for semiconductor lasers to meet system requirements better than erbium-doped fiber amplifiers (see Chap. 26).

Research Thrusts

Development of new InGaAsP lasers continues, with primary emphasis on the requirements of future communication systems. Important efforts include

- Exploring improvements possible with quantum-well and strained-layer structures.

- Ultra-narrow-linewidth lasers, for coherent communications.

- Lasers which exhibit low "chirp," a shift in wavelength caused by a change in drive current that effectively increases the linewidth.

- Multiwavelength sources, either monolithic or discrete, for wavelength-division multiplexing. The lasers should have narrow wavelength ranges and precisely controllable wavelengths, to avoid signal overlap and noise. One intriguing possibility is arrays of independently modulatable stripes emitting at different wavelengths.

- Optoelectronic integrated circuits, in which the laser and drive circuitry are integrated on a single chip.

- Wavelength-tunable lasers, for both coherent communications and wavelength-division multiplexing. Goals include very narrow linewidth and broad tunability; technologies include external-cavity lasers and various other structures.

- Vertical-cavity lasers, possibly as multiwavelength sources for wavelength-division multiplexing and optical interconnections.
- Optical amplifiers; semiconductor laser amplifiers are one possibility, but erbium-doped fiber amplifiers at 1.55 μm also may require InGaAsP lasers emitting at 1.48 μm as pump sources.
- Semiconductor laser materials for wavelengths longer than 1.7 μm.

Variations and Types Covered

Long-wavelength III–V lasers are less diverse than short-wavelength lasers because they share common applications in fiber-optic communications. The principal types are

- Fabry-Perot lasers emitting at 1.2, 1.3, and 1.55 μm, for fibers in which wavelength dispersion is not a critical issue
- Distributed-feedback (DFB) lasers with very narrow linewidth at 1.3 and 1.55 μm
- Semiconductor laser amplifiers for 1.3 and 1.55 μm, which lack resonant cavities and amplify signals from external sources
- High-power 1.3- and 1.55-μm lasers for measurement and research applications
- 1.48-μm lasers to pump erbium-doped fiber amplifiers (see Chap. 26)
- Ultra-short-pulse lasers, for novel transmission systems

The rest of this chapter concentrates on commercial technology, while indicating what new technologies are likely to emerge in the near future.

Beam Characteristics

Wavelengths and Output Power. InGaAsP/InP lasers can operate at wavelengths between about 1.1 and 1.65 μm. In practice, however, only a few wavelengths are available commercially:

- 1200 nm, at continuous power levels of 1 to 50 mW, primarily for communications, but not in common use.
- 1300 or 1310 nm, at continuous power levels from 0.1 to 100 mW, in widespread use for communications and measurement. Laboratory continuous-wave power levels have exceeded 200 mW (Oshiba et al., 1987).

- 1480 nm, at continuous power levels to 100 mW, for pumping erbium-doped fiber amplifiers. Laboratory continuous-wave power levels have reached 180 mW (Oshiba and Tamura, 1990).

- 1550 nm, at continuous power levels from 0.5 to 100 mW, for communications and measurement. Continuous-wave powers to 140 mW have been demonstrated in multiple-quantum-well, buried-heterostructure lasers with Fabry-Perot cavities (Tsang et al., 1990); distributed-feedback lasers have lower powers; the highest demonstrated is 77 mW (Kakimoto et al., 1989).

As for other semiconductor lasers, specified wavelengths are nominal values, with actual wavelengths lying in a range of ±20 to 40 nm. Thus there is little practical difference between lasers with specified wavelengths of 1300 and 1310 nm. Users who need better defined wavelengths must make special arrangements with the manufacturer or select lasers based on their own measurements.

External-cavity lasers can be tuned in wavelength. Commercial types have typical tuning ranges of 40 nm around center wavelengths of 1300 and 1550 nm; laboratory versions have been tuned between 1440 and 1640 nm (Lidgard et al., 1990).

Lasers with nominal continuous powers to about 10 mW (the upper limit for commercial distributed-feedback lasers) are used for fiber-optic communications. Those with higher powers generally are used in measurement equipment, such as optical time-domain reflectometers, which test fiber-optic communication systems. (The major exception is 1.48-μm lasers used to pump erbium fiber amplifiers.) Some measurement lasers operate only in pulsed mode.

Output power specifications often depend on packaging, and sometimes can be misleading. Some light is always lost in coupling laser output to optical fibers, especially single-mode fibers, so power from a fiber pigtail is less than from the same chip packaged to emit through an output window. This should be evident on data sheets.

Efficiency. The overall efficiency of a laser diode depends on the threshold current and the slope efficiency, the fraction of each added milliwatt of electrical input converted into light. Because of the threshold effect, the efficiency increases with operating power.

Typical threshold currents for commercial InGaAsP lasers are 20 to 30 mA, but a few have thresholds as low as 12 mA or as high as 100 mA. Threshold currents tend to be slightly lower at 1.3 μm than at 1550 nm. Considerably lower threshold currents have been reported in laboratory lasers. Forward voltages are low—typically 1.0 volt (V) for 1550 nm lasers, 1.1 V for 1300 nm, and 1.2 V for 1200 nm.

Differential efficiencies typically are 10 to 20 percent but can be considerably lower. Overall efficiencies can reach about 20 percent for the highest-power lasers, but typically are under 10 percent for outputs below 10 mW. Power dissipation is modest—25 to 50 mW in a few-milliwatt laser. Efficiency decreases and threshold current increases with temperature. Absorption limits 1.55-μm lasers to lower efficiencies than 1.3-μm lasers (Kakimoto et al., 1988).

Temporal Characteristics. Most InGaAsP lasers emit continuous-wave at room temperature. However, some are designed to generate pulses of up to about 100 mW for applications such as measurements of long lengths of optical fiber. Some pulsed InGaAsP lasers produce 1-microsecond (μs) pulses with duty cycles in the 1 percent range; others generate subnanosecond pulses.

Continuous-wave InGaAsP lasers with Fabry-Perot cavities typically have subnanosecond rise and fall times and modulation bandwidths in the 1-gigahertz (GHz) range. Distributed-feedback lasers typically have rise and fall times of 0.2 to 0.5 ns. The fastest digital fiber-optic systems in service have data rates over 2 gigabits per second. A few commercial InGaAsP lasers have much faster response, and modulation bandwidths in the 10-GHz range. Modulation bandwidths to 17 GHz have been reported (Uomi et al., 1989), and recent studies indicate that the upper limits for modulation bandwidth at 1.3 μm should be 24 GHz for distributed-feedback lasers and 40 GHz for Fabry-Perot lasers (Eom et al., 1990).

Modelocked InGaAsP lasers can produce extremely short pulses. A monolithic colliding-pulse-modelocked ring laser has generated 950-femtosecond (fs) pulses at over 100 GHz (Chen et al., 1991). Such short-pulse lasers are not available commercially, but they could prove important in the use of special pulses called *solitons* to transmit signals long distances at very high data rates. Soliton pulses represent stable propagation modes in an optical fiber. Nonlinear perturbations of pulse duration and wavelength range offset each other to re-create the original pulse shape, diminished in intensity by fiber attenuation, at certain points along the fiber (Marcuse, 1988, pp. 90–98). Optical amplifiers can maintain soliton intensity, allowing transmission over distances to 10,000 kilometers (km) (Mollenauer et al., 1990).

Spectral Bandwidth. InGaAsP lasers with Fabry-Perot cavities typically have bandwidths specified at 2 to 5 nm, covering multiple longitudinal modes. The bandwidths of "single-frequency" distributed-feedback lasers are much smaller, typically specified at 0.1 nm. If a distributed-feedback laser is operated continuous-wave, a typical instantaneous spectral bandwidth is tens of megahertz, but linewidths as low as 250 kHz have been reported for a multiple-quantum-well distributed-feedback lasers with 1500-μm-long cavity (Yamazaki et al., 1990). The narrowest linewidths available are about 100 kHz, produced by external-cavity lasers.

Direct modulation of a narrow-line InGaAsP laser causes additional broadening of its wavelength range because variations in electron density change the refractive index of the semiconductor. This changes the effective length of the laser cavity and thus the resonant wavelength, which thus shifts, or chirps, during a pulse. Chirp is rarely specified, but it functionally increases the bandwidth of any directly modulated laser, and can cause dispersion penalties in fiber-optic systems. Distributed-feedback structures per se do not prevent chirp, but certain types can limit it to as little as 0.4 nm (Uomi et al., 1990). Chirp can be avoided by external modulation of the beam from a laser operated at constant drive current.

Frequency and Wavelength Stability. The center wavelength of a diode laser depends on temperature, and typically increases about 0.3 to 0.5 nm/°C for InGaAsP. Each longitudinal mode of the laser makes this shift; the laser also can shift between longitudinal modes as the temperature changes. Increasing the drive current heats the laser, so the wavelength shifts to slightly longer values as output increases.

Modulation can increase a laser's spectral bandwidth and shift its operating wavelength. In addition to the wavelength chirp mentioned above, modulation can make the laser "hop" between longitudinal modes. This effect is mostly a problem for lasers operating in a single longitudinal mode. Active temperature stabilization may help limit changes.

Amplitude Noise. Manufacturers typically do not specify noise figures because these figures depend strongly on the operating environment. Diode lasers suffer from noise caused by internal mode competition and by stray reflections fed back into the laser cavity.

Some InGaAsP lasers are packaged with optical isolators, which block stray reflections and greatly reduce noise levels. An isolator, shown in Fig. 20.2, consists of a pair of polarizing filters separated by a Faraday rotator, which rotates the polarization of light passing through it by 45° in the same direction regardless of which way light passes through it. Typical isolation ratios are around 30 decibels (dB).

Beam Quality, Polarization, and Modes. Like other diode lasers, InGaAsP lasers have small emitting areas, 1 μm high by a few micrometers wide, and produce rapidly diverging beams. Divergence is somewhat larger in the plane perpendicular to the thin active layer than it is in the plane of the active layer, but the difference is not as

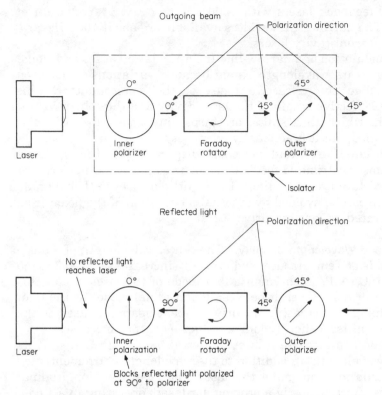

Figure 20.2 Optical isolator prevents reflected light from returning to a diode laser. Linearly polarized light from the laser is rotated 45° by the Faraday rotator before passing through another polarizer oriented at 45°. Reflected laser light passes through the outer polarizer and then is rotated an additional 45° by the Faraday rotator, making its polarization orthogonal to the inner polarizer, which blocks it from the laser.

large as that for shorter-wavelength semiconductor lasers. Cylindrical optics can compensate for the astigmatic beam shape, but typically the output is coupled into an optical fiber.

Commercial InGaAsP lasers usually oscillate in a single transverse mode; they are specified as operating in one or multiple longitudinal modes. Some lasers normally operate in multiple longitudinal modes, as shown in Fig. 20.3a. However, some nominally "single-mode" Fabry-Perot lasers emit in a single longitudinal mode with a steady drive current but shift into multimode operation when modulated. Others may have one dominant mode, but still have significant emission on other longitudinal modes, as shown in Fig. 20.3b. (One hint of this comes from comparing specifications; a "single-mode" la-

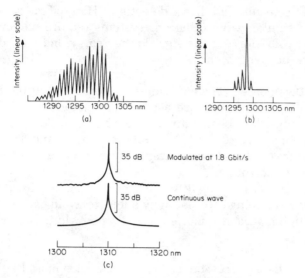

Figure 20.3 Longitudinal-mode structures of InGaAsP lasers: (*a*) laser specified for multiple longitudinal modes; (*b*) laser said to be single-mode despite some emission on secondary modes; (*c*) true single-longitudinal-mode output from a distributed-feedback laser.

ser with 4-nm spectral linewidth probably is not single-mode when modulated.)

Lasers with a mechanism that narrows linewidth, such as distributed feedback or an external cavity, oscillate reliably in a single longitudinal mode even when they are modulated, as shown in Fig. 20.3*c*.

Diode laser polarization normally is not specified, but as power increases, emission tends to be increasingly polarized with electric field in the plane of the active layer.

Coherence Length. Manufacturers usually do not specify coherence lengths for InGaAsP lasers. Typical values would be on the order of meters for a laser emitting in a single longitudinal mode and a few millimeters for a laser emitting multiple longitudinal modes. Noise considerations favor multiple-longitudinal-mode lasers for transmission through multimode fibers and single-longitudinal-mode lasers for single-mode fiber systems.

The term *coherent* is used in a different sense in some developmental fiber-optic communication systems. In these systems, light from the transmitter is mixed at the detector with light from another laser similar in wavelength, yielding a difference-frequency signal. That difference-frequency signal then is amplified, allowing detection of weaker signals than conventional direct detection. (Heterodyne radio receivers work in a similar way.) Coherent systems require narrow-linewidth lasers to avoid blurring the signals they carry, but they do not depend on coherence length per se.

Beam Diameter and Divergence.   The emitting areas of InGaAsP lasers typically are oval-shaped and measure a few micrometers wide and about a micrometer high. The beams diverge rapidly from that point. Typical beam divergences are 30° by 40° for 1300-nm lasers and 35° by 45° for 1550-nm lasers; the longer wavelengths accentuate diffraction effects from the small aperture.

In practice, the output of most InGaAsP lasers is coupled into optical fibers, generally after being collimated by a lens to avoid excessive loss. This and the high divergence make beam diameter largely irrelevant.

Suitability for Use with Laser Accessories.   The accessories most likely to be used with InGaAsP lasers are external modulators and optical amplifiers.

External modulators can be used to impose a signal on the beam from a diode laser operated at a steady drive current to avoid the wavelength chirp caused by direct laser modulation. At this point, most interest focuses on waveguide modulators in materials such as lithium niobate.

Optical amplifiers can increase the strength of signals, either before they enter a fiber-optic cable or after they have been attenuated by passage through a long fiber. Two types are available: semiconductor laser amplifiers, which are InGaAsP lasers of the appropriate wavelength without reflective facets, and fiber amplifiers doped with rare earths (see Chap. 26).

Diode lasers have been modelocked with external devices to generate short pulses at high repetition rates, but they also can be fabricated with modelocking components integral to the chip (Chen et al., 1991).

Harmonic generation has attracted little interest because InGaAsP powers are modest and better sources are already available near the second harmonics of 1300 and 1550 nm.

Operating Requirements

Input Power.   Diode lasers require only modest input powers by laser standards. InGaAsP lasers operate at voltages of 1.1 to 1.3 V for wavelengths of 1.55 to 1.2 μm, respectively. Typical operating currents for outputs to several milliwatts are about 50 mA. The signal which modulates the laser is a current added to a bias current which brings the laser near (or sometimes slightly above) the threshold for laser operation. The drive circuitry must be able to respond rapidly enough for the data rate of the signal.

InGaAsP diode lasers are very vulnerable to damage from voltage and current surges because of their low operating voltages and fast rise times. Thus careful control of drive power is essential.

Cooling Requirements. InGaAsP lasers normally operate at or near room temperature. All require heat sinking to remove excess heat generated during operation. Thermoelectric coolers are often used to stabilize temperature because both output wavelength and power vary significantly with temperature. Some high-performance fiber-optic lasers are supplied with integral thermoelectric coolers.

Consumables. Diode lasers are self-contained and need no consumables.

Required Accessories. A diode laser must be packaged to be useful. Some InGaAsP lasers are sold mounted on *chip carriers*, but most are housed in packages which mate to fiber-optic connectors, couple light out through standard fiber pigtails, or are similar to standard transistor cases.

Figure 20.4 shows some standard packages for long-wavelength semiconductor lasers. Virtually all packages (excluding chip carriers) include several key elements:

- A housing to simplify handling of the tiny chip
- A heat sink to help remove excess heat, or for connection to a thermoelectric cooler
- Optics or an optical fiber to direct the beam
- Electrical connections between the chip and the outside world
- A photodiode which monitors output from the rear facet of the laser, allowing feedback-loop control of laser output

Some InGaAsP laser packages may include other components, such as a thermoelectric cooler to stabilize operating temperature or electronic drive circuitry to power the laser, most likely to be found in units packaged for use as fiber-optic transmitters.

Diode lasers are vulnerable to damage from current surges, so care must be taken in selecting drive electronics for lasers which come

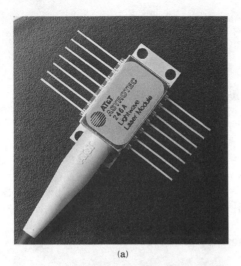

(a)

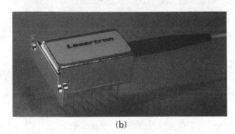

(b)

Figure 20.4 Two standard pack-ages for semiconductor lasers. *(a) (Courtesy of AT&T.) (b) (Cour-tesy of Lasertron Inc.)*

without them, and in avoiding exposing the lasers to electrostatic dis-charges. Thermoelectric coolers may be needed to stabilize operating temperature in sensitive applications.

Operating Conditions and Temperature. Long-wavelength III–V semi-conductor lasers usually operate at room temperature. Specified oper-ating temperatures range from 10 to 30°C to −40 to +70°C.

Laser power and wavelength vary significantly with temperature, so temperature stabilization may be needed. InGaAsP lasers are more sensitive than GaAlAs lasers to increases in threshold current and de-creases in operating efficiency caused by higher temperatures. Oper-ation at stable lower temperatures can extend expected lifetime.

InGaAsP lasers need no warm-up, and respond to direct electrical modulation with rise times of a nanosecond or less. Only a few high-power lasers are limited in duty cycle to prevent excessive heating.

Mechanical Considerations. InGaAsP laser chips are small, generally less than a millimeter in all dimensions. They usually are packaged in standard electronic packages, so they can be readily incorporated into fiber-optic transmitters. The packages provide electrical contacts for the laser, drive circuitry, and monitoring photodiode; a path (usually through an optical fiber or fiber connector) for the laser beam to travel; and a means of removing heat generated by laser operation. Except for packaging limitations, diode lasers should resist normal mechanical shocks.

External-cavity lasers are an exception because the laser cavity includes a collimating lens and reflective grating as well as the laser chip, as shown in Fig. 20.5. Models now available have cavities centimeters long and come in boxes measuring several centimeters on each side. External-cavity lasers are still in development and are not packaged for field use.

Safety. InGaAsP lasers pose no electrical hazards because of their low operating voltages and currents. The eye hazards depend on wavelength, operating power, and beam-handling optics.

The eye transmits the near-infrared wavelengths of InGaAsP lasers only weakly, limiting retinal hazards. Transmission to the retina is only about 10 percent at 1.2 and 1.3 μm and virtually zero at 1.55 μm (Sliney and Wolbarsht, 1980, p. 118). Safety codes recognize this difference, classing wavelengths less than 1.4 μm as retinal hazards, while applying less stringent restrictions to longer wavelengths. Wavelengths longer than 1.4 μm can damage the cornea, but only at higher power levels, and are sometimes called "eye-safe" because they cannot reach the retina. (Some diode laser makers do not appear to recognize this difference, because safety warnings in their sales literature do not differentiate between 1.3- and 1.55-μm lasers.)

Most 1.2- and 1.3-μm lasers emit only a few milliwatts, making

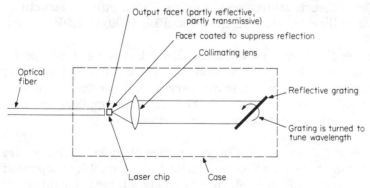

Figure 20.5 An external-cavity diode laser which includes beam expansion lens and tuning grating, in addition to the laser chip itself. In this version, the laser output is coupled directly into an optical fiber.

them fairly safe except for prolonged staring into the beam. An added safety margin comes from the rapid divergence of uncollimated diode laser beams and the fairly rapid divergence of beams emerging from optical fibers. In practice, InGaAsP laser beams are rarely collimated except when they are being coupled into optical fibers, so stray beams are unlikely. Thus hazardous exposures are unlikely except close to the output aperture or fiber end, and even there the hazards would come from long-term exposure. Note, however, that some 1.3-μm lasers emit tens of milliwatts or more, and safety goggles are advisable if their output is not contained entirely within optical fibers. The same considerations apply to 1.55-μm lasers, although that wavelength gives added safety margins.

Fluorescent screens or infrared viewers can render near-infrared beams visible to the eye, letting users align them in optical systems without removing infrared-blocking safety goggles.

Special Considerations.   Laser diodes are very vulnerable to damage from surge currents, so care should be taken to avoid exposing them to electrostatic discharge during handling. Precautions include grounded wrist straps for users, grounded work surfaces, and antistatic floors.

Reliability and Maintenance

Lifetime.   InGaAsP lasers tend to be longer-lived than GaAlAs diode lasers and are less subject to important degradation mechanisms. For example, the power levels needed to cause catastrophic degradation of the laser facet are about a factor of 10 higher in InGaAsP (Dutta and Zipfel, 1988). Many failures occur early, and devices likely to fail can be identified by burn-in tests at the factory. Most manufacturers do not quote lifetimes, but researchers estimate that InGaAsP lasers with buried-heterostructure and similar index-guided structures should have million-hour lifetimes (Yu and Li, 1990, p. 5.56).

Maintenance and Adjustments Needed.   Diode lasers need no special maintenance except for operation under proper conditions (including temperature), avoidance of static discharges, and careful cleaning of exposed optical surfaces if needed. However, some applications may require stabilization of temperature and output power.

Failure Modes and Causes.   The failure modes of InGaAsP lasers are similar to those of GaAlAs lasers described in Chap. 19, except that catastrophic facet damage is much less likely. "Infant mortality" of freshly made devices appears to be higher in InGaAsP, particularly at 1.55 μm, because of difficulties in making the material, but these can be caught by manufacturer screening. Most premature failures not caused by factory defects can be avoided by care in handling and operation.

Possible Repairs. If a diode laser fails, any operational problems which caused the failure must be corrected and the laser replaced.

Commercial Devices

Standard Configurations. Long-wavelength III–V diode lasers are sold both in general-purpose configurations and in forms designed specifically for fiber-optic applications. The most important types include

- General-purpose lasers in transistor-type cans
- Lasers delivering up to several milliwatts in packages with integral fiber pigtails
- Lasers delivering up to several milliwatts in packages with integral fiber connectors
- Dual in-line packages and similar modules for mounting in fiber-optic transmitters
- Chips mounted on carriers with electrical contacts, for incorporation into other equipment
- Lasers delivering up to 100 mW (often pulsed) for fiber-optic measurement systems
- Packaged external-cavity lasers, with tunable narrow-line output
- Semiconductor laser amplifiers at 1300 and 1550 nm, with facets coated to suppress reflection and laser oscillation

Options. The major options with commercial InGaAsP lasers are in the choice of packaging.

Pricing. Typical prices of InGaAsP lasers range from a few hundred dollars for 1300-nm Fabry-Perot lasers delivering a few milliwatts to several thousand dollars for the most sophisticated lasers, such as distributed-feedback 1550-nm lasers or external-cavity lasers. These prices are far above those of GaAlAs lasers for compact disks, reflecting the smaller scale of production, the need for very precise packaging to couple laser output into optical fibers, and higher production costs.

Suppliers. The market for InGaAsP lasers is competitive, with important suppliers in Japan, the United States, and Europe. Virtually all companies offer 1.3-μm Fabry-Perot lasers. Smaller numbers offer distributed-feedback lasers at 1.3 or 1.55 μm or Fabry-Perot lasers emitting at 1.2 and 1.55 μm. A number of companies purchase lasers from chip manufacturers and package them into forms useful in the laboratory or for certain applications in communications.

A few state-of-the-art products are available from at most a handful of suppliers. At this writing, this includes semiconductor laser amplifiers, external-cavity lasers, and 1.48-μm InGaAsP lasers for pumping erbium-doped fiber amplifiers. Those markets are small, and so far exclusively in research and development applications.

Applications

The only important applications of 1.2-, 1.3-, and 1.55-μm semiconductor lasers are in fiber-optic communications. Low-power lasers are used as signal sources; higher-power lasers are used in measurement equipment which sends pulses down a fiber-optic cable to test its optical characteristics.

The 1980s brought widespread acceptance of fiber-optic transmission between telephone switching offices and for backbone networks that spanned the globe, including submarine cables and transcontinental cables. These systems have been built with single-mode optical fiber having zero dispersion at 1.3 μm and laser sources operating at that wavelength.

Fiber attenuation is lower at 1.55 μm, but standard single-mode fibers have large wavelength dispersion at that wavelength, which can severely limit data rates possible with Fabry-Perot lasers. One way to limit the dispersion is with a laser that emits a narrow range of wavelengths, such as a distributed-feedback laser. Another is to use a different type of fiber with zero-dispersion wavelength shifted to 1.55 μm, and some new systems are being planned using such fibers. A strong push for development of 1.55-μm distributed-feedback lasers comes from the large number of installed cables with standard single-mode fiber. Interest in increasing the capacity of existing fiber systems also stimulated development of 1.3-μm distributed-feedback lasers.

Another way to increase the capacity of existing systems is by wavelength-division multiplexing, a technique which transmits separate signals at different wavelengths through the same fiber. For example, 1.2- and 1.55-μm lasers might be added to a system already operating at 1.3 μm.

In the long term, developers are trying to expand fiber-optic technology in two different directions—on one hand expanding transmission capacities and distances and on the other, developing lower-cost networks to carry advanced telecommunication services into homes and offices. High-capacity technology in development includes coherent transmission, which requires narrow-line lasers such as external-cavity types, and dense wavelength-division multiplexing, which requires narrow-line lasers which should be tunable in wavelength. High-capacity systems also require optical amplifiers to stretch transmission distances, including both semiconductor laser types and erbium-doped amplifiers, which are pumped with diode lasers emitting at 980 or 1480 nm. Extension of fiber technology to homes on a broad basis will require either significant reductions in laser prices or development of new architectures which reduce the need for active components such as lasers in people's homes. Optical amplifiers also may play an important role in allowing distribution of signals to many subscribers.

Bibliography

G. P. Agrawal and N. K. Dutta: *Long-wavelength Semiconductor Lasers*, Van Nostrand Reinhold, New York, 1986 (monograph).

J. E. Bowers and M. A. Pollack: "Semiconductor lasers for telecommunications," in Stewart E. Miller and Ivan P. Kaminow (eds.), *Optical Fiber Telecommunications II*, Academic Press, Boston, 1988, pp. 509–568 (review paper).

Y. K. Chen et al.: "Generation of subpicosecond transform-limited optical pulses for optical fiber communications," paper TuN4 in *Technical Digest, Optical Fiber Communications Conference*, San Diego, February 1991, Optical Society of America, Washington, D.C. (research report).

A. E. Drakin et al.: "InGaSbAs injection lasers," *IEEE Journal of Quantum Electronics* QE-23(6):1089–1094, 1987 (research report).

N. K. Dutta and C. L. Zipfel: "Reliability of lasers and LEDs," in Stewart E. Miller and Ivan P. Kaminow (eds.), *Optical Fiber Telecommunications II*, Academic Press, Boston, 1988, pp. 671–687 (review paper).

S. J. Eglash and H. K. Choi: "Efficient GaInAsSb/AlGaAsSb diode lasers emitting at 2.29 micrometers," *Applied Physics Letters* 57(13):1292–1294, September 24, 1990 (research report).

J. Eom, C. B. Su, and R. B. Lauer: "Observed differences in relationship of damping rates to resonance frequencies for Fabry-Perot and DFB lasers: implications for laser dynamics and bandwidth," paper CLF1 in *Technical Digest, Conference on Lasers and Electro-Optics*, Anaheim, Calif., May 21–25, 1990, Optical Society of America, Washington, D.C. (research report).

Luis Figueroa: "High-power semiconductor lasers," in Peter K. Cheo (ed.), *Handbook of Solid-State Lasers*, Marcel Dekker, New York, 1989, pp. 113–226 (review of high-power semiconductor lasers).

Luis Figueroa: "Semiconductor lasers," in Kai Chang (ed.), *Handbook of Microwave and Optical Components*, vol. 3, *Optical Components*, Wiley-Interscience, New York, 1990 (review of semiconductor lasers, concentrating on high-power types).

Yoshiji Horikoshi: "Semiconductor lasers with wavelengths exceeding 2 μm," in W. T. Tsang (ed.), *Semiconductors and Semimetals*, vol. 22, *Lightwave Communications Technology, Part C, Semiconductor Injection Lasers II, Light Emitting Diodes*, Academic Press, Orlando, Fla., 1985, pp. 93–151 (review article).

Syoichi Kakimoto et al.: "High output power and high temperature operation of 1.5 μm DFB-PPIBH laser diodes," *IEEE Journal of Quantum Electronics* 25(6):1288–1293, 1989 (research report).

Syoichi Kakimoto et al.: "Wavelength dependence of characteristics of 1.2–1.55 μm InGaAsP/InP P-substrate buried crescent laser diodes," *IEEE Journal of Quantum Electronics* 23(1):29–35, 1988 (research report).

Eli Kapon: "Semiconductor diode lasers: Theory and Techniques," in Peter K. Cheo (ed.), *Handbook of Solid-State Lasers*, Marcel Dekker, New York, 1989, pp. 1–112 (review paper).

T. Katsuyama and H. Matsumura: *Infrared Optical Fibers*, Adam Hilger, Bristol (U.K.) and Philadelphia, 1989 (review of nonsilica fibers).

Thomas L. Koch and Uziel Koren: "Semiconductor lasers for coherent optical fiber communications," *Journal of Lightwave Technology* 8(3):274–293, 1990 (review paper).

A. Lidgard et al.: "External-cavity InGaAs/InP graded index multiquantum well laser with a 200 nm tuning range," *Applied Physics Letters* 56(9):816–817, February 26, 1990 (research report).

Dietrich Marcuse: "Selected topics in the theory of telecommunications fibers," in Stewart E. Miller and Ivan P. Kaminow (eds.), *Optical Fiber Telecommunications II*, Academic Press, Boston, 1988, pp. 55–119 (see discussion of solitons on pp. 90–98).

Ramon U. Martinelli and Thomas J. Zamerowski: "InGaAs/InAsPSb diode lasers with output wavelengths at 2.52 μm," *Applied Physics Letters* 56(2):125–127, January 8, 1990 (research report).

L. F. Mollenauer et al.: "Experimental study of soliton transmission over more than 10,000 km in dispersion-shifted fiber," *Optics Letters* 15(21):1203–1205, November 1, 1990 (research report).

R. J. Nelson and N. K. Dutta: "Review of InGaAsP/InP laser structures and comparison of their performance," in W. T. Tsang (ed.), *Semiconductors and Semimetals*, vol. 22, *Lightwave Communications Technology, Part C, Semiconductor Injection Lasers II, Light Emitting Diodes*, Academic Press, Orlando, Fla., 1985, pp. 1–59 (review article).

Saeko Oshiba and Yasuaki Tamura: "Recent progress in high-power GaInAsP lasers," *Journal of Lightwave Technology* 8(9):1350–1356, September 1990 (review paper).

Saeko Oshiba et al.: "High-power output over 200 mW of 1.3 μm GaInAsP VIPS lasers," *IEEE Journal of Quantum Electronics* QE-23(6):738–743, 1987 (review paper).

David Sliney and Myron Wolbarsht: *Safety with Lasers and Other Optical Sources*, Plenum, New York, 1980 (comprehensive handbook).

T. Tanbun-Ek et al.: "High power output 1.48–1.51 μm continuously graded index separate confinement strained quantum well lasers," *Applied Physics Letters* 57(3):224–226, July 16, 1990 (research report).

H. Temkin et al.: "In GaAs/InP quantum well lasers with sub-mA threshold current," *Applied Physics Letters* 57(16):1610–1612, October 15, 1990 (research report).

P. J. A. Thijs, T. van Dongen, and B. H. Verbeek, paper WJ2 in *Technical Digest, Conference on Optical Fiber Communication*, San Francisco, January 22–26, 1990, Optical Society of America, Washington, D.C. (research report).

W. T. Tsang et al.: "Low threshold and high power output 1.5 μm InGaAs/InGaAsP separate confinement multiple quantum well laser grown by chemical beam epitaxy," *Applied Physics Letters* 57(20):2065–2067, November 12, 1990 (research report).

K. Uomi et al.: "Ultralow chirp and high-speed 1.55-μm multiquantum well quarter-wave shifted DFB lasers," *IEEE Photonics Technology Letters* 2(4):229–230, 1990 (research report).

K. Uomi, H. Nakano, and N. Chinone: "Ultrahigh-speed 1.55 μm multiquantum well quarter-wave shifted DFB PIQ-BH lasers with bandwidth of 17 GHz," *Electronics Letters* 25:668–669, 1989 (research report).

Hiroyuki Yamazaki et al.: "250 kHz linewidth operation in long cavity 1.5 μm multiple quantum well DFB-LDs with reduced linewidth enhancement factor," Postdeadline Paper 33 at *Optical Fiber Communication Conference*, San Francisco, January 22–26, 1990, *Postdeadline Paper Summaries*, Optical Society of America (research report).

Paul Kit Lai Yu and Kenneth Li: "Optical sources for fibers," in Frederick C. Allard (ed.), *Fiber Optics Handbook for Engineers and Scientists*, McGraw-Hill, New York, 1990, chap. 5 (overview).

Lead Salt Semiconductor Diode Lasers

A third family of semiconductor diode lasers emit at longer wavelengths than GaAs- or InP-based lasers and have quite distinct applications. They are called *lead salt* lasers because many (but not all) of the semiconductor compounds contain lead. They emit wavelengths between about 3.3 and 29 micrometers (μm), and most of their applications are in infrared spectroscopy.

Jack F. Butler and colleagues at the Massachusetts Institute of Technology (MIT) Lincoln Laboratory made the first lead salt diode laser of lead telluride in 1963 (Butler et al., 1964). That demonstration came only a year after the first diode lasers of any kind were reported, but for many years it stimulated little interest because of its long wavelength. Fiberoptic and military system needs for compact near-infrared sources instead pushed development of III–V diode lasers.

Interest in high-resolution infrared spectroscopy, which developed during the 1970s, has been the major impetus behind lead salt laser development. Most work has focused on spectroscopic applications, including measurement and monitoring of various compounds, but there has been some interest in long-wavelength fiber-optic communications and the testing of infrared detectors. Serious technological limits remain, however, including the need for cryogenic cooling and limited output powers from narrow-line lasers. Performance has been improved by adapting technology originally developed for shorter-wavelength diode lasers, and continuing work in that area should pay more dividends.

Internal Workings

Like III–V semiconductor lasers, lead salt lasers emit light when current carriers recombine at the junction of p- and n-type semiconductors. The basic physics are described in Chap. 18, but like III–V diode lasers, lead salt lasers have their own special characteristics.

The most obvious difference is in the materials which make up the diode. Lead salt lasers consist of elements with two or six valence electrons, drawn mostly from columns IVB and VI of the periodic table, including lead, tin, europium, sulfur, selenium, and tellurium; III–V materials consist of electrons with three or five valence electrons. Most lead salt lasers are composed of ternary compounds such as lead-tin telluride ($Pb_xSn_{1-x}Te$) or quaternary compounds such as $Pb_xEu_{1-x}Se_yTe_{1-y}$, in which emission wavelength depends on the proportions of the elements. The combination of valence states 2 and 6 in lead salt lasers leads to bonds somewhat more ionic than in III–V compounds and gives lead salt lasers a rock salt crystalline structure.

The gap between conduction and valence bands in lead salt lasers is much smaller in lead salt lasers than in III–V materials, less than half an electron volt. This makes the wavelengths of lead salt lasers much longer, from 3.3 to 29 μm. The nominal center wavelength of each laser depends on its composition.

Like other semiconductor lasers, the wavelength of lead salt lasers changes with temperature. While such shifts usually are a nuisance at shorter wavelengths, they can be valuable at the midinfrared wavelengths of lead salt lasers. Molecular spectra are particularly complex in the infrared, and the ability to generate a narrow laser line and tune its wavelength can be very valuable to spectroscopists. Thus lead salt lasers are designed to render their wavelengths temperature-sensitive and are packaged so that their output can be temperature-tuned.

Internal Structure

Like other diode lasers, lead salt diode lasers are tiny chips of semiconductor, with electrical and optical properties dependent on internal structure. Lead salt lasers typically have cavities 250 μm long, and the chips themselves are about 250 μm on each side.

Developers of lead salt lasers have adapted some structures and fabrication techniques originally developed for shorter-wavelength lasers, as mentioned in Chap. 18. Commercial lasers now use narrow-stripe double-heterostructure designs, which have replaced earlier broad-area lasers. One example is the structure shown in Fig. 21.1, a diffused junction laser with a 20-μm-wide mesa isolated by a pair of grooves etched into the chip. Another is the buried-heterostructure laser shown in Fig. 21.2, fabricated by molecular beam expitaxy with a stripe 4 μm wide (Feit et al., 1990), which is used at shorter wavelengths. Other structures have been used to make lead salt lasers in

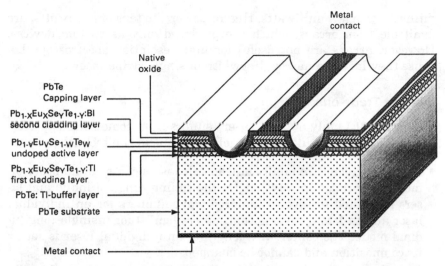

Figure 21.1 Double-channel PbEuSeTe laser with 20-μm mesa. (*Courtesy of Laser Photonics Inc.*)

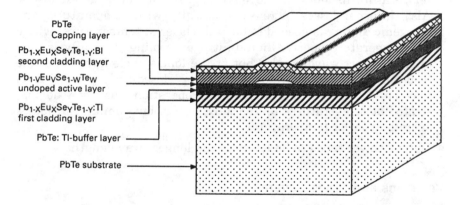

Figure 21.2 Buried-heterostructure PbEuSeTe laser with 4-μm stripe. (*Courtesy of Laser Photonics Inc.*)

the laboratory but are not used in standard products, including distributed-feedback, distributed-Bragg-reflection, and cleaved-coupled-cavity (C^3) lasers.

The cavity lengths of lead salt lasers are so short that the resonance lines are quite narrow in a Fabry-Perot cavity. Linewidths as narrow as 22 kilohertz (kHz) have been measured with Fabry-Perot cavities (Freed et al., 1983). The inherently narrow linewidths are valuable for spectroscopic applications.

Output powers of lead salt lasers typically are a fraction of a milliwatt to a few milliwatts. Higher powers, to tens of milliwatts, are available from arrays, which are produced only as custom devices. However, arrays are not useful for high-resolution spectroscopy because they lack the narrow linewidth of single-stripe lasers.

Inherent Trade-offs

The lead salt family of diode lasers suffers from some serious performance limitations which are largely inherent in the technology.

- Although the entire family of lead salt lasers can span an order-of-magnitude range in wavelength, the tuning range of individual lasers is limited to several hundred wavenumbers for an individual laser and to 1 to 3 inverse centimeters (cm^{-1}) for a single longitudinal mode. The center wavelength of an individual laser is set by its composition and cannot be changed.

- As wavelength increases, cooling requirements become increasingly stringent. One reason is that blackbody emission is strong at room temperature in much of the wavelength range of lead salt lasers. Also, internal losses increase drastically with temperature, and cryogenic cooling is needed to keep the laser at temperatures where it can operate properly. Liquid-nitrogen cooling generally is adequate at short wavelengths, but at the long-wavelength end of the lead salt laser range, liquid-helium or closed-cycle refrigeration may be necessary. Continuous-wave operation has been demonstrated at temperatures to 195 kelvins (K); but commercial lasers operate at lower temperatures.

- Output power tends to be lower at the longest wavelengths.

Variations and Types Covered

The basic variable in the lead salt family of semiconductor lasers is chemical composition, which determines the central emission wavelength. The basic compound consists of elements from two groups: the metals lead and tin and the nonmetals (often called *chalcogenides*) sulfur, selenium, and tellurium. The rare earth metals europium and ytterbium can be incorporated in lead salt lasers. The compounds that are commercially most important are PbEuSeTe, PbSSe, PbSnTe, and PbSnSe, which are shown in Fig. 21.3 with their respective wavelength ranges.

Beam Characteristics

Wavelength and Output Power. Center wavelengths of lead salt lasers range from 3.3 to 29 μm. Infrared spectroscopists, the main users of

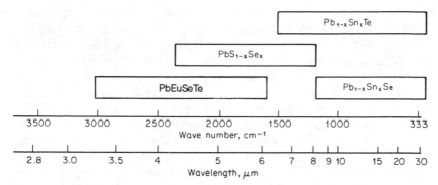

Figure 21.3 Wavelength ranges for lead salt lasers of various compositions.

lead salt lasers, typically measure this range in wavenumber or in-verse centimeters, and in those terms the range is about 3000 to 350 cm^{-1}. Individual lead salt lasers can be tuned over 50 to 200 cm^{-1}. Typical operating ranges for the major compositions offered commer-cially are shown in Fig. 21.3. Each laser is tunable over only a small part of the entire range for each compound; the entire range is covered by many lasers of different compositions.

Laser wavelength can be tuned by adjusting hydrostatic pressure, magnetic field, or temperature, all of which affect the laser's effective cavity length. Temperature tuning is the most straightforward and the most common approach. Coarse tuning is by adjusting diode tem-perature directly in the range 15 to 120 K. Fine-tuning relies on mak-ing small changes in drive current, which indirectly changes device temperature because an increase in current increases heat dissipated within the crystal.

In practice, tuning is complicated by hopping between different lon-gitudinal modes, 1 to 3 cm^{-1} apart and each about 0.0003 cm^{-1} wide. The gain curve of a lead salt laser allows oscillation on more than one longitudinal mode, but normally a single mode dominates. As temper-ature changes, it causes both the peak of the gain curve and the wave-lengths of the longitudinal modes of the Fabry-Perot cavity to shift, but at different rates. Eventually, the laser shifts to another dominant longitudinal mode, causing an abrupt change in wavelength. For com-mercial lasers, the tuning range of an individual longitudinal mode is typically 1 cm^{-1}; broader ranges have been demonstrated in the lab-oratory.

Output power from commercial lead salt diode lasers ranges from a fraction of a milliwatt to a few milliwatts, with powers limited at the longest wavelengths. Tens of milliwatts can be produced by custom multistripe arrays, at the cost of broader spectral width, and array powers have reached 100 mW (Linden, 1984).

Efficiency. Typically the energy in the drive current which powers commercial lead salt diode lasers is converted to laser output with an efficiency well under 1 percent. The efficiency of the laser itself is a minor consideration in practice because most of the energy is used by the system that cools the semiconductor to the desired operating temperature. While active cooling removes any excess heat up to reasonable power levels, it also consumes much more power than the laser itself. (Liquid-nitrogen dewars can be used for some lead salt lasers, avoiding the need for active coolers.) The main interest in improving the efficiency of lead salt lasers is not to reduce power consumption per se, but rather to make lasers capable of operating at high temperatures.

Temporal Characteristics. Like other diode lasers, lead salt types rapidly respond to changes in electric drive current and can produce pulsed or continuous output. Pulses can range from nanoseconds to much longer durations.

In practice, lead salt lasers are operated to meet the requirements of spectroscopic applications, which discourage rapid pulsing because it could distort measurement results. For spectroscopic measurements, a slow modulation of the drive current can be superimposed on a steady bias current to sweep the laser output over a range of wavelengths. Commercial instruments offer such modulation as a standard feature at rates of 50 Hz to 10 kHz. Intensity modulation is incidental to the wavelength shift, and spectroscopic instruments must accommodate changes in output power. Tunable lead salt lasers may be modulated at megahertz frequencies for some spectroscopic applications.

Modulation of the intensity of lead salt lasers for spectroscopic measurements normally is done externally with a chopper, to avoid drive-current-induced changes in wavelength.

Spectral Bandwidth. Lead salt lasers normally emit in a few narrow longitudinal modes which lie within the gain curve. Typically each longitudinal mode is 0.0003 cm^{-1} (9 MHz) wide, and mode spacing is 1 to 3 cm^{-1}. Buried-heterostructure lasers developed recently operate in a single longitudinal mode at drive currents 1.2 to 4.5 times the threshold current (Feit et al., 1989).

For high-resolution measurements with multimode lasers, a monochromator can separate longitudinal modes, so all but the desired wavelength can be filtered out, at the cost of reducing total power.

Frequency Stability. The seemingly smooth spectral tuning curve of a multimode lead salt laser hides more complex characteristics, as shown in Fig. 21.4. The tuning process changes the peak emission wavelength faster than the wavelength from any individual mode. As current increases, emission from a given mode first increases, then reaches a peak, and finally decreases, all while shifting gradually in wavelength. This can cause mode-switching instabilities.

Otherwise, frequency stability of lead salt lasers is excellent, with the main concern being stability of the temperature controller and drive-current regulator. Frequency stability of a packaged laser source spectrometer is specified at 0.0003 cm^{-1} for 1-second intervals, and 0.001 cm^{-1} over 30 minutes.

Amplitude Noise. Because the main applications of lead salt lasers are in spectroscopy, amplitude stability is less important than frequency

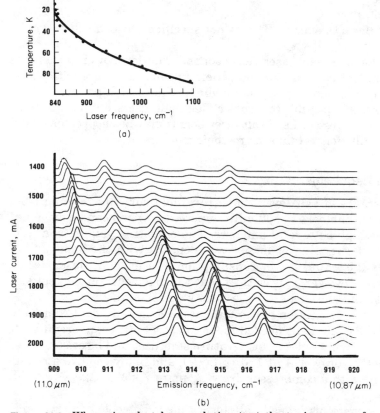

Figure 21.4 When viewed at low resolution (*top*) the tuning curve of a multimode lead salt laser looks smooth. However, a closer look (*bottom*) shows that the distribution of energy among modes is changing along with the wavelengths of the modes themselves. (*Courtesy of Laser Photonics Inc.*)

stability. However, as long as the laser is not near a mode-hopping point, amplitude stability is good: specified values are ±0.3 percent for 1 second and ±2 percent for 30 minutes.

Beam Quality and Polarization. Lead salt lasers emit a broad, rapidly diverging beam that may have multiple lobes. Typically a short-focus lens or mirror near the laser collects the output beam and collimates it before it can spread to a large diameter. Linear beam polarization typically is greater than 80 percent.

Beam Diameter and Divergence. Like other diode lasers, lead salt lasers have small light-emitting regions and short cavities, leading to large beam divergence. In practical instruments, a mirror or lens collimates the beam, producing a cone of light with less divergence than an $f/1$ lens.

Stability of Beam Direction. This is not specified for lead salt lasers.

Suitability for Use with Laser Accessories. The low-power continuous output from lead salt lasers makes them unsuitable for use with accessories such as harmonic generators which require high powers. It is at least conceptually possible to modelock lead salt lasers, but no one seems particularly interested. Lead salt laser beams may be chopped or modulated externally for certain spectroscopic measurements.

Operating Requirements

Input Power. Lead salt lasers have threshold currents ranging from under one milliampere to 500 milliamperes (mA) and can be driven at currents as high as several amperes. The operating voltage is a fraction of a volt, dropping at longer wavelengths with the bandgap energy. Thus the actual power consumption of a lead salt laser is less than a watt. However, it is critical that the drive current be kept stable because fluctuations alter the laser's wavelength.

The power consumption of the laser itself is insignificant compared to that of the equipment needed for it to operate. A commercial laser source assembly includes a temperature stabilizer and laser controller which together draw 1 A at 110 volts (V), and many require a refrigeration unit with a compressor drawing up to 2 kW at 220 V.

Cooling Requirements. Lead salt lasers must be cooled to cryogenic temperatures to stabilize operating wavelength, control heat dissipation in the laser, and reduce blackbody emission in the laser wavelength band. Dewars filled with liquid helium were used in early experiments, but now spectroscopic systems rely on closed-cycle mechanical refrigerators or dewars of liquid nitrogen. The coolers can operate between 12 K and room temperature and can be stabilized to 0.0003 K; in practice they are used for lasers which operate between 20 and 60 K. Liquid-nitrogen cooling is much simpler, and is used for lasers designed to operate between 80 and 120 K.

Consumables. No consumables are needed to operate lead salt lasers unless open-cycle or liquid-nitrogen cooling is used.

Required Accessories. Lead salt lasers are usable only when incorporated into a system or subsystem because of the many accessories they require. These include

- Cooling equipment
- Temperature controller
- Driver with current stabilization
- Shutter or chopper for modulation
- Collimating optics
- Current sweep generator
- Reference and sample cells
- Detection and measurement equipment

All this equipment is not needed for every application, but some of it—cooling, temperature control, and driver—is needed in all cases.

Operating Conditions and Temperature. Lead salt laser instruments are designed for normal laboratory conditions, although the lasers themselves must be operated well below room temperature. (The lasers can be stored at room temperature without significant degradation.) Some instruments have been operated under field conditions or in balloons for atmospheric measurements. The need for low-temperature operation leads to a cooldown delay before the laser can be used.

Mechanical Considerations. Unpackaged lead salt lasers are a fraction of a millimeter in all dimensions, but normally the lasers are mounted on a package a couple of centimeters long and about a centimeter high and wide. This package, in turn, is mounted in a benchtop laboratory instrument and requires either a liquid-nitrogen dewar or a 70-kilogram (kg) refrigeration unit for cooling.

Safety.   Lead salt lasers are innocuous. Their wavelength does not penetrate the eye, and their low-power output falls into Class 1 of the federal laser safety codes, which requires no special precautions or warning labels. No high voltages are needed, and the cold temperatures of the compact laser unit are safely inside the system.

Reliability and Maintenance

Lifetime.   Cryogenic temperatures are a benign environment for diode lasers, which should have a long lifetime if used as recommended. However, excessive currents could damage the laser. It is reasonable to expect the laser to outlive some other system components, such as the refrigeration subsystem, typically specified as capable of operating for 10,000 hours (1 year) without maintenance.

Maintenance and Adjustments Needed.   Cooling systems require periodic attention, and liquid-nitrogen dewars require replenishment. The lasers themselves are maintenance-free.

Mechanical Durability.   Lead salt laser instruments are designed for normal laboratory environments but can be ruggedized for field use such as airborne or balloon systems. The lasers themselves are durable.

Failure Modes and Causes and Possible Repairs.   Failure of a lead salt laser usually is caused by progressive changes in device resistance, leading to irreparable degradation of the semiconductor device. However, the laser is far from the most expensive system component and can be replaced easily. Cryogenic refrigeration systems rely on the compression and expansion of helium; leaks are possible but not too common. Most refrigerators run for about 15,000 hours without problems between rebuilds.

Commercial Devices

Standard Configurations.   Lead salt lasers normally are offered in three forms:

- As laser modules, packaged lasers which require cooling and other accessories

- As "laser source assemblies," which contain the laser, cooling equipment, electronic driver, and stabilization circuits for temperature control and drive electronics

- As part of complete systems such as laser spectrometers, which include the laser source assembly, plus sample cells, optics, chopper, monochromator, detector assembly, and possibly signal-processing equipment

Designs are modular, and the laser modules are designed to plug into the source assemblies, which in turn can plug into spectrometers. Because each laser has only limited wavelength coverage, several lasers may be used with one source assembly or spectrometer. Generally the systems can accommodate four lasers, which can be switched as the operating wavelength changes. Other lasers can be kept in reserve for broader wavelength coverage.

Options. The choice of options for a lead salt laser depends on how the laser is purchased. For a laser source assembly, collimating optics and certain controls are optional. For a spectrometer, optics, measurement hardware, and signal-processing capabilities are options. In practice, the choice of individual lasers is functionally an option because specific lasers must be chosen to cover the desired wavelengths.

Special Notes. Lead salt lasers emitting near 16 μm are subject to special export controls because they can be used to measure the proportions of different uranium isotopes in uranium hexafluoride. This capability is important in uranium enrichment, making the technology sensitive from the standpoint of nuclear proliferation, and hence subject to stringent export controls.

Pricing. Standard lead salt lasers currently have list prices of about $2000 to $2600, with the more expensive types capable of higher-temperature operation. Laser source assemblies with refrigeration systems are about $53,000; with liquid-nitrogen dewar cooling the assemblies are only $22,000. Detector assemblies also can be expensive, with prices in the $15,000 range. Complete spectroscopic systems, including a variety of equipment for general-purpose laboratory use, can run $100,000 or more.

Suppliers. The small market for lead salt lasers and systems built around them has long been dominated by a single company in the United States, which has some competition from overseas.

Applications

High-resolution infrared spectroscopy is the only commercially important application of lead salt diode lasers. Figure 21.5 compares the resolution of a lead salt laser system with that of a conventional in-

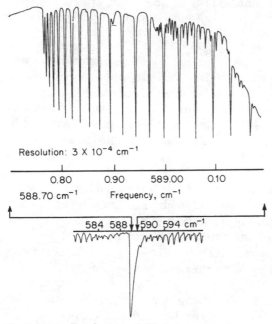

Figure 21.5 Resolution of diode-laser spectroscopy (*top*) compared with that of a conventional infrared spectrometer (*bottom*). (*Courtesy of Laser Photonics Inc.*)

frared spectrometer. Such high resolution is not needed for all spectroscopic problems, and sometimes may supply far more detail than the user needs, wants, or can interpret. Most uses of high-resolution lead salt laser spectroscopy remain in the laboratory, but a few are in industrial settings with demanding requirements for process control or other measurements. Examples include measurement of water content in integrated-circuit packages, monitoring plastics, checking halogen-lamp quality, measuring automotive emissions, or monitoring hydrogen fluoride content in UF_6. Tunable diode lasers are promising for applications in environmental monitoring and medical diagnosis by studying the ratio of carbon-13 to carbon-12 in a person's breath. Lead salt lasers may find other applications if ultra-low-loss mid-infrared fibers are developed successfully, but formidable technical problems remain.

Bibliography

G. P. Agrawal and N. K. Dutta: *Long-Wavelength Semiconductor Lasers*, Van Nostrand Reinhold, New York, 1986, pp. 432–460 (review).

J. F. Butler et al.: "PbTe diode laser," *Applied Physics Letters* 5:75, 1964 (research report).

R. S. Eng, J. F. Butler, and K. J. Linden: "Tunable diode laser spectroscopy: An invited review," *Optical Engineering* 19:945–960, November–December 1980 (review).

Z. Feit et al.: "PbEuSeTe buried heterostructure lasers grown by molecular beam epitaxy," *Journal of Vacuum Science and Technology B* 8(2):200–204, March–April 1990 (research report).

Charles Freed, Joseph W. Bielinsi, and Wayne Lo: "Fundamental linewidth in solitary, ultranarrow output $PbS_{1-x}Se_x$ diode lasers," *Applied Physics Letters* 43:629–631, October 1, 1983 (research report).

Yoshiji Horikoshi: "Semiconductor lasers with wavelengths exceeding 2 μm," in W. T. Tsang (ed.), *Semiconductors and Semimetals*, vol. 22, *Lightwave Communications Technology, Part C, Semiconductor Injection Lasers II, Light Emitting Diodes*, Academic Press, Orlando, Fla., 1985, pp. 93–151 (review article).

A. Katzir et al.: "Tunable lead salt lasers," in Peter K. Cheo (ed.), *Handbook of Solid-State Lasers*, Marcel Dekker, New York, 1989, pp. 227–347 (scholarly review).

Kurt J. Linden: "Diode laser array with high power in the 4–5 μm infrared region," *Optical Engineering* 23(5):685–686, September–October 1984 (research report)

Neodymium Lasers

The neodymium laser is the most common member of a family grouped together as *solid-state* lasers, a term which in the laser world does not encompass semiconductor devices. Qualitatively, neodymium lasers operate much like ruby (the first working lasers) and other solid-state types. Atoms present in impurity-level concentrations—roughly 1 percent—in a crystalline or glass host material are excited optically by light from an external source, producing a population inversion in a rod of the laser material. The rod is mounted in an optical cavity which provides the optical feedback needed for laser action. The result is a simple and versatile laser which has become a standard tool for diverse applications.

Neodymium lasers are more a family of devices than a single type. The neodymium may be incorporated into various host materials, either single crystals or glasses of different compositions. Accessories can shift the output wavelength from the near-infrared into the visible or ultraviolet. The choice of optical pump source—semiconductor laser, pulsed flash lamp, or continuous arc lamp—strongly influences laser characteristics. Neodymium lasers can generate continuous beams of a few milliwatts to over a kilowatt, short pulses with peak powers in the gigawatt range, or pulsed beams with average powers in the kilowatt range. Their applications range from esoteric laboratory research to materials working on industrial production lines. No single laser can do all those tasks, but the broad spectrum of commercial products does cover the entire range.

Active Medium

Strictly speaking, the active medium in a neodymium laser is triply ionized neodymium in a crystal or glass matrix. In a crystal, the neodymium is essentially an impurity, which takes the place of another element of roughly the same ion size (typically yttrium, another rare earth element). The energy-level structure, laser wavelength, and other optical properties of the neodymium ions are influenced by the host material. Table 22.1 summarizes some important characteristics of major host materials; many other hosts have been studied.

Glasses and common crystalline hosts are doped with about 1 percent neodymium by weight, giving a neodymium concentration on the order of 10^{20} atoms per cubic centimeter, considered the optimum for laser action. Crystals also can have neodymium as an integral component of the lattice, such as neodymium pentaphosphate, NdP_5O_{14} (Chinn, 1982), but so far practical problems have offset the theoretical advantages of these materials, and they remain in the laboratory.

The most common host for neodymium is yttrium aluminum garnet (YAG), a synthetic crystal with a garnetlike structure and the chemical formula $Y_3Al_5O_{12}$ that the laser world knows by the acronym YAG. YAG is hard and brittle, and its growth is best characterized as a black art, but it has desirable optical, mechanical, and thermal properties. Although not an ideal laser material, it is the best for many neodymium laser applications. Its strongest advantage is its thermal characteristics, which allow an Nd–YAG laser to produce a good-quality continuous beam, something difficult at room temperature for most other solid-state laser materials.

Many other crystal hosts have been tested for neodymium, but only a handful have proved useful. The best-known alternative to YAG is yttrium lithium fluoride ($YLiF_4$), known by the acronym YLF. It has lower thermal conductivity and is not as hard as YAG, but still dissipates heat well enough to operate continuously at room temperature. YLF exhibits less thermal lensing than YAG, making it more desirable for some applications. The gain cross section of Nd–YLF is half that of Nd–YAG, allowing higher energy storage densities and gener-

TABLE 22.1   Optical Characteristics of Major Nd Hosts for 1-μm Line

Material	Wavelength, nm	Cross section, $\times 10^{-20}$ cm	Linewidth, nm	Lifetime, μs
YLF*	1047	37	—	480
	1053	26		
Phosphate glass	1054	4.0–4.2	28	290–330
GSGG	1061	11	—	222
Silicate glass	1061–1062	2.7–2.9	19–22	340
YAG	1064	34	0.45	244

*YLF is a birefringent material with different refractive indexes for light of different linear polarizations; wavelength depends on the polarization.

ation of higher-energy Q-switched pulses from oscillators. YLF is birefringent, with different refractive indexes for light of different polarizations, so it can oscillate at 1047 or 1053 nanometers (nm), depending on polarization. Both pulsed and continuous Nd–YLF lasers are offered commercially.

Other crystals which have received considerable attention as neodymium hosts are gadolinium scandium gallium garnet (GSGG; $Gd_3Sc_2Ga_3O_{12}$) and yttrium aluminate (YALO; $YAlO_3$, sometimes called YAP). GSGG usually is codoped with chromium, which can absorb pump light and transfer it to neodymium ions, increasing laser efficiency as described below. It is available commercially. YALO offers some advantages over YAG, but they have been difficult to realize in practice. Nd-doped YALO is available commercially at this writing, but Nd–YALO lasers are not.

Problems with crystal growth, including its speed and the uniformity of the material, limit the size of YAG, YLF, and GSGG rods. YAG grows at about 0.5 millimeter per hour (mm/h), so it takes several weeks to produce a typical boule 10 to 15 centimeters (cm) long. In practice, the largest Nd–YAG rods are about 1 cm in diameter and 15 cm long; typical rods are several millimeters in diameter and several centimeters long.

Neodymium-doped glass can be produced in much larger sizes and is the material of choice when high pulse energies are needed. Glass rods can be made tens of centimeters long; disks and slabs which measure tens of centimeters in two directions can be made for use in amplifiers or in slab-geometry lasers. Segmented elliptical disks 46 by 85 cm are used in the final amplifier stage of the Nova fusion laser at the Lawrence Livermore National Laboratory; segmentation along the long axis prevents premature depletion of the population inversion by amplified stimulated emission (Martin et al., 1981).

In addition to being available in larger chunks, glass has other advantages for generating high-energy pulses. It can be doped with higher neodymium concentrations than YAG, up to several percent. It also has a smaller cross section for stimulated emission, so it can store more energy per unit volume than YAG. Glass has a much broader emission bandwidth than crystalline hosts, aiding in production of short pulses and energy storage.

The primary problem with glass is its poor thermal qualities. Its thermal conductivity is much less than that of YAG, and thermal gradients in the glass can degrade beam quality. Average powers to 900 watts (W) have been produced by a glass laser using a moving slab 34 by 37 by 0.5 cm (Basu et al., 1990). However, most glass lasers operate at low repetition rates. One extreme is Livermore's gigantic Nova laser, which produces less than one pulse per hour.

Energy Transfer

All neodymium lasers rely on light from an external source to raise neodymium atoms to an excited energy level. The primary transitions involved are shown in Fig. 22.1. Ground-state neodymium ions absorb most strongly at pump bands near 0.73 and 0.8 micrometers (μm), but also absorb some at other wavelengths. Once in the higher energy level, they decay to a metastable level, producing a population inversion between the $^4F_{3/2}$ and $^4I_{11/2}$ states. The latter level decays by a fast nonradiative process to the ground state. Neodymium is a four-level laser and, as theory predicts, it is more efficient than the three-layer ruby system described in Chap. 23, and can operate continuous-wave.

Neodymium laser materials can be codoped with other ions which also absorb pump energy and then transfer it to the neodymium ions. The effect is to absorb more pump energy, thus potentially enhancing laser efficiency. This technique has been studied extensively in the laboratory, but its only practical use is in Nd,Cr–GSGG, which is codoped with chromium.

The strongest neodymium line is near 1.06 μm, with precise wavelength dependent on the host, as indicated in Table 22.1. If oscillation on this line is suppressed, energy-level splitting allows weaker emission on over a dozen other lines, which are not shown in Fig. 22.1. Most are seldom used because their wavelengths are close to the primary line; the only important exception is at 1.3 μm. The actual energy-level kinetics are quite complex.

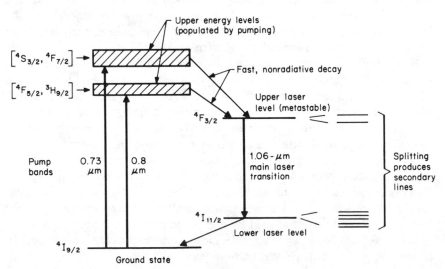

Figure 22.1 Energy levels in Nd–YAG, simplified.

Until the mid-1980s, all commercial neodymium lasers were pumped by broad-spectrum light sources, continuous tungsten arc lamps, or pulsed flashlamps. These sources offer high intensity, but much of their emission is not absorbed by the neodymium ions. A much more efficient approach is to excite the 0.8-μm pump band with a GaAlAs semiconductor laser (see Chap. 19), although output power cannot yet approach lamp-pumped levels.

For lamp pumping, hollow reflective cavities such as shown in Fig. 22.2 transfer pump light from one or more pump lamps to the laser rod. The first solid-state lasers were pumped by helical flashlamps placed around the laser rod, as shown in Fig. 22.2a, but this approach has been supplanted by linear lamps. One modern approach (Fig. 22.2b) is to place the linear lamp next to the laser rod in a *closed-coupling* configuration. Another (Fig. 22.2c) is to put the lamp and laser rod at the two foci of an ellipse, so the geometry efficiently focuses the pump light onto the rod. Two lamps and a rod can be put into a dual elliptical cavity, shown in Fig. 22.2d, which in cross section looks like two overlapping ellipses with the rod at the shared focus. High-power oscillators or amplifiers may be surrounded by arrays of many flashlamps.

Diode laser pumping can be from the end of the rod, as shown in Fig. 22.3, or from the side. In end pumping, standard at lower power levels, optics focus the rapidly diverging beam from the diode laser to fill as much space as possible in the crystal, which strongly absorbs at the diode laser wavelength near 810 nm. End pumping can generate an ultra-high-quality output beam. Multiple diode lasers can be arranged along the side of the rod for side pumping at higher powers. Both

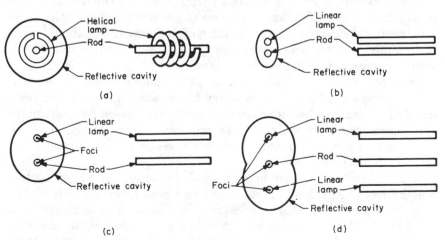

Figure 22.2 Representative cavity configurations for flashlamp pumping: (a) helical lamp, (b) closely coupled lamp, (c) elliptical cavity, (d) dual elliptical cavity.

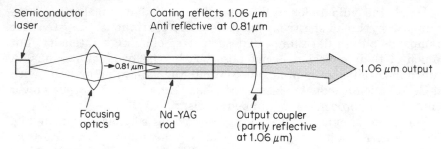

Figure 22.3 End pumping of an Nd–YAG rod with an 810-nm diode laser. Virtually all of the pump light is absorbed in the laser rod, so only the 1.06-μm beam emerges from the output coupler.

single-element and multistripe diode lasers can be used in either approach. Normally the laser rod (Nd–YAG or Nd–YLF are the usual choices) is small.

Internal Structure

Three simple resonator designs for lamp-pumped lasers are shown in Fig. 22.4; a simple diode-pumped laser cavity is shown in Fig. 22.3. Many lasers have more complex designs which incorporate intracavity polarizers, beamsplitters, or other schemes to couple light out of the cavity.

Solid-state laser cavities are designed to pump and extract energy from as much as possible of the laser medium, typically a rod 6 or 9 mm in diameter and several centimeters long. Thus the resonator mirrors should confine light to oscillate through most of the rod, as shown in Fig. 22.4. The cavity design also must account for the subtle "thermal lensing" caused by uneven heating of the laser rod. It should maximize both output power and beam quality. For many applications, stable, narrow-line output is important.

Some resonators are designed to be as compact as possible, particularly for applications such as military laser rangefinders, where portability is critical. Laboratory lasers usually leave room for users to insert accessories between the laser rod and mirrors, either to take advantage of the high powers available there or for beam-control devices such as Q switches.

Neodymium lasers can be arranged in series in an oscillator-amplifier configuration to produce pulses whose laser power is higher than that possible with a single-laser oscillator. The oscillator is a conventional laser, but the amplifier is a rod or disk without cavity mirrors, pumped by a separate pump lamp. Multiple amplifiers may be used where extremely high laser powers are required, such as in

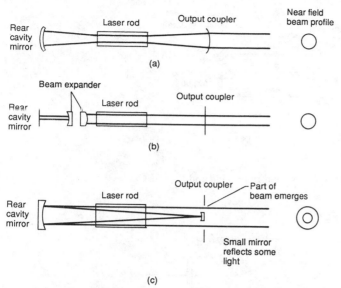

Figure 22.4 Three simple resonant cavities used with lamp-pumped Nd–YAG lasers: (a) simple stable resonator to generate TEM_{00} beam; (b) stable resonator with internal telescope, which helps generate low-divergence beam; (c) unstable resonator with output coupled around a small mirror, producing a near-field beam with a null at its center. In the far field all three beams have central bright regions.

laser fusion experiments. The Nova fusion laser at the Lawrence Livermore National Laboratory has 20 amplifier stages. At high powers, optical isolators may be inserted between amplifier stages to prevent any light from going backward toward the oscillator. In high-energy, low-repetition-rate systems the amplifiers typically are glass rather than crystalline because of the greater size and energy storage capabilities of glass.

Solid-state laser materials are not always used in the form of cylindrical rods. High-energy amplifiers such as those in Livermore's Nova laser use the disk geometry shown in Fig. 22.5. Use of thin disks rather than thick chunks of laser glass minimize thermal problems with the material.

Oscillators can be constructed with slabs of laser material, either glass or YAG (Jones, 1989). In the zigzag slab laser shown in Fig. 22.6, the beam bounces back and forth along the length of the slab, confined by total internal reflection. Slab lasers have a number of attractions, including a fairly large pumped volume and a potential to compensate for some thermal problems within the material, but they have yet to realize their full potential (Koechner, 1988, pp. 391–397).

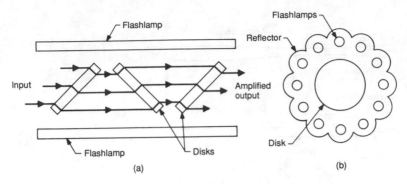

Figure 22.5   Disk amplifier, shown in (a) side and (b) end views.

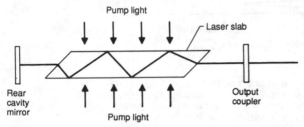

Figure 22.6   In a zigzag slab laser, the laser beam bounces back and forth between top and bottom of the slab, confined by total internal reflection. This design requires very flat top and bottom surfaces.

Continuous-wave Nd–YAG slab lasers have specified average powers to 1800 watts (W). Continual motion of a glass slab so the beam passes through different regions has allowed generation of average powers about half that level (Basu et al., 1990).

Another alternative is a monolithic ring configuration, in which the beam makes a loop through the solid-state medium in the oscillator. This has been demonstrated with Nd–YAG, with total internal reflection confining the beam within the crystal; application of a magnetic field can limit oscillation to one direction (Kane and Byer, 1985).

Inherent Trade-offs

The major trade-offs with neodymium lasers deal with output power. Peak power in an emitted pulse tends to decrease as repetition rate increases beyond a certain point. In general, the higher the peak power, the longer the interval needed between pulses to dissipate excess heat that otherwise might distort the optical properties of the rod and degrade the quality of the output beam. Thermal problems are much more severe in glass than in Nd–YAG or Nd–YLF.

Saturation effects limit the amount of energy that can be stored in and extracted from neodymium laser rods. A resonator which oscillates in a single transverse mode extracts only a small fraction of the energy in a normal rod because it creates a narrow-diameter beam. More power can be extracted in a multimode beam, but the beam quality is poorer. Some designers strike a balance by using an unstable resonator, which extracts energy from most of the rod and generates a beam of reasonable quality.

Research Directions

Many aspects of neodymium laser technology are mature. While Nd–YAG lasers are subject to continued engineering improvements, the material itself is well developed. The most important advances in Nd–YAG technology are likely to be in development of improved pumping geometries such as slab lasers, and in better pump sources, notably higher-power diode lasers (see Chap. 19). Neodymium-glass lasers seem nearly as mature, although efforts to develop fiber amplifiers at 1.3 μm (see Chap. 26) may pay some dividends for larger-scale lasers. Some progress may be possible on Nd–YAG lines other than 1.06 μm, if they offer important advantages for other applications.

Research continues on other potential neodymium hosts. Nd–YLF lasers are gaining acceptance, and work continues to improve their performance. Development of multiply doped materials could pay dividends in higher efficiency by absorbing more pump energy than neodymium ions alone; the leading candidate, Nd,Cr–GSGG, is available commercially. Work also continues on extremely high-power lasers for fusion applications, although much of that effort is concentrating on solid-state materials with active ions other than neodymium.

Progress is being reported on novel materials which serve as both neodymium hosts and nonlinear harmonic generators. Their attraction is the potential to internally generate visible light by frequency-doubling the strong near-infrared lines of neodymium. One example is lithium niobate doped with neodymium and magnesium oxide (added to increase optical damage threshold). Neodymium's fundamental wavelength in $LiNbO_3$ is 1.094 μm; the second harmonic is 547 nm in the green. Researchers have demonstrated flashlamp-pumped versions producing 400-microjoule (μJ) green pulses (Gong et al., 1990) and diode-pumped versions operating continuous-wave (Cordova-Plaza et al., 1988).

Variations and Types Covered

Neodymium lasers are versatile and can operate in many configurations and host materials. Commercial types range from small diode-pumped models emitting several milliwatts continuous-wave to materials-working systems generating over 1 kilowatt (kW). Research versions cover an even broader range, from tiny fiber lasers to building-sized lasers used for fusion research at a handful of major laboratories.

Pumping with diode lasers has become an important technology in the past few years because it offers much higher overall efficiency than lamp pumping. The higher efficiency means there is less waste heat to remove, allowing smaller packages. On the other hand, diode lasers are expensive and limited to much lower powers than lamps, so diode-pumped lasers are limited in power level and relatively expensive. It is reasonable to expect diode lasers costs to decline and powers to increase, but at least for the near future, diode pumping will be limited to the low end of the neodymium power range.

Flashlamp and arc lamp pumping remains the conventional approach to obtaining higher powers. Flashlamps can produce higher peak powers, but arc lamps are required for continuous-wave operation; arc lamps also are used for modelocked lasers and some Q-switched lasers.

The most common neodymium host is YAG, but Nd-glass, Nd–YLF, and Nd,Cr–GSGG lasers also are available. Only Nd–YAG and Nd–YLF lasers operate continuous-wave. Many other materials have been tried, but at this writing they are not offered in commercial lasers.

Some of the variations among neodymium lasers most important to users are essentially matters of packaging. Harmonic generators may be packaged with the laser to produce visible or near-ultraviolet light. Q-switched and short-pulse modelocked neodymium lasers also are available. Although these are actually enhancements to operation of the basic 1.06-μm neodymium laser, they often are packaged as distinct models operating at wavelengths or pulse durations different from those in standard neodymium lasers. These accessories and the changes they can produce are described in more detail below.

Beam Characteristics

Wavelength and Output Power. The wavelength of neodymium lasers normally is quoted as 1.06 μm. This is a good working value, but three factors can cause differences, ranging from slight to major:

- Interactions between the neodymium ion and the host material can shift wavelength by about 1 percent.

- The neodymium laser can operate on weak transitions which normally are overwhelmed by the 1.06-μm line.

- Harmonic generation can multiply the frequency by a factor of 2, 3, or 4, thus dividing the wavelength by the same amount and shifting emission into the visible or ultraviolet.

Principal wavelengths of the major neodymium materials are

- Nd–YLF: 1.047 or 1.053 μm, depending on polarization (1.047 μm is the usual choice)
- Nd-phosphate glass: 1.054 μm
- Nd,Cr–GSGG: 1.061 μm
- Nd-silicate glass: 1.062 μm
- Nd–YAG: 1.064 μm

For most laser applications, there is no significant difference among these wavelengths. However, care must be taken in matching wavelengths of oscillators and amplifiers made of different materials. Thus Nd–YLF oscillators are used with phosphate glass amplifiers; and Nd–YAG oscillators, with silicate glass amplifiers.

Nd–YAG has significant gain at several wavelengths between 1.052 and 1.444 μm, but normally these are overwhelmed by 1.064-μm emission. The strongest line outside the 1.06-μm region is at 1.3188 μm, and a few companies offer Nd–YAG laser operation at that wavelength, either as an option for 1.064-μm lasers or as a separate product. The birefringence of Nd–YLF splits that transition into two wavelengths, 1.313 and 1.321 μm. Commercial lasers operating on this 1.3-μm transition are available only in Nd–YAG and Nd–YLF.

The second, third, and fourth harmonics of the neodymium laser's fundamental frequency are produced by passing the fundamental wavelength through nonlinear crystals. This generates wavelengths of 532, 335, and 266 nm for Nd–YAG, and slightly different wavelengths for other neodymium hosts. Some energy is lost in harmonic generation, which becomes more efficient as power density in the nonlinear material increases. Both pulsed and continuous neodymium lasers generate sufficiently high power densities for harmonic generation, but with continuous-wave lasers the harmonic generator usually is placed inside the laser cavity, between the laser rod and the output coupler. Harmonic generators may be packaged with the laser. Raman shifting, sum-and-difference frequency mixing, and parametric conversion also can shift neodymium laser wavelengths, but these normally are done by separate accessories.

A wide range of output powers is available form Nd–YAG, Nd–YLF, and Nd-glass lasers, depending on configuration, pumping source, and wavelength; representative figures (derived from information published in industry directories) are listed in Table 22.2. Steady or av-

TABLE 22.2   Typical Output Powers for Neodymium Lasers Derived from Specifications

Type	Pump source	Wavelength, nm	Output powers (average or CW), W
Pulsed quadrupled Nd-glass	Flashlamp	263–266	≤0.04
Pulsed quadrupled Nd–YAG	Flashlamp	266	0.001–2
Pulsed tripled Nd-glass	Flashlamp	351–355	≤0.07
Pulsed tripled Nd–YAG	Flashlamp	355	0.001–20
Pulsed doubled Nd–YLF	Diode	523	0.00002–0.01
Pulsed doubled Nd–YLF	Flashlamp	523–527	1–3
CW Doubled Nd–YAG	Diode	532	0.001–0.08
Pulsed doubled Nd–YAG	Diode	532	0.000001–0.0004
Pulsed doubled Nd–YAG	Flashlamp	532	0.01–50
CW doubled Nd–YAG	Arc lamp	532	0.3–4*
Pulsed doubled Nd-glass	Flashlamp	532	≤0.15
CW Nd–YLF	Diode	1047	0.035–0.25
CW Nd–YLF	Diode	1053	0.015–0.1
CW Nd–YLF	Arc lamp	1053	≤20
Pulsed Nd–YLF	Flashlamp	1053	≤20
Pulsed Nd-phosphate glass	Flashlamp	1054	0.01–1
CW Nd,Cr–GSGG	Diode	1061	≤0.1
Pulsed Nd,Cr–GSGG	Flashlamp	1061	≤1
CW Nd–YAG	Diode	1064	0.002–3
Pulsed Nd-silicate glass	Flashlamp	1064	0.1–100
CW Nd–YAG	Arc lamp	1064	0.5–1800
Pulsed Nd–YAG	Flashlamp	1064	≤1500 W
CW Nd–YLF	Diode	1313	≤0.1
CW Nd–YLF	Lamp	1313	≤3
CW Nd–YAG	Diode	1319	≤to 0.25
Pulsed Nd–YAG	Flashlamp	1319	≤20
CW Nd–YAG	Arc lamp	1319	0.05–40
CW Nd–YLF	Diode	1321	0.015–0.1
CW Nd–YAG	Diode	1335	0.012–0.1
Pulsed Nd,Cr–GSGG	Flashlamp	1335	>1

*Average power for modelocked second-harmonic output; continuous-wave (CW) power would be lower.

erage powers in the kilowatt range can be obtained either from pulsed or continuous-wave lasers, with the highest powers from multirod, oscillator-amplifier, and slab designs.

Peak powers available from neodymium lasers are much higher, and depend on both the pulse energy and duration. Neodymium lasers can operate in a variety of pulsed modes, depending both on excitation and on control of the energy in the laser cavity:

- "Long" pulses, with duration dependent on the length of the pulse from the excitation source, and the fluorescence lifetime of the medium (typically 200 to 500 μs). For flashlamps this can reach milliseconds.

- Q-switched pulses (with pulsed or continuous excitation), lasting from a few nanoseconds to hundreds of nanoseconds.

- Cavity-dumped pulses (with pulsed or continuous excitation), from a few to tens of nanoseconds; now rarely used in commercial lasers.

- Modelocked pulses (generally with continuous excitation), usually from tens to hundreds of picoseconds.

The shorter the pulse, the higher the peak power for a given laser rod and pump source; however, shrinking pulse length beyond a certain point can reduce the pulse energy. Adding pulse control can reduce output power; for example, one YAG laser can deliver 24 watts (W) continuous-wave (CW), 22 W average power when modelocked, 13 W average power when Q-switched, and 3 W average power when modelocked and Q-switched. Pulse lengths are described in more detail below and in Table 22.3 (based on data from industry directories).

Because glass can be made in larger sizes and has lower gain than Nd–YAG, glass lasers can store more energy and produce more energetic pulses. However, its thermal problems limit repetition rate and

TABLE 22.3 Duration and Repetition Rates Available from Pulsed Neodymium Lasers Operating at 1.06 μm (Not All Available Commercially)

Type	Modulation	Excitation source	Typical repetition rate	Typical pulse length
Nd-glass	Lamp only	Flashlamp	0–3 Hz	0.1–10 ms
Nd-glass	Q switched	Flashlamp	0–3 Hz	3–30 ns
Nd-glass	Modelocked	Flashlamp	Pulse trains*	5–20 ps
Nd–YAG	Lamp only	Flashlamp	0–200 Hz†	0.1–10 ms
Nd–YAG	Q switched	Flashlamp	0–200 Hz†	3–30 ns
Nd–YAG	Cavity dump	Flashlamp	0–200 Hz†	1–3 ns
Nd–YAG	Modelocked	Flashlamp	Pulse trains*	20–200 ps
Nd–YAG	Q switch	Arc lamp	0–100 kHz	40–700 ns
Nd–YAG	Q switch	Diode	0–15 kHz	20–30 ns
Nd–YAG	Modelocked	Arc lamp	50–500 MHz	20–200 ps
Nd–YAG	Modelocked and Q-switched	Arc lamp	Varies‡	20–200 ps§
Nd-YLF	Modelocked	Arc lamp	Same as YAG	About half YAG
Nd–YLF¶	Modelocked	Diode	160 MHz	7 ps
Nd–YLF/ glass	Modelocked and chirped pulse amplification	Lamp	0–2 Hz	1–5 ps

*Series of 30–200 ps pulses, separated by 2–10 ns, lasting duration of flashlamp pulse, on the order of a millisecond.
†Depends on flashlamp pulse rate.
‡Depends on Q-switch rate; modelocked pulses.
§In pulse trains lasting duration of Q-switched pulses (100–700 ns).
¶Laboratory results (Juhasz et al., 1990).

thus lead to low average power, as shown in Table 22.2. Nd–YLF also can store more energy than Nd–YAG because of its lower cross section, but it cannot be made in pieces as large as glass. Nd–YAG remains the material of choice for the highest average powers, but at lower powers Nd–YLF can generate nearly as much power as Nd–YAG.

Efficiency.   Overall efficiency of a commercial lamp-pumped neodymium laser, measured as laser power out divided by electrical power in, typically is 0.1 to over 1 percent. Diode-pumped lasers can be considerably more efficient because neodymium absorbs the 0.8-μm pump beam much more efficiently than the broadband output of a lamp.

A hierarchy of optical losses contributes to the low efficiency of lamp-powered lasers. They account for the following typical fractions of the input energy to a continuous lamp (Koechner, 1988, p. 336):

- Only half the input energy emerges as pump light at 0.3 to 1.5 μm; the lamp dissipates the other 50 percent as heat.

- Only 8 percent of the original energy is absorbed by the laser rod. The laser cavity absorbs 30 percent, the coolant and flowtubes absorb 7 percent, and the lamp reabsorbs 5 percent.

- Only 2.6 percent of the lamp input energy goes into stimulated emission from the laser rod. Five percent is lost as heat and 0.4 percent is lost as fluorescence.

- Optical losses reduce stimulated emission to 2 percent of the lamp input energy; 0.6 percent is optical loss. Some energy is lost in driving the pump lamp and operating other parts of the laser, reducing typical overall efficiency to the 1 percent range. The fractions differ somewhat with design details.

Diode pumping is inherently more efficient. The high-power diode lasers and arrays used for Nd-laser pumping convert 10 to over 30 percent of electrical input into laser output, a smaller fraction than lamps. However, a much larger fraction of the pump light is converted into laser light; slope efficiency of that optical conversion can reach 40 percent. Thus diode-pumped Nd–YAG lasers can deliver optical output at 1.06 μm that exceeds 10 percent of the electrical input.

Harmonic generation is an inefficient process, especially at lower powers, so neodymium lasers are less efficient when operating at their second, third, or fourth harmonics than at 1 μm.

Temporal Characteristics of Output.   Neodymium lasers can operate in several pulsed modes or, in the case of YAG and YLF, deliver a continuous beam. Representative pulse lengths for various operating modes at the 1-μm wavelength are listed in Table 22.3. Harmonic generation reduces pulse length because the nonlinear process depends strongly on intensity and thus is far more effective at the peak of the pulse than when intensity is rising or falling.

As indicated in Table 22.3, pulse duration and repetition rate depend on both the pump source and the pulse switching method. With flashlamp pumping, the lamp pulse rate determines the repetition rate of Q-switched or cavity dumped lasers. Modelocked lasers emit series of short pulses, with spacing equal to the round-trip time of the laser cavity. Many of these pulses may be produced during a single flashlamp pulse, and Q switches can select single modelocked pulses.

Glass lasers have lower repetition rates than Nd–YAG or Nd–YLF, and commercial models typically produce no more than a few pulses per second. Repetition rates may be as low as one pulse every few minutes, or single-shot for large experimental lasers.

Because they have broader spectral bandwidth than Nd–YAG, both Nd-glass and Nd–YLF can generate shorter modelocked pulses. The shortest pulses are produced by a technique called *chirped pulse amplification*, which is similar to that used to generate femtosecond pulses from a dye laser. A modelocked pulse, typically from an Nd–YLF laser, is passed through an optical fiber, where nonlinear effects spread its spectrum over a broader range than the laser generates. Then that pulse is compressed in duration by special optics, and amplified in a broad-bandwidth Nd-glass laser. The resulting pulses are in the 1- to 3-picosecond (ps) range, and when amplified can reach the terawatt (10^{12} W) level in commercial lasers (although their brief duration means that pulse energy is on the order of only a joule). Researchers have demonstrated 3×10^{12} W pulses and are working on a 10×10^{12} W laser (Perry, 1988).

Spectral Bandwidth. Many manufacturers do not list spectral bandwidth in their specifications for neodymium lasers, and for most purposes the lasers can be considered monochromatic. The natural linewidths of the major neodymium hosts are shown in Table 22.1.

Commercial lamp-pumped Nd–YAG lasers without line-narrowing accessories generally have linewidths of about 1 to 5 parts in 10^4 at 1.06 μm, or 1 to 5 inverse centimeters (cm^{-1}). An intracavity etalon can reduce linewidth below 0.2 cm^{-1}, and an additional electronic line-narrowing accessory can reduce linewidth to 0.02 cm^{-1}. Narrower bandwidths, to about 0.003 cm^{-1}, are possible by injection seeding an amplifier with the output of a low-power narrow-line laser.

Neodymium-glass and Nd–YLF lasers have broader linewidths. A typical specification is 5-nm linewidth (about 5 parts in 10^3) for a pulsed glass laser. Nd–YLF linewidths are closer to those of Nd–YAG than Nd-glass.

Extremely narrow linewidths, as narrow as 5 kilohertz (kHz) over a 1-ms interval, are available from commercial continuously diode-pumped Nd–YAG lasers. Jitter makes the linewidth considerably larger over reasonable periods (e.g., it is 130 kHz over 1 second), but these values are still small compared to most other lasers. Diode-pumped lasers benefit from their mechanical simplicity, which avoids the vibrations which can degrade stability of a lamp-pumped oscillator.

Amplitude Noise. The type and magnitude of amplitude noise depend on the pump source. Pulsed lamp-pumped lasers suffer pulse-to-pulse variations in peak power, including relaxation oscillations which produce sharp output spikes when the laser starts oscillating. Continuous lamp-pumped lasers have noise caused by fluctuations in pump lamp intensity, often at a harmonic of the 60-Hz line current which is rectified to power the lamp. Both types are subject to mechanical noise caused by vibration from equipment such as cooling water pumps.

The magnitude of amplitude fluctuations are comparable for pulsed and continuous lamp-pumped lasers, typically 1 to 10 percent, with higher-power and multimode lasers at the upper end. Nd–YLF lasers tend to be more stable than Nd–YAG in similar configurations. Active stabilization, available as an option, can improve stability by a factor of 2 to 4. Harmonic generation tends to magnify amplitude fluctuations.

Continuous diode-pumped lasers have lower amplitude noise; a typical rating is 0.2 to 1 percent root-mean-square (rms) noise at 10 Hz to 10 MHz. However, Q switching introduces noise; pulse-to-pulse variation for Q-switched diode-pumped Nd–YAG lasers from the same manufacturer is specified at ±10 percent.

Beam Quality, Polarization, and Modes. Neodymium lasers are available with output in a single (TEM_{00}) mode, in multiple transverse modes, or in modes determined by unstable resonators. Unstable resonators can be used only with high-gain lasers, but they have become popular because they offer a combination of good beam quality with high power and efficient extraction of energy deposited in the laser rod.

Lamp-pumped neodymium lasers normally operate in multiple longitudinal modes. Diode-pumped lasers, with much shorter cavities, have broader longitudinal mode spacing and can more easily be limited to oscillation on a single longitudinal mode.

Nd–YLF is birefringent, so its output is inherently polarized. Commercial Nd–YAG and Nd-glass lasers are available with unpolarized output, or with output polarized by polarizing components in the laser cavity. Polarizing the output reduces power somewhat, but a polarized beam may be required in some cases, notably for harmonic generation or other nonlinear processes.

Beam quality can be degraded by operating at repetition rates or power levels too high for thermal stresses within the laser rod to be dissipated between pulses. The power levels at which these problems occur depend on the material. Thermal gradients can cause beam distortion and depolarization effects.

Coherence Length.   Coherence length of lamp-pumped neodymium lasers generally is not specified. Output of Nd–YAG and Nd–YLF lasers is coherent enough for harmonic generation, sum-and-difference frequency mixing, and other nonlinear processes which require coherent light. The nominal linewidth of Nd–YAG corresponds to a coherence length of about a centimeter, but line-narrowing options can extend it to meters.

Diode-pumped neodymium lasers can have extremely narrow linewidths, which give them very long coherence lengths. The calculated coherence length for a diode-pumped laser with 5-kHz bandwidth is 60 kilometers (km).

Beam Diameter and Divergence.   Divergence of lamp-pumped neodymium laser beams is a fraction of a milliradian to about 10 milliradians (mrad). Divergences typically are smaller for second-, third-, and fourth-harmonic beams. Nd–YAG and Nd–YLF beams generally are 1 to 10 mm, with larger diameters possible with beam expanders. Glass lasers, which can be made with larger cross-sectional areas, can have larger beams, particularly from high-power amplifiers at the end of a chain. Most beams are circular, but glass and slab lasers and amplifiers may be exceptions.

Typical beam divergences from diode-pumped neodymium lasers are 1 to 10 mrad at 1.06 μm and as small as 0.6 mrad at 532 nm. Beam diameters are 0.2 to 2 mm, an indication of the small size of the rods used in diode-pumped lasers.

Stability of Beam Direction.   Beam direction generally is stable to within a milliradian for lamp-pumped lasers, but this quantity is rarely specified. Diode-pumped lasers offer better stability; one manufacturer claims beam wander no more than 10 microradians (μrad) between pulses. Active stabilization is available for some modelocked lasers.

Suitability for Use with Laser Accessories.   The neodymium laser is so well suited for use with laser accessories that many models have space within their optical cavities for them. Other models are sold packaged with such accessories as Q switches and harmonic generators.

Harmonic generators are widely used with neodymium lasers. Frequency doublers generate second and fourth harmonics. The third harmonic is produced by mixing the second harmonic with the fundamental. A single laser can generate all four wavelengths, but output from any one aperture normally is filtered to restrict emission to one wavelength. Pulsed Nd–YAG, Nd–YLF, and Nd-glass lasers generate high peak powers, so the harmonic generators normally are placed outside the laser cavity. Continuous-wave lasers have lower peak powers, so harmonic generators normally are placed inside the laser cavity. Despite their low powers, diode-pumped lasers can generate sufficient intracavity powers for harmonic generation.

Continuously pumped YAG lasers can be used with Q switches, modelockers, pulse selectors, cavity dumpers, and/or pulse compressors to generate pulses of various durations and peak powers. Pulsed Nd–YAG and Nd–YLF lasers also can be used with Q switches, pulse selectors, modelockers, and cavity dumpers to generate short pulses. Nd-glass lasers can be used with the same accessories, but their thermal limitations make them a poor choice for Q switching at rates faster than a few pulses per minute. On the other hand, the broader spectral bandwidth of Nd-glass permits much shorter modelocked pulses than does Nd–YAG at reasonable repetition rates, as long as pulse energy is limited.

The shortest pulses are produced by the amplification and compression of frequency-broadened or chirped pulses. Nd–YLF or Nd-glass are used in the oscillator because they have broader bandwidths than does Nd–YAG. Oscillator pulse length is limited to the several-picosecond range, but amplification (usually in Nd-glass), pulse broadening, and compression can reduce the pulse duration to as little as 30 fs. Pulses as short as one picosecond can be generated with powers in the terawatt (10^{12} W) range from commercial lasers using Nd–YLF oscillators, Nd-glass amplifiers, and pulse compression (Maine and Mourou, 1988; Olson and Weston, 1990).

Operating Requirements

Input Power. Small lamp-pumped neodymium lasers can plug into an ordinary wall outlet and draw about 500 W from a 117-volt (V) supply. Large industrial models draw tens of kilowatts from three-phase 220-V or 440-V power lines. Low-power military range finders based on YAG lasers can operate from a 24- or 28-V battery pack, with the number of shots and degree of portability depending primarily on battery size.

Diode pumping requires less electrical power to produce the pump light than does lamp pumping, as little as several watts of electricity.

However, many models require extra electrical power for forced-air cooling. Most operate from standard 117-V electrical outlets and have separate power supply modules which produce the direct current needed to drive the diode lasers. Battery operation is possible.

Cooling. Purely convective air cooling suffices for very low-power lamp- or diode-pumped YAG lasers used where portability is a paramount concern. However, this limits operation to low repetition rates for lamp-pumped and low powers for diode-pumped lasers. Most military range finders and target designators employ closed-cycle cooling. Typical diode-pumped YAG lasers rely on forced-air cooling, but some are water-cooled. Some low-power low-repetition-rate lamp-pumped lasers also use forced-air cooling.

Most laboratory and industrial lamp-pumped lasers used open-cycle water cooling. A typical Nd–YAG laser with one-watt average power requires about 3 liters (L) [0.8 gallons (gal)] of tap water per minute, while a massive industrial system delivering a few hundred watts needs about 60 L (16 gal) per minute. Some high-power lasers have closed-cycle water-cooling systems which transfer heat to an open-cycle water cooler.

Consumables. Solid-state neodymium lasers are self-contained units which do not require replenishment of supplies, except when a continuous supply of cooling water is needed. The only components which might be considered consumable are pump lamps, which require replacement at intervals of a few hundred hours or several million shots. Diode lasers have much longer operating lifetimes in normal use, but they can fail catastrophically if exposed to power surges.

Required Accessories. Neodymium lasers generally are marketed as self-contained units which operate directly from available electrical power and are basically ready to begin operation. Some systems include components such as Q switches, harmonic generators, or modelockers, which could be considered accessories. The following accessories also may be useful with neodymium lasers for certain applications:

- Amplifiers to boost pulse power and energy
- Beam dumps to absorb power at the fundamental wavelength after harmonic generation
- Beam expanders to enlarge the output beam, particularly before amplification
- Cavity dumpers
- Collimators to align laser optics

- Infrared beam viewers or viewing cards
- Injection seeding for single-longitudinal-mode operation
- Optical benches, for stable mounting of accessories and cavity optics
- Polarizers and polarization rotators, for use with harmonic generators
- Pulse compression systems
- Pulse diagnostic equipment
- Raman wavelength shifters
- Shutters and beam isolators, to prevent undesired feedback in large amplifiers.
- Temperature-control system to enhance laser stability
- Video beam diagnostics

Some of these accessories may be supplied as standard equipment or options. Generally they are required only for certain applications. For example, harmonic generators are needed to generate green or ultraviolet beams for dye pumping, but not for materials-working with the 1.06-μm fundamental wavelength.

Operating Conditions and Temperature.   Most Nd–YAG, Nd–YLF, and Nd-glass lasers are designed for normal conditions in a laboratory or factory. Temperature ranges of lamp-pumped lasers normally are not specified, presumably because most industrial models require active cooling with flowing water. Operating temperatures are specified for diode-pumped lasers, typically 10 to 35°C, because the pump lasers are quite sensitive to temperature. Military range finders and target designators are designed to meet the broad range of conditions encountered in military operations.

Mechanical Considerations.   The smallest neodymium lasers are diode-pumped versions, such as the one shown in Fig. 22.7. One company offers a series of compact lasers in which the pumping diodes and Nd–YAG rod are packaged in a 0.37-kilogram (kg) [12.9-ounce (oz)] cylindrical head 108 mm [4.25 inches (in)] long and 45 mm (1.75 in) in diameter and the electronics and power supply are in a separate 1.9 kg (67-oz) box measuring 228 by 102 by 76 mm (9 by 4 by 3 in). Another manufacturer delivers pump energy through a fiber-optic cable, allowing a more compact 0.17-kg (6-oz) cylindrical head which contains only the neodymium crystal rod and is only 19 mm (0.75 in) in diameter and 91 mm (3.5 in) long. However, the box containing power supply, controls, and pump laser is larger and heavier.

Lamp-pumped military laser range finders and target designators

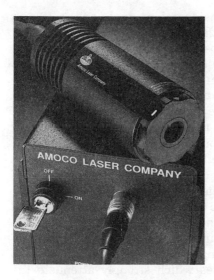

Figure 22.7 A diode-pumped Nd–YAG laser with pump laser in the 108-mm-long laser head and power supply and controls in a separate box. (*courtesy of Amoco Laser Co.*)

can be smaller than a pair of binoculars. A modest 18-pound (lb) (8-kg) package measuring 14 by 5.4 by 5.5 in (35.6 by 13.7 by 14 cm) can deliver average power of several watts with open-cycle water cooling.

At higher powers, the laser head (such as the one shown in Fig. 22.8) normally is around a meter long, and separate from the controls and power supply; the box containing the power supply can be bulky. High-power industrial lasers may have power supplies the height of a person, bulky cooling equipment, and laser "heads" big enough to cover a desktop. Experimental Nd-glass lasers used in fusion research occupy entire laboratories.

Safety. Pulsed neodymium lasers probably have been responsible for more eye injuries than any other type, both because of their widespread use and because of the particular hazards they pose.

Although the human eye cannot see the 1.06-μm fundamental wavelength of the neodymium laser, that light can pass through the eyeball to damage the retina in the same way as visible light (Sliney

Figure 22.8 A commercial 50-W *Q*-switched continuously pumped Nd–YAG laser head on a 1-m rail. (*Courtesy of Lasermetrics Inc.*)

and Wolbarsht, 1980). The same considerations apply to the 1.3-μm neodymium line. The risk of being zapped in the eye by an unseen beam is increased by the fact that most neodymium lasers used in the laboratory produce short, intense pulses. A number of eye injuries have occurred when a single neodymium laser pulse was reflected unexpectedly from an optical component into the eye of an unwary person. The fact that the beam cannot be seen increases the likelihood of such reflections. The best protection against such accidents is safety goggles which effectively block all optical paths to the eye, transmitting most visible light but strongly blocking the near-infrared neodymium lines. Because one pulse can do serious damage, it is important to block even peripheral paths.

Harmonics of neodymium lasers also come in short, intense pulses and present serious eye hazards. The second harmonic at 532 nm is visible green light, and can penetrate the eye in the same way as the fundamental wavelength. Although the wavelength is visible, frequency-doubled Nd–YAG lasers require safety goggles which strongly attenuate the green part of the spectrum, and should make the beam invisible.

Hazards from the 355-nm third harmonic and the 256-nm fourth harmonic are different from those at longer wavelengths and from each other. Both pose eye hazards similar to those for excimer lasers described in Chap. 13. The fourth harmonic also can cause a sunburn-like effect on the skin. Ultraviolet-blocking safety goggles can provide eye protection against the third and fourth harmonics.

Users should be aware that some light at longer wavelength may remain after harmonic generation unless the optical system is explicitly designed to get rid of it. Some commercial lasers come with optional beam dumps for this purpose. Multiwavelength laser output can pose a severe eye hazard because it is difficult to design safety goggles to block two or more wavelengths.

Continuous-wave neodymium lasers pose serious eye hazards, but the power levels are sufficiently low that an instantaneous exposure (as from a scanning beam) is less likely to cause permanent damage. Some diode-pumped lasers produce continuous beams on the order of a milliwatt at the second-harmonic wavelength, but most neodymium lasers generate considerably higher powers. Eye protection is necessary when working with any neodymium laser.

Lamp pumping poses two additional hazards. All lamps require high drive voltages, posing serious electrical hazards if any high voltages are exposed. Overloaded or worn-out flashlamps can explode, producing potentially dangerous shrapnel, but a properly designed laser cavity should contain the pieces.

Special Considerations. Neodymium and ruby lasers are similar enough that a few companies offer a single laser system which can be adapted for use with either, and sometimes with other materials such as alexandrite. Nd–YAG, Nd–YLF, and Nd-glass are sufficiently similar that rods of one material may work in cavities designed for the others, if operating conditions do not exceed material requirements (e.g., if repetition rate is not too high for Nd-glass).

Lamp-pumped lasers generally should not be left operating for long periods if they are not being used because lamp lifetimes are limited. (This warning does not apply to leaving flashlamps in "simmer" mode, where they are biased below threshold but not firing pulses.) Some warm-up time may be needed, depending on the laser model and application. Some time also may be needed to set up and align sensitive accessories such as line-narrowing and resonator optics, harmonic generators, Q switches, and modelockers.

Reliability and Maintenance

Lifetime. Complete neodymium laser systems can last a decade or more, and some have outlived their manufacturers by several years. However, some components must be replaced during that period.

The shortest-lived component in a lamp-pumped neodymium laser is the pump lamp. Continuous lamps have rated operating lifetimes of only a few hundred hours. Flashlamps typically are rated for a few million to tens of million shots. To put that figure in perspective, at a dozen pulses a second—not uncommon for Nd–YAG—a flashlamp will fire a million times in 24 hours of operation. Some pulsed high-voltage components also have comparatively short lifetimes.

Diode lasers used as pump sources have much longer lifetimes. Typical rated figures are 20,000 hours and up at maximum output, but the pump lasers are rarely operated at that figure. However, pump diodes are vulnerable to electric surges and static discharges.

Lifetime of the laser rod and optics generally are limited only by abuse or poor maintenance. Potential problems include mechanical damage, or optical damage produced by operation at excessive power levels or with dirty surfaces.

Maintenance and Adjustments Needed. The most frequent maintenance needed for lamp-pumped neodymium lasers is lamp replacement. Most commercial lasers are designed to make such replacement simple. Exposed optical surfaces must be cleaned periodically to prevent accumulation of dirt, which can cause excess absorption of light leading to optical damage. High-voltage electronics may require periodic maintenance. Cooling system filters require periodic cleaning at intervals on the order of 1000 hours. Except for cleaning, most of these maintenance requirements do not apply to diode-pumped lasers.

Cavity optics and optical accessories require periodic alignment and adjustment to be certain that the laser is producing the desired output. A visible alignment laser may be needed for major adjustments but is not critical for routine periodic alignment.

Mechanical Durability. Neodymium lasers can be packaged well enough to meet stringent military specifications for portable field equipment, and both lamp- and diode-pumped types have been space-qualified. Industrial lamp-pumped neodymium lasers generally are packaged to operate under factory-floor conditions, and diode-pumped lasers are inherently durable because they have sealed heads and lack fragile pump lamps. However, users should remember that pump lamps are made of fragile glass, and that laser rods are brittle. Some laboratory lasers have exposed optics, and bumping a mirror can knock the laser cavity out of alignment.

Failure Modes and Possible Repairs.   Several things can go wrong with neodymium lasers, some easy to fix and others requiring more extensive repairs:

- Failure or degradation of the pump lamp, requiring replacement.
- Failure or degradation of diode-laser pumps, requiring replacement.
- Misalignment of the optical cavity, reducing output power and/or degrading optical quality, requiring realignment.
- Optical damage to the laser rod or other optical components. This can occur rapidly if the laser is operated at excessive power, if dirt on a surface absorbs enough laser energy to trigger damage, or if reflections from external components return too much energy to the laser. Damage or long-term degradation often can be repaired by repolishing and recoating the affected surface.
- Faults in high-voltage electronics, which can reduce power delivered to the pump lamp, reducing laser output.
- Explosion of flashlamps, which can damage the laser rod and other components in the pump cavity. This is not common, but can occur at very high power levels. During the first 3 years the Lawrence Livermore National Laboratory operated the Shiva fusion laser, 40 of 2000 flashlamps exploded, and repairs operated for about 3 percent of Shiva operating costs (Martin et al., 1981).

Generally neodymium and other solid-state lasers are assembled from separate components and subassemblies rather than built as an integral unit (like a gas-laser tube). This makes it reasonable to refurbish an old laser, and a few companies offer such services. Most solid-state laser makers maintain an inventory of spare parts, and a few companies continue to offer spare parts for lasers made by defunct companies.

Commercial Devices

Standard Configurations. Neodymium lasers generally are offered as complete units, including laser rod, pump source, pump cavity, cavity optics, power supply, cooling system (if required), and controls. Some lasers are packaged in complete laser machining systems, which include beam-delivery systems, parts-handling equipment, beam enclosures, and computer controls. The difference between lasers and systems is obvious from sales literature.

There are several standard types of neodymium lasers offered. Important types include

- High-power materials-working lasers, pumped by arc lamps or flashlamps. They operate at the 1.06-μm fundamental wavelength where output power is highest.

- Moderate-power lamp-pumped industrial lasers for applications such as resistor trimming.

- Pulsed neodymium lasers for industrial marking, usually at 1.06 μm.

- Laboratory lamp-pumped lasers delivering extremely high peak powers in very short pulses for research. Operation may be at the fundamental wavelength or a harmonic.

- Laboratory lamp-pumped lasers which generate a continuous beam at the fundamental or second harmonic. These lasers may be modelocked to generate short pulses, and the modelocked pulses may be used to generate harmonics.

- General-purpose lamp-pumped lasers, including both pulsed and continuous types.

- Diode-pumped lasers emitting up to a few watts at 1.06 μm. Second-harmonic output is available but at lower power. Normally the compact laser head is separate from the power supply.

Neodymium lasers often are packaged with harmonic generators, Q switches, or modelockers and are marketed as devices generating short pulses or wavelengths shorter than 1.06 μm. There are significant differences between industrial lasers, usually designed for specific applications, and some laboratory lasers which are designed as general-purpose devices which include room for accessories. However, some laboratory lasers are packaged for specific uses.

Options. The usual range of options for neodymium lasers includes harmonic generation, Q switching, modelocking, line-narrowing optics, frequency mixing, Raman wavelength-shifting optics, injection seeding to generate narrow-line output, and pulse compression. Most lamp-pumped general-purpose and laboratory neodymium lasers provide ways to accommodate accessories inside or outside the laser resonator cavity; some diode-pumped laser heads accommodate accessories on the outside of the head. External amplifiers can be added to raise the power of pulsed lasers to high levels. Some options may be packaged as standard features.

Special Notes. The versatility of Nd–YAG lasers has led to a wide range of commercial devices. Nd–YLF has reached the market only in the past few years, but it is gaining acceptance. Crystalline neodymium materials are difficult to grow, and the companies which produce them are separate from laser manufacturers, who buy crystals or rods and build lasers around them.

Neodymium laser designs tend to be modularized, allowing companies to offer a broad range of standard products which are assembled from a small stock of standardized components. This increases flexibility and range of user choice.

Pricing. Single diode-pumped Nd–YAG or Nd–YLF lasers can be purchased for $3000 to about $30,000. The least-expensive lamp-pumped Nd–YAG lasers sell for about $3000, but prices of high-power and high-performance systems sell for well over $100,000. Lamp-pumped Nd-glass and Nd–YLF are comparable in price to Nd–YAG.

Suppliers. The neodymium laser market as a whole is highly competitive. The 1991 edition of one industry directory lists specifications of diode-pumped lasers from eight suppliers. The list of companies offering lamp-pumped neodymium lasers is considerably longer. Some companies specialize in certain market niches, and few manufacturers of diode-pumped lasers offer lamp-pumped models. The range of suppliers for a few specialized, high-performance neodymium laser systems may be limited.

Applications

Neodymium lasers have found many applications in science, civilian industry, medicine, and military equipment. A main reason is their ability to generate high power at wavelengths near 1 µm with reasonable efficiency in a convenient manner. Harmonic generation can shift the wavelength into the green and ultraviolet, and Q switching and modelocking can generate short pulses with high power. Diode laser pumping is quite efficient, and produces a beam of much better quality than is available from the pump lasers. All in all, neodymium has proved to be an exceptionally valuable and versatile laser system.

The major industrial uses of Nd–YAG and Nd-glass lasers have been in materials working. Neodymium lasers complement CO_2 lasers because the two have different wavelengths and output characteristics. The high peak powers and short wavelength of neodymium lasers make them a better choice for drilling, particularly of the many metals which strongly reflect the 10-µm wavelength of CO_2. For similar reasons, neodymium lasers are often better choices for spot welding and laser marking of codes and symbols on certain materials. Neodymium lasers also are widely used in electronics fabrication, for applications such as resistor trimming or mask or memory repair where laser energy must be focused precisely onto a small spot. Neodymium lasers are used in some automatic soldering systems. On the other hand, the higher continuous output of CO_2 lasers and the strong absorption of some materials (notably titanium and nonmetals) at 10 µm make CO_2 the better choice for many types of continuous cutting. The detailed trade-offs are complex and beyond the scope of this book (e.g., see Duley, 1983).

Neodymium lasers also have found some medical uses, in treatments where their short pulses and short wavelength offer advantages over the CO_2 lasers preferred for many types of laser surgery. Pulsed neodymium lasers have become the standard tool for treating a common complication of cataract surgery, in which a membrane inside the eye becomes cloudy after the natural lens is replaced by a plastic implant. A tightly focused, short pulse from a neodymium laser can cut the membrane, restoring normal vision. The 1-µm wavelength of neodymium lasers also can be carried through standard optical fibers, unlike CO_2 lasers, allowing their use with endoscopes for gall-bladder surgery and for treatment of gastrointestinal bleeding. The 532-nm second harmonic is being studied for use in surgery to reduce pressure that spinal disks are applying to nerves.

The power and versatility of neodymium lasers have led to many applications in scientific research. Doubled, tripled, and quadrupled neodymium lasers are widely used to pump dye lasers. The high peak powers available with modelocked and Q-switched lasers are valuable in many research applications, including plasma physics, materials interactions, and nonlinear optics. The ability to generate extremely high peak powers has made neodymium lasers the most common type used in inertial confinement fusion research, in which laser pulses irradiate and compress a target, heating it so that some atomic nuclei in the target fuse and generate nuclear energy.

Neodymium lasers have found some measurement applications. Neodymium laser pulses can vaporize small samples for Raman spectroscopy and mass spectroscopy (Moenke-Blankenburg, 1989). Doubled neodymium lasers can generate short pulses for holographic interferometry and measurements of deformation under stress. Other measurement applications are in research and development.

The advent of low-power diode-pumped neodymium lasers has led to research on a new range of applications in optical storage, display, reprographics, communications, and information processing. Inexpensive green diode-pumped solid-state lasers could replace rare gas ion lasers in some applications that require tens of milliwatts of visible light, but at this writing diode-pumped neodymium lasers remain comparatively costly because the pump diodes themselves are expensive. The 1.3-μm neodymium line may prove useful in fiber-optic communications or in measurement of optical fiber properties.

The largest single use of Nd–YAG lasers probably is as military range finders and target designators. Comparatively little of this technology has been transferred to the civilian sector, although military agencies have sponsored much fundamental research. Nd–YAG lasers are not ideal because their 1-μm wavelength can be obstructed by smoke and fog, and because their pulses pose a serious eye hazard to friendly troops in training exercises. Other lasers have been studied, including CO_2 and such eye-safe materials as erbium-doped glass. However, Nd–YAG range finders and designators have become standard battlefield equipment, which the Department of Defense is buying in quantity.

Military researchers are working on potential applications of diode-pumped neodymium lasers, including high-repetition-rate range finders and intersatellite communications.

Bibliography

Norman P. Barnes et al.: "Comparison of Nd 1.06 and 1.33 μm operation in various hosts," *IEEE Journal of Quantum Electronics* QE-23(9):1434–1451, 1987 (research review).

Santanu Basu, Robert L. Byer, and Josef R. Unternahrer: "Slab lasers move to increase power," *Laser Focus World* 26(4):131–141, April 1990 (status report).

S. R. Chinn: "Stoichiometric lasers," in M. J. Weber (ed.), *CRC Handbook of Laser Science and Technology*, vol. 1, CRC Press, Boca Raton, Fla., 1982, pp. 147–169 (review article).

A. Cordova-Plaza et al.: "Nd:MgO:LiNbO₃ continuous wave laser pumped by a laser diode," *Optics Letters* 13(3):209–211, 1988 (research report).

W. W. Duley: *Laser Processing and Analysis of Materials*, Plenum, New York, 1983 (overview of technology).

M. Gong et al.: "Nd:Mg:LiNbO$_3$ self-frequency-doubled laser pumped by a flashlamp at room temperature," *Electronics Letters* 26(25):2062–2063, December 6, 1990.

Tim Gray and Curt Frederickson: "Pumping Nd:YAG lasers: lamp or diode array?" *Lasers & Optronics* 8(11):40–48, November 1990.

William B. Jones Jr.: "Slab geometry lasers," in Peter K. Cheo (ed.), *Handbook of Solid-State Lasers*, Marcel Dekker, New York, 1989 (review).

T. Juhasz, S. T. Lai, and M. A. Pessot: "Efficient short-pulse generation from a diode-pumped Nd:YLF laser with a piezoelectrically induced diffraction modulator," *Optics Letters* 15(24):1458–1460, December 15, 1990 (research report).

Thomas J. Kane and Robert L. Byer: "Monolithic, unidirectional, single mode Nd:YAG ring laser," *Optics Letters* 10(2):65–67, 1985 (research report).

Walter Koechner: *Solid-State Laser Engineering*, 2d ed., Springer-Verlag, Berlin and New York, 1988 (reference book).

P. Maine and G. Mourou: "Amplification of 1-nsec pulses in Nd:glass followed by compression to 1 psec," *Optics Letters* 13(6):467–469 June 1988 (research report).

W. E. Martin et al.: "Solid-state disk amplifiers for laser fusion systems," *IEEE Journal of Quantum Electronics* QE-17(9):1744–1755, 1981 (system description).

Gordon McFadden and David M. Filgas: "Designing Nd–YAG lasers for higher power," *Lasers & Optronics* 8(8):33–40, August 1990 (review).

John McMahon: "Solid-state lasers," in Kai Chang (ed.), *Handbook of Microwave and Optical Components*, vol. 3, *Optical Components*, Wiley-Interscience, New York, 1990, pp. 323–398 (review).

Lieselotte Moenke-Blankenburg: *Laser Micro Analysis*, Wiley-Interscience, New York, 1989 (monograph).

Peter F. Moulton: "Paramagnetic ion lasers," in M. J. Weber (ed.), *CRC Handbook of Laser Science and Technology*, vol. 1, CRC Press, Boca Raton, Fla., 1982, pp. 21–147 (review article).

Ron Olson and Jeremy Weston: "Designing custom solidstate lasers," *Lasers & Optronics* 8(5):73–80, May 1990 (review paper).

Michael D. Perry: "High-intensity, short-pulse lasers," *Energy & Technology Review*, November 1988, pp. 9–15 (published by Lawrence Livermore National Laboratory, Livermore, Calif.) (review).

David Sliney and Myron Wolbarsht: *Safety with Lasers and Other Optical Sources*, Plenum, New York, 1980 (reference).

S. E. Stokowski: "Glass lasers," in M. J. Weber (ed.), *CRC Handbook of Laser Science and Technology*, vol. 1, CRC Press, Boca Raton, Fla., 1982, pp. 215–264 (review article).

D. C. Winburn: *Practical Laser Safety*, 2d ed., Marcel Dekker, New York, 1990 (safety guidelines).

Ruby Lasers

The ruby laser has proved surprisingly long-lived. Thirty years ago, it was the first laser demonstrated by Theodore H. Maiman (Maiman, 1960), who soon helped make the first commercial versions. Ruby lasers remain on the market today, although limited to a few applications, while most other types envisioned by laser pioneers are little more than laboratory curiosities.

Ruby emits pulses at 694.3 nanometers (nm) in the deep-red when pumped by a flashlamp. Continuous-wave operation has been demonstrated in the laboratory, but is difficult to achieve. Ruby was the first laser used for materials working, but has been replaced by other types except for a few special applications. It remains a valuable source of high-power red pulses for pulsed holography, interferometry, nondestructive testing, plasma measurements, and research.

Internal Workings

Ruby lasers are similar to the pulsed neodymium lasers described in Chap. 22, although repetition rates are much lower and ruby cannot be diode-pumped. In some cases, ruby lasers can operate in the same cavities as neodymium lasers, if suitable resonator optics are used.

Active Medium. Ruby laser rods are grown from sapphire (Al_2O_3) doped with about 0.01 to 0.5 percent chromium to from a synthetic ruby crystal colored red or pink with about 10^{19} chromium atoms per cubic centimeter. Rods are up to 20 centimeters (cm) long and 3 to 25 millimeters (mm) in diameter. Ruby resists optical damage at normal power levels if its surface is clean, and conducts heat better than does glass or Nd–YAG. However, it is a three-level laser system, which severely limits performance. It also suffers from self-absorption in unpumped regions.

Energy Transfer.   Visible photons from the pump lamp raise Cr^{3+} ions to one of the two excited levels shown in Fig. 23.1, which decay in about 100 nanoseconds (ns) to a pair of metastable levels with room-temperature lifetimes about 3 milliseconds (ms). If emission is allowed on both lines, the 694.3-nm transition from the E state will dominate.

High pump powers are needed to produce a population inversion in the three-level ruby system because the lower level is the ground state. This leads to a high laser threshold and low efficiency because the population must be inverted relative to the ground state, but energy storage is high, allowing generation of high-power pulses. Excess pump energy remains in the rod as heat, limiting repetition rate to a few hertz except for very small rods.

Internal Structure

Maiman silvered the ends of the rod in his first laser, but modern ruby lasers use separate mirrors, one totally reflecting and one partly transparent. The mirrors usually are flat, or slightly concave to compensate for thermal lensing in ruby, and form a stable resonator. Ruby lasers work well in oscillator-amplifier configurations, which can produce high-energy pulses with better beam quality than can single-stage oscillators.

Ruby lasers use the same kinds of pump cavities as lamp-pumped neodymium lasers, as shown in Fig. 22.1. The higher pump power requirements make efficient coupling and reflection more critical for ruby. Cooling is so important that some designs have water flowing

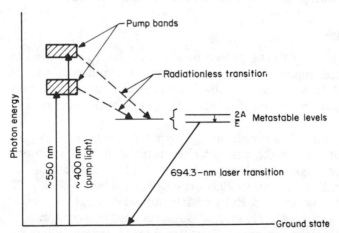

Figure 23.1   Energy levels of Cr^{3+} in ruby, showing pump and laser transitions. The ground state is the lower laser level in this three-level laser, thus limiting efficiency.

through the coupling cavity, either filling it or flowing annularly around the rod and flashlamps.

Beam Characteristics

Wavelength and Output Power. Ruby lasers emit at 694.3 nm. Oscillators can produce millisecond pulses of 50 to 100 joules (J), and oscillator-amplifier configurations can emit well over 100 J, both in multiple transverse modes; Q-switching or limiting output to TEM_{00} mode reduces power considerably. Average power is limited by low repetition rates to around 100 watts (W) for a single-rod oscillator. Typical wall-plug efficiency of ruby lasers is 0.1 to 1 percent.

Temporal Characteristics. Without pulse-control accessories, ruby laser pulse duration depends on the pump-lamp pulse and ranges between 0.3 ms to a few milliseconds. Pulse energy is highest in this mode, but average power during the pulse is no more than tens of kilowatts. Internal effects generate strong spikes during these long pulses, so instantaneous power can vary considerably. Some engineers consider this spiking desirable for drilling holes.

Q switching limits pulses from a single oscillator to a few joules, but by compressing length to 10 to 35 ns, it can raise peak power to the 100-megawatt (MW) range. Oscillator-amplifiers can have peak powers of 1 gigawatt (GW). Ruby lasers also can be modelocked, producing trains of 20 to 30 pulses of 1 millijoule (mJ), each about 3 to 4 picoseconds (ps); although peak powers reach hundreds of megawatts, modelocking is not common because of poor pulse-to-pulse stability.

Spectral Bandwidth. The ruby laser transition is about 330 GHz or 0.53 nm wide, but bandwidth can be limited by about 30 megahertz (MHz) by using etalons in the laser cavity for holographic lasers.

Beam Quality, Coherence, and Modes. Ruby laser beams range from near-diffraction-limited TEM_{00} output for holography to multiple transverse-mode beams for brute-force power. Operation at high repetition rates or with inadequate cooling degrades beam quality. Holographic ruby lasers have coherence lengths to 10 m, but high-power models have much shorter coherence lengths. Beam diameters range from 1 to 25 mm, with divergence from 0.25 to 10 milliradians (mrad).

Suitability for Use with Laser Accessories. Ruby lasers work well with Q switches and can be modelocked; Q-switched pulses can be readily frequency-doubled to 347 nm in rubidium hydrogen arsenate (RDA; RbH_2AsO_4) with about 30 percent conversion efficiency.

Operating Requirements

Input Power.   Ruby lasers typically operate from single-phase 110-volt (V) ac lines, but some require 220 V ac. Charging the capacitor bank which powers the flashlamps may require 10 to 20 amperes (A), but current requirements drop after charging is complete.

Cooling.   Ruby lasers require active cooling. Heat does not damage the crystal, but it can raise laser threshold, decrease output power, and degrade beam quality. Typical designs flow deionized water through the laser head and through a heat exchanger which transfers the excess heat to air or flowing tap water. Deionized water is needed because it passes over electrical connections.

Consumables.   Flashlamps must be replaced periodically, but repetition rates are so low that the period is long. Closed-cycle lasers consume water filters and deionizer cartridges. Focusing lenses for drilling lasers are functionally consumables because they become coated with debris.

Operating Conditions.   Ruby lasers are built for laboratory or industrial environments. Stability is important for holographic lasers because of stringent demands on beam quality.

Safety.   Ruby lasers present a serious eye hazard because their high-power pulses can penetrate the eye and permanently damage the retina. As with Nd–YAG lasers, a single pulse can permanently impair vision. Ruby pulses can carry enough energy to burn the skin.

The 5- to 10-kilovolt (kV) potentials applied across flashlamps carry large enough currents to electrocute a person. Because the laser operates in pulsed mode and the power supply charges between pulses, many components may retain large charges after the laser is turned off. Most current commercial lasers automatically dump capacitor charge when power is off or the case is opened, but do not take this feature for granted because many older lasers still exist. Flashlamps can explode if driven beyond their rated input, but a sealed pump cavity should contain any shrapnel. Special care should be taken with old high-voltage equipment, especially if it does not appear to have been used for a long time.

Special Considerations.   Ruby lasers are the sort of large and costly equipment that may be moved to a storeroom and forgotten for many years. Old ruby lasers may still be useful, especially where budgets are limited. However, long-neglected lasers should be examined carefully before firing them. Some high-voltage components (e.g., electrolytic capacitors) may degenerate with age, and could fail catastrophically. Old lasers may not meet modern electrical or optical safety requirements. The best course is to have such equipment checked by someone familiar with its operation. Industry directories list companies which offer repair services.

Reliability and Maintenance

Ruby lasers are reliable when given proper care, and some have lasted many years longer than the companies which made them. Keeping optics clean is critical for high-power operation, because energy absorption by dirt or dust can damage optical surfaces. Silver coatings used in some pump cavities can tarnish and must be repolished or reapplied periodically to maintain high reflectivity. The resonant cavity may require periodic realignment to maintain maximum output; many ruby laser manufacturers offer optional helium-neon lasers and autocollimators for that purpose.

Commercial Devices

Standard Configurations. Ruby lasers may be designed for general-purpose use or for specific applications. The most common specialized versions are built for holography, where factors such as coherence length, beam quality, and pulse timing are crucial. Ruby lasers also have been packaged for hole-drilling applications.

Typical ruby laser heads are 1 to 2 m long with an internal rail or other stabilizing structure. The head and optics usually are sealed in high-power lasers but may be exposed in lower power systems to allow user adjustment. In many high-power lasers, optical elements are connected by hollow plastic or aluminum tubes to keep dust away from the beam and optics. Oscillator heads typically weigh 20 to 50 kg. The power supply and controls are packaged separately and usually are much larger and heavier. Total shipping weights range from about 100 kg for a small oscillator to nearly a ton for a large oscillator-amplifier.

Solid-state laser manufacturers often use modular designs and may assemble their ruby lasers from components which also are used in neodymium lasers.

Options. Common ruby laser options include He–Ne alignment laser, Q switches, autocollimators, pulse counters, double- or triple-pulse output, TEM_{00} output, etalons, remote control, integration with other system components, and choice of cooling system.

Pricing.   Ruby laser prices in 1990 ranged from $15,000 to $70,000.

Suppliers.   The ruby laser market is small, and the several companies that offer them also produce neodymium lasers. The basic technology and designs are well established, but recent years have seen the addition of computer controls.

Applications

The largest use of ruby lasers has been in holography. A Q switch can produce one, two, three, or more highly coherent pulses during a single flashlamp pump pulse. These 10- to 30-ns pulses can record holograms to freeze the motion of moving objects. Double pulses can record a pair of pulses on the same plate, making it possible to observe deformation in techniques widely used in nondestructive testing (Abramson, 1981).

Thomson scattering of ruby laser pulses can record the density and temperature of electrons in high-energy plasmas. Ruby lasers also can perform atmospheric ranging, scattering and lidar measurements where the high peak power and red wavelength are important. Other solid-state lasers have largely supplanted ruby in some older applications such as hole drilling, military range finding, and military target designators. However, ruby's wavelength and its spiking behavior in long-pulse mode may make it attractive for some specialized applications.

Ruby lasers are being investigated for tattoo removal, because the ruby wavelength is strongly absorbed by dyes used in some tattoos. Ruby lasers also are used in general research, including plasma production, studies of liquid flow, and fluorescence spectroscopy.

Bibliography

Nils Abramson: *The Making and Evaluation of Holograms*, Academic Press, New York, 1981 (reference).
L. Allen: *Essentials of Lasers*, Pergamon Press, Oxford, 1969 (introductory book on lasers).
Yvonne A. Carts: "Ruby lasers shine on," *Laser Focus World 26*(7):83–91, July 1990.
Walter Koechner: *Solid-State Laser Engineering*, 2d ed., Springer-Verlag, Berlin and New York, 1988 (review of all solid-state lasers).
Theodore H. Maiman: "Stimulated optical radiation in ruby," *Nature 187*:493, 1960 (research report).

Tunable Vibronic Solid-State Lasers

One of the newer additions to the commercial laser world is a family of solid-state lasers with tunable wavelength. Their tunability originates from their operation on "vibronic" transitions in which the active species (an atom in a solid host) changes both electronic and vibrational states. Dye lasers operate on similar transitions. Solid-state tunable vibronic lasers are much easier to use than dye lasers, which has stimulated their rapid commercial development. However, tunable solid-state lasers cannot yet generate all wavelengths available from dye lasers. Their fundamental wavelengths are in the near infrared; visible and near ultraviolet light can be produced only by harmonic generation or other nonlinear processes.

The first tunable solid-state laser to reach the market was alexandrite, chromium-doped $BeAl_2O_4$, which can be tuned between 701 and 826 nanometers (nm) at room temperature, although not across the entire range under the same conditions. It has since been joined by titanium-doped sapphire (Al_2O_3), which is tunable from 660 to 1180 nm, the broadest tuning range of any single conventional laser medium. (Dye lasers can be tuned across a broader range only by switching dyes.) Titanium-sapphire lasers gained rapid acceptance at the end of the 1980s and are displacing dye lasers from the market for near-infrared tunable lasers. Indeed, one company which makes both dye and Ti-sapphire has advertised that "dye is dead" for the near-infrared. Other vibronic solid-state lasers are in development or offered commercially on a small scale, including cobalt-doped magnesium fluoride and chromium doped in other hosts.

Thus far, most tunable solid-state lasers have been used in research and development, particularly in spectroscopy and other studies where tunability is essential. Tunable near-infrared lasers are receiving serious attention for some medical applications, where their wavelengths may fulfill specific needs. Military researchers have investigated alexandrite lasers for use as antisensor weapons because of their ability to generate high-power tunable pulses. Titanium-sapphire lasers have been studied for potential use as laser radars.

Introduction and Description

As in most other solid-state lasers, the active species in vibronic lasers is an impurity in a crystalline (or glass) host pumped by light from a flashlamp or another laser. The tunability comes from the energy-level structures. In conventional narrow-line solid-state lasers such as ruby or neodymium, the laser levels are discrete energy states (shown as isolated horizontal lines in Fig. 24.1). Vibronic laser transitions occur between electronic energy levels which are spread into bands by vibrational sublevels. The active species can change its vibrational state as it changes its electronic state, so the transition can take place over a range of energies. In an untuned cavity, a vibronic laser can emit at a range of wavelengths; a tunable cavity selects a narrow part of the band in which the laser has positive gain.

Tunability might seem an ideal addition to the features of a solid-state laser, but life is not that kind. With tunability comes a trade-off.

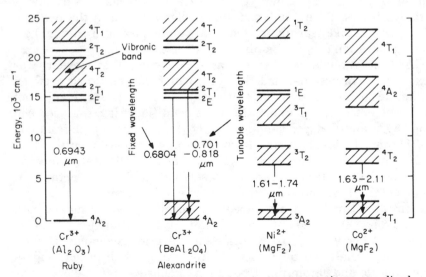

Figure 24.1 Energy levels and laser transitions in a representative narrow-line laser (ruby) compared with those in three vibronic lasers, alexandrite, Ni–MgF$_2$, and Co–MgF$_2$. Note that transitions between discrete energy levels (lines in the figure) in ruby and alexandrite produce narrow-line emission, while transitions between vibronic bands allow tunable emission.

The product of the excited-state lifetime and the cross section for stimulated emission in a solid-state laser is roughly proportional to the inverse of the gain bandwidth. Thus, the broader the gain bandwidth, the lower the product of lifetime and cross section. This, in turn, means that no tunable solid-state laser can match *both* the high cross-section and the long excited-state lifetime of a narrow-line laser such as Nd–YAG.

The balance between lifetime and cross section differs among materials and strongly affects the choice of pump source. Titanium-sapphire has a large cross section, comparable to Nd–YAG, but its excited-state lifetime is only 3.2 microseconds (μs), too short for pumping with a standard flashlamp. Most Ti-sapphire lasers are pumped with short-pulse or continuous-wave lasers. Alexandrite has a longer excited-state lifetime, but a lower cross section, so it can be pumped with standard high-brightness flashlamps. Most other tunable solid-state lasers work best with laser pumping.

Developers are continuing efforts to improve vibronic solid-state lasers and to develop new materials. The technology is new enough that more progress can be expected, although some performance limitations may not be easy to surmount. Existing materials have an important head start because it takes time and money to grow laser crystals, characterize their performance, and design lasers around them.

Internal Workings and Structure

Table 24.1 lists important vibronic solid-state lasers. All are four-level laser systems in optically pumped crystals. Many rely on chromium,

TABLE 24.1   Types of Vibronic Solid-State Lasers

Type	Pump source	Operation	Wavelength, nm
Alexandrite	Arc lamp	CW	730–810
Alexandrite	Flashlamp	Pulsed	701–858*
Ce–YLF	KrF excimer laser	Pulsed	309–325
Co–MgF$_2$	1320-nm Nd–YAG	Pulsed†	1750–2500‡
Cr–LiCaAlF$_6$	Laser or lamp	Pulsed or CW	720–840+
Cr–LiSrAlF$_6$	Laser or lamp	Pulsed or CW	760–920+
Emerald (Cr-doped)	Laser	Pulsed or CW	720–842
Forsterite (Cr-doped)	Laser	Pulsed or CW	1167–1345
Thulium-YAG	Laser	CW	1870–2160
Ti-sapphire	Usually laser	Pulsed or CW	660–1180

*Wavelengths longer than 826 nm possible only at elevated temperatures.

†At room temperature; continuous-wave operation is possible at cryogenic temperatures.

‡At room temperature; wavelengths as short as 1500 nm are possible at cryogenic temperatures.

the active medium in nontunable ruby lasers, but operate on a different transition than ruby.

The four-level transitions work in essentially the same way. Pump light excites the active species to a vibrationally excited level of an excited electronic state. The atom then releases some vibrational energy (as what is called a "phonon" or unit of vibrational energy), dropping or relaxing to the bottom of the band, which is the upper laser level. The laser transition is to a vibrationally excited sublevel of the ground electronic state, as shown in Fig. 24.1, which relaxes back to the ground state by releasing vibrational energy.

If the laser cavity does not include wavelength-selective optics, oscillation is at a range of wavelengths near the peak of the gain curve. Tuning optics limit the range of wavelengths which can oscillate in the cavity. The wavelength range depends on temperature because the vibrational sublevels of the ground electronic state—the lower laser level—are thermally populated. Higher temperatures populate higher vibrational levels, preventing population inversions of those levels. This restricts the tuning range by making it impossible to sustain laser action at the shorter wavelengths (which involve transitions to the lower vibronic levels which are thermally populated). As a result, vibronic laser wavelengths become longer at higher temperatures. For alexandrite, the shift is typically 0.2 to 0.3 nm/°C.

The optimum laser-cavity design differs among vibronic lasers. Flashlamp-pumped alexandrite lasers use cylindrical rods several centimeters long housed in cavities similar to those used for Nd–YAG lasers. Design details differ because alexandrite has lower gain and is tunable in wavelength, unlike fixed-wavelength Nd–YAG. On the other hand, Ti-sapphire lasers are designed for laser pumping and strongly resemble tunable dye lasers. Indeed, "Ti-dye" lasers are available in which an external laser can pump either a small Ti-sapphire crystal or a dye jet. Continuous-wave Ti-sapphire lasers are manufactured with both linear standing-wave cavities and with ring cavities similar to those for continuous-wave dye lasers. Because the material has higher gain than alexandrite, smaller crystals can be used. Titanium-sapphire also has high enough gain to operate with an unstable resonator (Rines and Moulton, 1990).

Alexandrite. Alexandrite is a beryllium-containing compound called *chrysoberyl* ($BeAl_2O_4$). It was first studied as a possible replacement for ruby as a chromium host, but laser performance was poor on the 680.4-nm fixed-wavelength transition. Later, laser action was discovered on the $^4T_2 \rightarrow {}^4A_2$ vibronic transition at longer wavelengths (Walling et al., 1979). That transition attracted more attention because it is both more efficient and tunable.

Alexandrite laser rods contain 0.01 to 0.4 percent chromium. The normal pump band is at 380 to 630 nm, similar to that of ruby, and compatible with conventional flashlamps. A peak near 680 nm allows pumping with semiconductor lasers at that wavelength (Scheps et al., 1990), although output power of red diode lasers is limited. Alexandrite lasers can operate continuous-wave when pumped by an arc lamp (Samelson et al., 1988) or by a 647-nm krypton laser (Sam, 1989), but typically operation is pulsed.

Optical pumping and subsequent fast relaxation populate a mixture of electronically excited levels, primarily a short-lived $^4T_2$ level and a longer-lived ^2E level with slightly lower energy. Together these two levels act as an upper laser level with lifetime and emission rate depending on temperature. This unusual combination causes laser gain to increase with heating above room temperature, rather than decrease as in most solid-state lasers. Thermal population of the lower laser level becomes sufficient to reduce gain well above room temperature.

The alexandrite laser transition terminates in the lower vibronic band of the ground electronic state $^4A_2$, as shown in Fig. 24.1. The laser transition goes to vibrational sublevels in the upper or middle part of the band which are depopulated by vibrational relaxation. Tuning of the cavity controls transition wavelength. At room temperature, the tuning range is 701 to 826 nm, but tuning to 858 nm in long-pulse mode is possible at 360°C (Kuper et al., 1990).

The emission cross section of alexandrite is only 6×10^{-21} square centimeters (cm^2) at room temperature, an order of magnitude less than neodymium-glass. This makes gain much lower than in neodymium lasers, so energy extraction efficiency is a much more important design consideration and pumping thresholds are relatively high. However, the low emission cross section allows alexandrite to store much more energy per unit volume than that stored in neodymium lasers. This combines with a long excited-state lifetime (260 μs) to allow high-power Q-switched pulses. Alexandrite can amplify pulses from other lasers, generating high peak powers.

Several years of effort overcame initial problems to produce large high-quality crystals of alexandrite, up to 15 cm long, from which rods up to 9.5 mm in diameter can be produced. The crystal is hard and has good thermal qualities, but the very high peak powers in Q-switched pulses can cause optical damage to rod surfaces.

Alexandrite is unusual in that its laser performance improves as it is heated above room temperature, as shown in Fig. 24.2. The stimulated emission cross section reaches a peak of about 3×10^{-20} cm^2, comparable to that of Nd-glass, at 250°C. This unusual improvement is a consequence of internal energy-transfer characteristics. Alexandrite also ben-

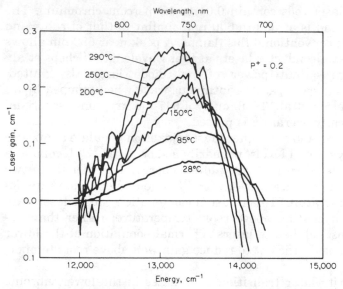

Figure 24.2 Gain of alexandrite as a function of wavelength for various temperatures, showing how gain increases and wavelength peak shifts slightly as temperature increases. (*From Shand and Jenssen, 1983, copyright 1983 IEEE. Reproduced with permission.*)

efits from good thermal properties which make it less subject to thermal lensing than most other solid-state laser materials.

Ti-Sapphire. Titanium-doped sapphire (Al_2O_3) has the broadest tuning range of any conventional laser, 660 to 1180 nm. It was first demonstrated in 1982 (Moulton, 1982); commercial continuous-wave models were available by 1988. They quickly began replacing dye lasers in the near-infrared because they are much easier to use and offer many of the same features, including tunability and the ability to generate subpicosecond modelocked pulses. The absence of a flowing dye solution greatly enhances stability, reducing noise by a factor of 10 and allowing linewidths as narrow as 1 kilohertz (kHz) in a ring configuration.

Sapphire has the same crystalline structure as ruby, which is Al_2O_3 doped with enough chromium to color it red or pink. Titanium-sapphire contains on the order of 0.1% titanium, added in the form of Ti_2O_3 to produce the desired Ti^{3+} ion, which replaces aluminum in the crystal lattice. (Ti^{4+} ions can cause detrimental absorption.) The Ti^{3+} ion is in the same class of $3d$ transition-metal ions that includes Cr^{3+}, Co^{2+}, and Ni^{2+}, active media in other vibronic solid-state lasers. Titanium-sapphire crystals are red in color. Like ruby, they are hard and durable, with good optical properties. The crystal can be grown readily, although care must be taken to avoid producing Ti^{4+}. Crystals can be grown up to 30 cm long, but the gain is so high that typical continuous-wave lasers use crystals less than 1 cm long.

The laser transition is between the ^2E excited state and the $^2T_2$ ground state. As shown in Fig. 24.3, optical pumping excites Ti^{3+} from the ground state to a vibrationally excited sublevel of the upper laser level. The ion then drops to a lower sublevel of the upper laser level before making the laser transition to a vibrationally excited sublevel of the ground state. Vibrational relaxation returns the ion to the ground state. The strong interaction between the titanium atom and host crystal, combined with the large difference in electron distribution between the two energy levels, leads to a broad transition linewidth.

The absorption and emission bands of Ti-sapphire overlap only slightly. Absorption peaks near 500 nm, at wavelengths readily available from argon ion, frequency-doubled neodymium, or copper vapor lasers. All three have been used as pump sources. GaAlAs diode lasers cannot directly pump Ti-sapphire because their emission is outside its absorption band. However, 173 milliwatts (mW) from a doubled diode-

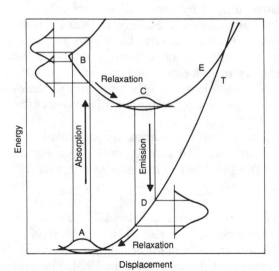

Figure 24.3 Ti-sapphire laser is optically excited to an upper vibronic level of the upper laser level, which then relaxes to the upper laser level. Laser emission leaves the ion in a vibrationally excited sublevel of the ground state. (*From Wall and Sanchez, 1990, reprinted with permission of Lincoln Laboratory, Massachusetts Institute of Technology, Lexington, Massachusetts.*)

pumped YAG laser was sufficient to generate 4.2 mW from a low-threshold Ti-sapphire laser (Harrison et al., 1991). Fluorescence peaks at 780 nm, but Ti-sapphire output is readily tunable, with power highest at 700 to 900 nm.

The high peak gain cross section of Ti-sapphire (ca. 3.5×10^{19} cm^2) is comparable to Nd–YAG and much higher than in other vibronic solid-state lasers. The high gain allows use of unstable resonators. However, high pump intensities are needed because radiative lifetime of the laser transition is only 3.8 μs. That is too short for pumping with a standard flashlamp, so Ti-sapphire lasers normally are pumped with either short-pulse or continuous-wave lasers. Sapphire crystal can withstand high pump intensities, one of its important advantages as a titanium host.

Titanium-sapphire lasers can be pumped with special flashlamps, which generate microsecond pulses, but lamps operated in that range generate excess short-wavelength light which can produce defects in the sapphire crystal. Lamp output can be filtered, or converted to longer wavelengths with a liquid dye fluorescence converter. Lamp pumping has generated pulses to 6.5 joules (Brown et al., 1990), but that requires high titanium doping and rods up to 2 cm in diameter and 17 cm long, which are costly and hard to produce.

Some Ti-sapphire lasers are pumped by short-pulse Q-switched lasers, but most are pumped by continuous-wave lasers and emit continuously. A variety of modelocking techniques have been demonstrated, and pulses under 100 femtoseconds (fs) have been produced. The colliding-pulse and synchronous modelocking techniques used with dye lasers do not work with Ti-sapphire, so other techniques have been developed, including self-modelocking using nonlinearities of the material.

Pulsed Ti-sapphire lasers can be used to generate second, third, and fourth harmonics. Tuning ranges are 340 to 470 nm for the second harmonic, 235 to 300 nm for the third harmonic, and as short as 207 nm for the fourth harmonic.

Co–MgF$_2$. Cobalt-doped magnesium fluoride is a broadly tunable laser which operates at room temperature between 1750 and 2500 nm, a wavelength range not readily available from other tunable lasers. Laser action from Co^{2+}–MgF_2 was first demonstrated in 1964, but that version required the cumbersome combination of cryogenic cooling and flashlamp pumping, so it was neglected for 20 years. Later investigation led to pumping with a 1.32-μm Nd–YAG laser and a pulsed Co–MgF$_2$ laser at room temperature (Welford and Moulton, 1988). Commercial versions were introduced soon after room-temperature operation was demonstrated.

Laser action is between the $^4T_2$ and $^4T_1$ states of Co^{2+}. At 80 kelvins (K), fluorescence extends from under 1.5 μm to 2.5 μm. A strong absorption peak between 1.2 and 1.4 μm makes the 1.32-μm Nd–YAG laser a good pump source (Moulton, 1985). The upper-state lifetime depends strongly on temperature, dropping from 1.65 ms at 77 K to 210 μs at 225 K and 36.5 μs at room temperature. Fluorescence intensity also drops as temperature increases. Gain cross section is very low at room temperature, only 9×10^{-22} cm^2 (Welford and Moulton, 1988).

These factors make Co–MgF$_2$ a more powerful laser at cryogenic temperatures, where it produces continuous-wave powers to 4 W (Moulton, 1985) and emits wavelengths as short as 1.5 μm. However, commercial interest centers on the more convenient room-temperature version, which emits lower power and operates only in pulsed mode.

Forsterite. Chromium-doped forsterite (Mg$_2$SiO$_4$) is a tunable laser that emits at 1167 to 1345 nm, wavelengths longer than those of Ti-sapphire. It is one a large family of chromium-doped vibronic lasers, but is unusual in that the light-emitting species is Cr^{4+}, not Cr^{3+}, as in alexandrite, ruby, and most other Cr-doped materials. Doping levels are relatively low—3×10^{18} to 6×10^{18} per cubic centimeter (cm^{-3}), a factor of 20 to 40 lower than neodymium in Nd–YAG. Some chromium ions replace Mg^{2+} as Cr^{3+}; others replace silicon as Cr^{4+}.

Forsterite has three principal pump bands, 350 to 550 nm, 600 to 850 nm, and 850 to 1200 nm. Pulsed and continuous-wave laser action is possible with laser pumping at 532, 578, 629, and 1064 nm. Slope efficiency has reached 23 percent in pulsed operation and 38 percent in continuouswave operation (Petricevic et al., 1990). The broad pump bands appear attractive for flashlamp pumping, but the short excited-state lifetime (only 3 μs) poses a problem. The peak gain cross section is 1.45×10^{-19} cm^2, high for vibronic lasers, but not as large as for Nd–YAG or Ti-sapphire.

Forsterite lasers have recently been modelocked, generating pulses as short as 25 ps at a 76-MHz repetition rate, with average power above 250 mW. They also have been Q-switched, producing 60- to 80-ns pulses. The high peak powers available in such pulses could facilitate generation of second harmonic at 575 to 675 nm in the orange and red parts of the spectrum, where dye lasers so far have been the only tunable sources. Continued encouraging results could lead to commercial products.

Other Chromium Hosts. The attractive qualities of the Cr^{3+} ion have led to many experiments with many potential laser hosts. Laser-pumped vibronic Cr^{3+} lasers have been demonstrated in over 20 hosts, but most have not proved practical. A major problem in many hosts is absorption by excited states, which reduces laser efficiency. Other problems include crystal absorption, poor optical quality, and vulnerability to optical damage from pump light.

Emerald [Cr-doped $Be_3Al_2(SiO_3)_6$] has attractive laser characteristics and can be tuned between 729 and 842 nm, roughly the same range as alexandrite (Shand and Lai., 1984). However, crystal growth has been difficult.

Two chromium hosts which seem particularly promising at this writing are the similar crystals, $LiSrAlF_6$ (LiSAF) (Payne et al., 1989) and $LiCaAlF_6$ (LiCAF) (Chase and Payne, 1990). As in most other vibronic Cr^{3+} laser media, the laser transition is between the $^4T_2$ and $^4A_2$ states. Tuning ranges are 720 to 840 nm for LiCAF and 780 to 920 nm for LiSAF.

The two materials are somewhat complementary. Chromium-doped LiSAF has emission and absorption spectra that are broader and at longer wavelengths than those of LiCAF; its absorption and emission cross sections also are about four times higher. However, LiCAF has a longer excited-state lifetime, 170 μs, in contrast to 67 μs for LiSAF. LiCAF also has higher projected slope efficiency, 67 percent, in contrast to 53 percent for LiSAF.

The chromium-doped crystals absorb strongly in the red and blue but are transparent at green wavelengths, giving them an emerald-green color. Pumping has been demonstrated with the 647-nm red line of krypton ion lasers, with flashlamps, and with 670-nm diode lasers. The crystals are not as rugged as oxide-based materials, and have lower thermal conductivity, but they also suffer less thermal lensing. LiCAF can tolerate large chromium dopings, adding design flexibility. Recent studies predict that rods of good optical quality 1 cm in diameter and 10 cm long should be available soon.

Unlike forsterite or Co–MgF_2, Cr–LiSAF and Cr–LiCAF do not promise unique wavelengths. Both operate in ranges covered by Ti-sapphire and partly by alexandrite. However, developers state that Cr–LiSAF offers a better balance between lifetime and emission cross section for Q switching and pulse amplification (Chase and Payne, 1990). Both Cr–LiSAF and Cr–LiCAF promise good beam quality at sizable power levels, although the materials are not as robust as Ti-sapphire or Nd–YAG. They remain in development.

Other Vibronic Lasers. Vibronic solid-state lasers have been demonstrated with a few other ions as active media, but progress has been limited. These include

- *Thulium:* Tm^{3+}-doped YAG is tunable between 1.87 and 2.16 µm; the similar Tm^{3+}–yttrium scandium gallium garnet (YSGG; $Y_3Sc_2Ga_3O_{12}$) laser is tunable between 1.85 and 2.14 µm (Stoneman and Esterowitz, 1990). Pumping is at 785 nm, which should allow diode laser pumping, or with a flashlamp.

- *Cerium:* Ce^{3+}-doped $LiYF_4$, pumped by a KrF excimer laser and tunable between 309 and 325 nm (Ehrlich et al., 1978).

- *Samarium:* Sm^{2+}-doped CaF_2, tunable between about 710 and 750 nm.

- *Vanadium:* V^{2+}-doped MgF_2, tunable near 1100 to 1200 nm, and V^{2+}-doped $CsCaF_3$, tunable around 1.3 µm.

- *Nickel:* Ni^{2+}-doped MgF_2, tunable in the 1.7- to 1.8-µm range.

- *Ytterbium:*   Yb-doped single-mode fibers, tunable between 1040 and 1170 nm, emitting continuously to over 50 mW when pumped by a diode laser (Gapontsev et al., 1991).

Research Trends and Types Covered

The impressive progress in tunable solid-state lasers during the 1980s reflects both the breakthrough discovery of new laser species (particularly titanium) and the steady improvement of already known laser species (notably chromium and cobalt). Breakthroughs, by their very nature, cannot be predicted, but refinement of existing lasers will continue.

One broad goal is to extend tuning of solid-state lasers throughout the visible, near-ultraviolet, and near-infrared, replacing dye lasers and extending the wavelength range available. Material absorption imposes a limit on short wavelengths, while the internal kinetics make solid-state lasers difficult to develop at wavelengths longer than 3 µm. Titanium-sapphire lasers have made impressive progress as continuous-wave sources in the near-infrared. Alexandrite and Co–MgF_2 are valuable pulsed sources. Harmonic generation and (for alexandrite) Raman shifting can generate other wavelengths from those lasers. Other materials such as forsterite and Cr–LiSAF also may be able to provide other wavelengths at either the fundamental frequency or harmonics.

Noteworthy progress has been made in extending short-pulse techniques developed for dye lasers to solid-state lasers with broad gain bandwidths. Modelocked Ti-sapphire lasers are coming onto the market as this book is being completed, and pulse durations have been pushed below 100 fs. Researchers say that the broad gain bandwidth of Ti-sapphire may allow pulse lengths as short as 5 fs. That would beat the 6-fs record for dye lasers. Other researchers are investigating modelocking of other tunable vibronic solid-state lasers.

The rest of this chapter concentrates on the three vibronic solid-state lasers now available commercially: Ti-sapphire, alexandrite, and $Co–MgF_2$. Other vibronic lasers may become available if development efforts prove successful.

Beam Characteristics

Wavelength and Output Power.   The tuning ranges and output powers of commercial vibronic lasers typically are more restricted than those which have been demonstrated in the laboratory, or built under special contracts. In the case of alexandrite, some large, high-power lasers have been custom-built for military and other government laboratories using technology which is not readily available. For example, a 40-W continuous-wave (CW) alexandrite laser was built for the Lawrence Livermore Laboratory and a 100-joule (J) pulsed alexandrite laser was built for the Air Force Space Technology Center.

- *Alexandrite:*   Commercial pulsed alexandrite lasers are tunable between 720 and 800 nm, generating average power to 20 W. Pulsed alexandrite lasers can generate the second harmonic at 360 to 400 nm, with average power up to a few watts. Output depends on wavelength, as shown in Fig. 24.4. Fundamental wavelength pulse energy is 0.1 to 0.5 J *Q*-switched and 1 to 3 J in 100-μs pulses. Raman shifting can generate additional wavelengths, as shown in Fig. 24.5.

- *Ti-sapphire:*   Specified tuning range for commercial Ti-sapphire lasers can be as broad as 670 to 1100 nm, but some have a more limited tuning range. Tuning across the entire range usually requires multiple sets of optics. Output powers reaches several watts continuous-wave. Power depends on operating wavelength, as shown in Fig. 24.6 for pumping with a multiline argon ion laser; it

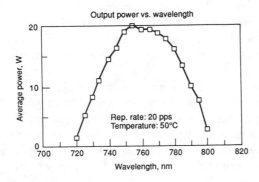

Figure 24.4 Output of an alexandrite laser as a function of wavelength. (*Courtesy of Light Age Inc.*)

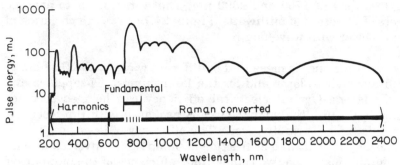

Figure 24.5 Raman shifting can generate additional tunable wavelengths from an alexandrite laser. (*Courtesy of Light Age Inc.*)

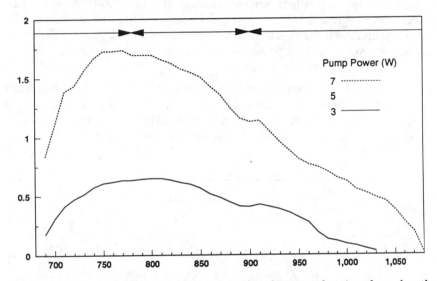

Figure 24.6 Output of a continuous-wave Ti-sapphire laser as a function of wavelength, for various pump powers from an argon ion laser. Arrows at top indicate tuning ranges of different cavity optics. (*Courtesy of Schwartz Electro-Optics.*)

also depends on pump power. Pulsed Ti-sapphire lasers can generate the second harmonic at 350 to 470 nm, the third harmonic at 235 to 300 nm, and the fourth harmonic near 210 nm. Laboratory versions have generated continuous powers above 40 W at the fundamental wavelength.

- *Co–MgF₂:* The tuning range of commercial Co–MgF$_2$ lasers, which operate at room temperature, is 1750 to 2500 nm. Shorter wavelengths are possible at cryogenic temperatures. Pulses, generated at a 10-Hz repetition rate, have peak power of 75 mJ at 2050 nm, dropping to 20 mJ at 1750 and 2500 nm. That corresponds to average powers of hundreds of milliwatts. Figure 24.7 shows the variation of output power with wavelength.

Efficiency. Efficiency is measured in different ways for the flashlamp-pumped alexandrite laser and for the laser-pumped Ti-sapphire and Co–MgF$_2$ lasers. Overall electrical efficiency can be measured for alexandrite because the pump source is packaged with the laser (although this figure is rarely given). The efficiency of laser-pumped lasers generally is measured as fraction of pump light converted into laser output. To make the two comparable, efficiency of the pump laser must be considered.

- *Alexandrite:* Overall electrical-to-optical efficiency is in the 1 percent range for alexandrite at its fundamental wavelength. Alexandrite has a high threshold because of its low gain, so efficiency increases for operation well above threshold. Slope efficiency, the increase in output power for an incremental increase in electrical input, can exceed 5 percent.

- *Ti-sapphire:* At the peak of its tuning curve (770 nm), output from a continuous Ti-sapphire laser may reach about 25 percent of pump power from a 7-W ion laser. Conversion efficiency is lower at lower pump powers and at wavelengths away from the peak.

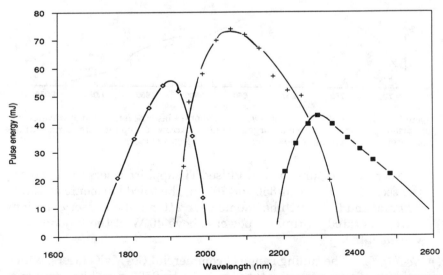

Figure 24.7 Output of a pulsed Co–MgF$_2$ laser as a function of wavelength; the three curves represent different sets of cavity optics. Pumping is with a 1.3-μm Nd–YAG laser. (*Courtesy of Schwartz Electro-Optics.*)

- *Co–MgF$_2$:* Conversion efficiency can reach 30 percent at the peak wavelength (2050 nm) for pumping with a pulsed 1.3-μm Nd–YAG laser at room temperature, with slope efficiency to 46 percent (Welford and Moulton, 1988). Efficiency is lower at other wavelengths.

Temporal Characteristics of Output.   The three major vibronic lasers have distinct temporal modes of operation:

- *Alexandrite:*   Although continuous-wave operation has been demonstrated in the laboratory, and custom models have been built, standard commercial alexandrite lasers are pulsed. They can generate 100- to 400-μs pulses in long-pulse mode, or Q-switched pulses of 30 to 250 ns. Repetition rates can exceed 100 Hz for commercial lasers; considerably higher rates have been demonstrated. Pulse stretching can extend the range of pulse durations. Subpicosecond pulses can be generated by amplifying the output of short-pulse Ti-sapphire lasers.

- *Ti-sapphire:*   Many Ti-sapphire lasers generate continuous beams. When pumped by a Q-switched Nd–YAG laser, they produce pulses with length and repetition rate set by the pump laser. A 10-Hz 15-ns Nd–YAG pump generates 10-ns Ti-sapphire pulses at the same repetition rate. Modelocking of Ti-sapphire lasers has been demonstrated experimentally, and products were emerging as this book was going to press. Some are expected to have durations of a couple of picoseconds, others may emit subpicosecond pulses.

- *Co–MgF$_2$:*   Standard Co–MgF$_2$ lasers produce 80-μs pulses at up to 20 Hz, with pulse duration set by the normal-mode pump laser.

Spectral Bandwidth.   The spectral bandwidths of vibronic lasers are controlled by the laser cavity and typically are lowest for Ti-sapphire lasers, which use the most sophisticated tuning optics.

- *Alexandrite lasers:*   Standard commercial pulsed models have linewidths of 0.2 to 0.5 nm; untuned lasers have broader bandwidths. Lasers with linewidths under 150 MHz are available on special order. Optical phase conjugation can correct distortions which limit narrowband pulse energy, allowing generation of a 1.8-J pulses with linewidth under 200 MHz in the laboratory (Rapoport et al., 1990). Injection locking of an alexandrite laser to a narrowband input also can generate narrowband output.

- *Ti-sapphire lasers:*   Spectral widths of continuous-wave Ti-sapphire lasers depend strongly on the laser cavity, often a type adapted from dye lasers. Commercial continuous-wave Ti-sapphire lasers with standing-wave cavities have linewidths of 2 to 450 GHz. Models with unidirectional ring cavities have linewidths of 0.5 to 40 MHz. Titanium-sapphire lacks the dye jets which cause instabilities in continuous-wave dye lasers, and ring lasers have produced linewidths as low as 1 kHz in the laboratory. Pulsed Ti-sapphire lasers much broader linewidths, about 0.5 nm, but this can be narrowed to about 0.001 nm by "seeding" the pulsed lasers with narrow-line light from a continuous-wave Ti-sapphire laser.

- $Co-MgF_2$: Linewidth is not specified in present commercial models, but the use of a birefringent tuner as the only line-narrowing element indicates that it is comparable to alexandrite lasers.

Beam Quality, Polarization, and Modes. Alexandrite lasers produce multimode or "near-TEM_{00}" beams when Q-switched and multimode pulses in long-pulse operation. Continuous Ti-sapphire lasers produce TEM_{00} beams with linear polarization, as do pulsed $Co-MgF_2$ lasers.

Beam Diameter and Divergence. Alexandrite lasers emit beams 3 to 6 mm in diameter with divergence of 1.2 to 5.0 milliradians (mrad), with beam quality 1.5 to 10 times the diffraction limit. Pulsed Ti-sapphire lasers have beam diameters 0.5 to 3 mm and divergences of 0.6 to 2.0 mrad; continuous Ti-sapphire lasers have beam diameters of 0.5 to 1 mm and divergences of 1 to 2 mrad. $Co-MgF_2$ lasers have 4.5-mrad divergence and 1.3-mm diameter.

Amplitude Stability. Continuous Ti-sapphire lasers have amplitude noise and short-term amplitude stability in the 1 to 2 percent range and long-term amplitude stability of about 5 percent. Pulse-to-pulse stability for alexandrite and pulsed Ti-sapphire is a few percent at the fundamental wavelength and roughly double at the second harmonic.

Suitability for Use with Laser Accessories. The principal accessories used with vibronic solid-state lasers are harmonic generators, Q switches, modelockers, and amplifiers.

- *Alexandrite:* Alexandrite lasers can be readily Q-switched, producing high peak power pulses. High pulse powers allow harmonic generation and Raman frequency shifting. Alexandrite is suitable for use in oscillator-amplifier configurations and has been modelocked both actively and passively in the laboratory. Raman shifting can generate additional wavelengths, as shown in Fig. 26.5.

- *Ti-sapphire:* Q-switching of Ti-sapphire has been demonstrated (N. Barnes et al., 1988), but the short lifetime of the upper laser level causes energy-storage losses, so it has found little practical use. However, short, powerful pulses can be generated by pumping with short, high-power pulses from an external laser, and Ti-sapphire can be modelocked (although not with some standard dye laser techniques). Pulse powers are sufficient to generate second, third, and fourth harmonics in suitable nonlinear materials. Master-oscillator power amplifier configurations have been demonstrated, but remain in development (J. Barnes et al., 1988); Ti-sapphire oscillators also can be used with alexandrite amplifiers at wavelengths where alexandrite has adequate gain. Titanium-sapphire lasers can be used as regenerative amplifiers for short pulses at repetition rates to 1 kHz (Vaillancourt et al., 1990).

- $Co–MgF_2$: Q-switching and modelocking of $Co–MgF_2$ are possible but have received limited attention because of the limited applications of the laser.

Operating Requirements

Vibronic lasers are quite new in the commercial realm; hence their operating requirements have not been well quantified. Some characteristics, such as input power requirements, are defined only sketchily or are not mentioned on most data sheets; others, such as lifetimes, maintenance needs, and durability, remain to be established.

Input Power.   Vibronic lasers require optical pumping by an internal flashlamp or an internal or external laser. Alexandrite lasers, which are flashlamp-pumped, require electrical connections similar to flashlamp-pumped Nd–YAG lasers of comparable power.

Titanium-sapphire lasers can be pumped by doubled Nd–YAG, doubled Nd–YLF, argon ion, copper vapor, or even flashlamp-pumped dye lasers. Normally the pump laser is external, but some models have dedicated internal pump lasers. Titanium-sapphire lasers with external pump lasers have minimal electrical power requirements, depending on their control systems. One model with an internal Nd–YAG pump laser draws up to 15 amperes (A) from a 115-volt (V) ac source.

$Co–MgF_2$ lasers are supplied with an integral pulsed 1.3-μm Nd–YAG pump laser. Electrical power requirements are comparable to those of small Nd–YAG lasers.

Cooling.   Typical vibronic lasers require active cooling, either flowing water, refrigeration, or forced air. All normally operate at room temperature. Some alexandrite lasers are supplied with air-cooled heat exchangers rather than water cooling.

Consumables and Maintenance Requirements. Pump lamps are the only consumables for vibronic lasers. They may be the lamps which pump the vibronic laser itself (e.g., alexandrite), or lamps which pump a dedicated Nd–YAG pump laser in Ti-sapphire or Co–MgF$_2$ lasers.

Required Accessories. Many Ti-sapphire lasers require separate pump lasers, but others are supplied with internal pump lasers.

Operating Conditions and Temperature. Tunable vibronic solid-state lasers are primarily laboratory instruments, although some models have been designed for field use. All commercial versions operate at room temperature.

Mechanical Considerations. Vibronic lasers are packaged for laboratory use. They range from small units 30 by 60 cm for benchtop use to larger units 0.5 m long with separate power supplies and coolers. Alexandrite lasers tend to be bulkier because they are flashlamp-pumped and contain their own lamps; many Ti-sapphire lasers and Co–MgF$_2$ lasers contain internal pump lasers.

Safety. Users of vibronic lasers at wavelengths shorter than 1.4 μm should follow the usual laser safety precautions for lasers with wavelengths able to penetrate to the retina of the eye. As with other tunable lasers, make sure that protective goggles block the wavelengths being used. Solid-state vibronic lasers are new enough that standard safety procedures for their use are not readily available; do not simply assume they are "just another" solid-state laser, or that Ti-sapphire deserves exactly the same treatment as dye lasers.

Users should be aware of the following special hazards and precautions:

- Short, powerful Q-switched pulses are especially dangerous because a single pulse can cause permanent eye damage. Users should be certain that all paths to the eye are blocked, including peripheral paths.

- Safety goggles are not readily available for specific vibronic lasers because all are new or in limited use. Users should be certain to check specifications to make sure that all wavelengths in use are blocked.

- Be aware of the possibility that stray light from the pump laser may reach the laboratory environment. In many cases, this pump light could pose greater hazards than the vibronic laser. If pump light does reach the laboratory environment, goggles which block the appropriate wavelengths should be worn.

- In flashlamp-pumped lasers, be alert to potential lamp hazards: high voltages applied to the lamps, charges retained within the power supply, and the risk of explosions of overloaded lamps.

- In laser-pumped lasers, be alert to the safety issues relating to the pump laser, whether it is separate or packaged with the vibronic laser.

- Users should beware of safety pitfalls at the near-infrared wavelengths of alexandrite and Ti-sapphire lasers. Retinal sensitivity drops off logarithmically beyond about 650 nm. Although 700 nm is often given as the long-wavelength end of the visible spectrum, the eye can sense longer wavelengths, but only very inefficiently. Thus what the eye perceives as a faint red beam may be a near-infrared beam with enough power to cause retinal damage despite its seeming "faintness."

The eye is much less vulnerable to wavelengths longer than 1.4 μm because these wavelengths cannot penetrate far enough into the eyeball to reach the retina. However, intense infrared radiation at such long wavelengths could burn the outer layers of the eye just as it can burn the skin. Long-term effects of exposure to infrared radiation at high to moderate intensities are not well understood.

Reliability and Maintenance. Because tunable vibronic solid-state lasers are relatively new, little data are available on their reliability. These lasers require standard laser maintenance, including cleaning optical surfaces exposed to high powers and replacing cooling-system filters periodically.

Commercial Devices

Standard Configurations. Titanium-sapphire and Co–MgF$_2$ lasers are packaged in small, tabletop units, typically about 30 by 60 cm, not counting any cooling equipment. Alexandrite and pulsed Ti-sapphire lasers may be somewhat larger and may include a bulkier cooling system and a remote controller.

Options. Important options for tunable solid-state lasers include special cavities, line-narrowing optics, injection seeding with external sources, computer-controlled tuning, harmonic generation, and cavity optics for specific wavelength ranges. Special options for alexandrite include generation of 500-ns to 5-μs pulses and Raman shifting of wavelength.

Pricing. Prices for tunable solid-state lasers start at around $20,000.

Suppliers. Some half-dozen companies now offer Ti-sapphire lasers, creating a strongly competitive, technology-driven market. A few companies offer alexandrite lasers, and at this writing only a single company offers Co–MgF$_2$ lasers. Suppliers range from small companies which specialize in tunable solid-state lasers to diversified laser manufacturers which offer many other types of lasers. Dye laser manufacturers are increasingly entering the Ti-sapphire market.

Patent coverage affects market entry. Titanium-sapphire was developed at the Massachusetts Institute of Technology (MIT) Lincoln Laboratory, which chose not to patent it, thus leaving the field open. Alexandrite was developed at the Allied Corporation, which secured a number of strong patents and controls access to the technology through its licensing policy.

Applications

The primary uses of solid-state vibronic lasers are in research and development. Their tunability makes them attractive for spectroscopy and other measurement applications which require tunable wavelengths, including laser radar and remote sensing. They also are used to investigate medical applications. As tunable sources with reasonable power levels, they can serve as "stand-ins" for other lasers which are not readily available with the precise wavelength. For example, much research on the feasibility of diode-laser pumping of others lasers is done with Ti-sapphire lasers.

Other applications under investigation include

- Shattering stones in the urinary tract or kidneys with alexandrite lasers.

- Undersea communications, by harmonic generation of the desired blue-green wavelength.

- Antisensor military laser systems using alexandrite. Its tunability makes it possible to shift wavelength and thus foil enemy sensors which detect laser irradiation at the neodymium wavelength.

- Photodynamic therapy for the treatment of cancer, in which patients are administered dyes selectively absorbed by cancer cells. Most research has been done with hematoporphyrin derivative, which absorbs in the red, but other dyes absorbing at longer wavelengths (produced by tunable solid-state lasers) may offer advantages.

- Remote sensing of atmospheric gases in the near-infrared; $Co-MgF_2$ wavelengths are particularly attractive.

- Control of tissue ablation and penetration depth, which depend on wavelength, for medical applications. $Co-MgF_2$ wavelengths are attractive because they cover major water-absorption lines, where strong tissue absorption causes rapid ablation with minimal damage to adjacent tissue.

- Range-resolved imaging with charge-coupled device (CCD) cameras using alexandrite lasers, which emit at the peak of silicon sensitivity.

Bibliography

James C. Barnes et al.: "Master oscillator power amplifier performance of Ti:Al$_2$O$_3$," *IEEE Journal of Quantum Electronics* 24(6):1029–1038, 1988 (research report).

Norman P. Barnes et al.: "A self-injection locked Q-switched, line-narrowed Ti:Al$_2$O$_3$ laser," *IEEE Journal of Quantum Electronics* 24(6):1021–1028, 1988 (research report).

A. J. W. Brown et al.: "High energy, short pulse flashlamp pumped Ti:Al$_2$O$_3$ laser," paper CWF13 in *Technical Digest CLEO '90*, Anaheim, Calif., May 21–25 1990, Optical Society of America, Washington, D.C. (research report).

Lloyd L. Chase and Stephen A. Payne: "New tunable solid-state lasers: Cr^{3+}:LiCaAlF$_6$ and Cr^{3+}:LiSrAlF$_6$," *Optics & Photonics News* 1(8):16–19, August 1990 (informal research report).

D. J. Ehrlich, P. F. Moulton, and R. M. Osgood, Jr.: "Ultraviolet solid-state Ce:YLF laser at 325 nm," *Optics Letters* 3:184, 1978 (research report).

V. P. Gapontsev et al.: "Laser diode pumped Yb-doped single-mode tunable fiber laser," paper WCl in *Technical Digest, Advanced Solid State Lasers Topical Meeting*, March 18–20, 1991, Hilton Head, S.C., Optical Society of America, Washington, D.C. (research report).

James Harrison et al.: "A low-threshold CW all solid state Ti:Al$_2$O$_3$ laser," in *Technical Digest, Advanced Solid State Lasers Conference*, March 1991, Hilton Head, S.C., Optical Society of America, Washington, D.C. (research report).

Walter Koechner: *Solid-State Laser Engineering*, 2d ed., Springer-Verlag, New York and Berlin, 1988 (overview of all solid-state lasers).

J. W. Kuper, T. Chin, and H. E. Aschoff: "Extended tuning range of alexandrite at elevated temperatures," paper MA2 in *Technical Digest, Advanced Solids State Lasers Meeting*, Salt Lake City, March 5–7, 1990, Optical Society of America, Washington, D.C. (research report).

Peter F. Moulton: "Ti-doped sapphire tunable solid state lasers," *Optics News* 8:9, November–December 1982 (overview).

Peter F. Moulton: "An investigation of the Co:MgF$_2$ laser system," *IEEE Journal of Quantum Electronics* QE-21(10):1582–1595, 1985 (review paper).

Peter F. Moulton: "Spectroscopic and laser characteristics of Ti:Al$_2$O$_3$," *Journal of the Optical Society of America* B3(1):125–133, 1986 (research report).

Stephen A. Payne et al.: "Laser performance of LiSrAlF$_6$:Cr^{3+}," *Journal of Applied Physics* 66(3):1051–1056, August 1, 1989 (research report).

V. Petricevic, A. Seas, and R. R. Alfano: "Forsterite laser tunes in near-IR," *Laser Focus World* 26(11):109–116, November 1990 (overview).

W. R. Rapoport et al.: "High brightness alexandrite laser," paper WD4 in *Technical Di-*

gest, *Advanced Solid State Lasers Meeting*, Salt Lake City, March 5–7, 1990, Optical Society of America, Washington, D.C. (research report).

Glen A. Rines and Peter F. Moulton: "Performance of gain-switched Ti:Al$_2$O$_3$ unstable resonator lasers," *Optics Letters* 15(8):434–436, April 15, 1990 (research report).

Richard C. Sam et al.: "Design and performance of a 250-Hz alexandrite laser," *IEEE Journal of Quantum Electronics QE-24*(6):1151–1166, 1988 (research report).

Richard C. Sam: "Alexandrite lasers," in Peter K. Cheo (ed.), *Handbook of Solid-State Lasers*, Marcel Dekker, New York, 1989, pp. 349–455 (review).

Harold Samelson et al.: "CW arc-lamp-pumped alexandrite lasers," *IEEE Journal of Quantum Electronics* 24(6):1141–1150, 1988 (research report).

Richard Scheps et al.: "Alexandrite laser pumped by semiconductor lasers," *Applied Physics Letters* 56(23):2288–2290, June 4, 1990 (research report).

Michael L. Shand and H. P. Jenssen: "Temperature dependence of excited-state absorption of alexandrite," *IEEE Journal of Quantum Electronics QE-19*:480–484, March 1983.

Michael L. Shand and S. T. Lai: "CW laser pumped emerald laser," *IEEE Journal of Quantum Electronics QE-20*(2):105–108, 1984 (research report).

M. D. Shinn et al.: "Progress in the development of LiCaAlF$_6$:Cr^{3+} laser crystals," paper MA1 in *Technical Digest, Advanced Solid State Lasers Meeting*, Salt Lake City, March 5–7, 1990. Optical Society of America, Washington, D.C. (research report).

R. C. Stoneman and L. Esterowitz: "Efficient, broadly tunable, laser-pumped Tm:YAG and Tm:YSGG cw lasers," *Optics Letters* 15(9):486–488, May 1, 1990 (research report).

G. Vaillancourt et al.: "Operation of a 1-kHz pulse-pumped Ti:sapphire regenerative amplifier," *Optics Letters* 15(6):317–319, March 15, 1990 (research report).

K. F. Wall and A. Sanchez: "Titanium sapphire lasers," *Lincoln Laboratory Journal* 3(3):447–462, 1990 (overview).

John C. Walling et al.: "Tunable laser performance in BeAl$_2$O$_4$Cr^{3+}," *Optics Letters 4*: 182, 1979 (research report).

John C. Walling: "Alexandrite lasers: physics & performance," *Laser Focus* 18(2):45–50, 1982 (overview).

John C. Walling: "Tunable paramagnetic-ion solid-state lasers," in L. F. Mollenauer and J. C. White (eds.), *Tunable Lasers*, Springer-Verlag, Berlin and New York, 1987, pp. 331–398 (review.

D. Welford and Peter F. Moulton: "Room-temperature operation of a Co:MgF$_2$ laser," *Optics Letters* 13(11):975–977, 1988 (research report).

Color-Center Lasers

Color-center lasers are types in which optical pumping of a crystal generates wavelength-tunable output in the near-infrared. They sometimes are called *F*-center lasers from the German word for color, *Farbe*. Some of their characteristics resemble those of continuous-wave dye or tunable vibronic solid-state lasers, but there also are important differences. Color-center lasers have an intrinsically narrow spectral linewidth that makes them attractive for spectroscopy. Experimental versions have operated near 0.4 micrometer (μm) and at 0.8 to 4.0 μm, but commercial models operate in a more restricted range, near 1.5 μm and beyond 2 μm.

Continuous-wave color-center lasers differ markedly from the family of flashlamp-pumped crystalline or glass solid-state lasers described in Chaps. 22 to 24 and 27. The active medium is a thin crystal, into which point defects have been introduced deliberately. The resulting microscopic defects in the crystals tend to absorb light, coloring the normally colorless crystals and earning the name color centers. Color centers are found in many types of normally colorless crystals. Laser action in color centers, first seen in 1965 (Fritz and Menke, 1965) is mostly in alkali halide crystals.

Like continuous-wave dye lasers, color-center lasers require pumping with another laser. The usual choices are ion, dye, or neodymium-YAG lasers. Normally operation is continuous, but color-center lasers also can be synchronously pumped by modelocked lasers. Pulsed operation of color-center lasers has been demonstrated experimentally with both laser and flashlamp pumps. However, only continuous-wave and modelocked versions are marketed.

Internal Workings

Active Medium. The active media in color-center lasers are crystals doped with selected impurities to generate the desired color-center (or

F-center) flaws. The crystals then are processed in ways that cause holes in the lattice (which are filled by free electrons instead of by negative ions) to come together with positively charged impurities, which generally are smaller than the positive ions in the lattice. In some defects, a single electron occupies each hole; in others two adjacent holes share a single electron, giving the F-center a positive charge. These color centers tend to absorb light in the visible or near-infrared, an effect used to pump color-center lasers.

The type of color center depends on the atomic configuration of the defect, and that, in turn, depends on the type of crystal and the nature of the impurity. (Only a few types of color center produce laser action.) Figure 25.1 shows the normal configuration of one type of color center, the $F_A(II)$ center, and its configuration after it has absorbed light. This type is formed only when lithium is added to crystals of potassium chloride or rubidium chloride. Figure 25.2 shows a second, sim-

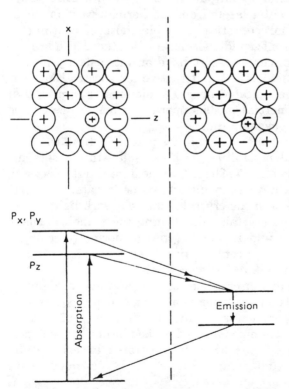

Figure 25.1   Structure of a $F_A(II)$ color center showing (a) normal (vacancy) configuration and (b) configuration in upper laser level. The energy-level structure is shown at bottom. (*From Mollenauer, 1982, with permission.*)

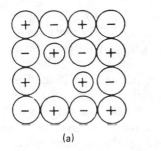 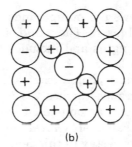

(a)   (b)

Figure 25.2 Structure of a $F_B(II)$ color center, showing (a) normal configuration and (b) excited configuration. (*From Mollenauer, 1982, with permission.*)

ilar type, the $F_B(II)$ center, formed in KCl or RbCl crystals doped with sodium. These two types can operate as lasers with emission wavelengths between 2.2 and 3.65 μm, although each center and crystal operates over more limited range. Related $Tl^0(I)$ color centers rely on doping alkali halides with thallium, producing neutral thallium defects in the crystals. These defects consist of a Tl impurity adjacent to an anion vacancy, with an electron bound to the complex. Lasers based on this defect have shorter emission wavelengths, tunable between about 1.4 and 1.7 μm (Gellerman et al., 1982).

Another type of color center, the F_2^+ center, can produce laser emission at visible and near-infrared wavelengths. The F_2^+ center forms when two adjacent holes share a single electron; it is positively charged because the region contains one electron less than it should to balance the charge of the positive ions.

Color centers can form only under restricted conditions, and laser operation is possible only in a more limited range. Commercial models operate at the 77 kelvin (K) temperature of liquid nitrogen, with the low temperature essential because the fluorescence quantum efficiency decreases as the temperature increases. The defects in the crystals are stable at temperatures as high as about −10°C. The type II color centers gradually dissociate if the crystals are stored at room temperature, but no permanent damage is done and the color centers can be regenerated. Type II centers can be regenerated even if the crystal is exposed to light at room temperature. F_2^+ centers are harder to produce and stabilize than $F_A(II)$ and $F_B(II)$ centers (Gellermann et al., 1982).

The first commercial color-center lasers relied on the stabler $F_A(II)$ and $F_B(II)$ centers. $Tl^0(I)$ lasers were introduced later.

Energy Transfer. Color-center lasers are pumped by an external laser with output wavelength falling into the color center's absorption band. Figure 25.3 shows the pump bands of $F_A(II)$ and $F_B(II)$ center crystals in commercial lasers. The pump band of the $Tl^0(I)$ center is at slightly longer wavelengths and includes the 1.06-μm output wavelength of the Nd-YAG laser.

Absorption of the pump photon excites the atoms at the defect site, causing them to rearrange themselves as shown in Figs. 25.1 and 25.2. The rearrangement, accompanied by a radiationless relaxation to a slightly lower overall energy level, puts the defect site in the upper laser level. The atoms stay in this rearranged state long enough to be stimulated to emit light on a transition to a lower energy level of the rearranged state. From that lower laser energy level, they then relax to the original configuration. As indicated at the bottom of Fig. 25.1, this corresponds to the standard four-level laser scheme, and it can operate quite efficiently.

The multitude of possible vibrational states in the crystal leads to homogeneous broadening of the laser transition. The broadening is large enough to permit tuning of the emission wavelength over a comparatively broad range, with total range about 15 to 20 percent of the central wavelength.

Internal Structure

Commercial color-center lasers are pumped longitudinally with a continuous beam (or modelocked pulse train) using the type of optical cavity shown in Fig. 25.4. The design bears a general resemblance to that used for many continuous-wave dye lasers. The beamsplitter reflects the pump wavelength through the laser crystal but transmits

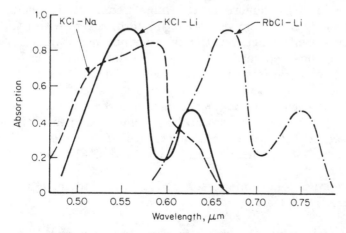

Figure 25.3 Absorption bands of three color-center crystals used in commercial lasers. (*Courtesy of Burleigh Instruments Inc.*)

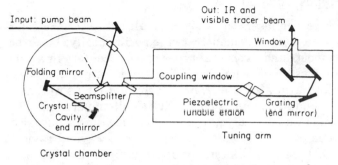

Input: pump beam

Out: IR and
visible tracer beam

Window

Folding mirror

Coupling window

Beamsplitter

Crystal

Cavity
end mirror

Piezoelectric
tunable etalon

Grating
(end mirror)

Tuning arm

Crystal chamber

Figure 25.4 Internal structure of commercial color-center laser, showing configuration with an intracavity etalon to limit oscillation to a single frequency. (*Courtesy of Burleigh Instruments Inc.*)

the color-center emission to the tuning cavity. Part of the pump beam also is transmitted by the beamsplitter and ultimately emerges through the output window along with the color-center laser beam.

The color-center crystal is housed in a chamber that cools it to 77 kelvins (K); the same chamber houses a pair of curved mirrors that direct the pump beam through the crystal. Wavelength is tuned by adjusting the angle of a grating in the tuning arm. The first-order reflected beam from the grating is fed back into the laser cavity; the zeroth-order beam is coupled out the output window as the laser beam. The grating design keeps direction of the output beam constant to within 0.25 milliradian (mrad) as wavelength is tuned by tilting the grating.

At low pump power, a color-center laser tends to emit a single longitudinal mode even without an intracavity etalon. However, as pump power increases, spatial hole burning within the crystal can produce a second longitudinal mode. Insertion of an etalon within the cavity can eliminate the second mode and assure reliable single-frequency operation.

Both crystal chamber and tuning arm are sealed, and when the laser is in operation they are evacuated to low pressures. In addition to helping avoid condensation on the crystal and other components at the low operating temperatures, this helps avoid atmospheric absorption. Water vapor in the atmosphere absorbs strongly at 2.5 to 3.0 µm, and other strong atmospheric bands lie at longer and shorter wavelengths; excessive absorption could reduce laser output or prevent lasing altogether.

The beam passes through only a small part of the crystal, a thickness of about 2 mm, and the crystal itself is oriented at Brewster's angle. The result is linear polarization of the pump and color-center beams that helps separate them at the beamsplitter that couples in the pump beam. The entire color-center laser is fairly compact, and the total length of the folded cavity is about half a meter.

Variations and Types Covered

Commercial color-center lasers are only a small subset of the color-center lasers that have been demonstrated experimentally. Laser action has been demonstrated in several types of color centers in a variety of alkali halides. Operation of F_2^+ center lasers has been demonstrated in LiF, NaF, KF, NaCl, KCl, KBr, and KI, as well as in sodium-doped KCl and RbCl and Li-doped KCl (Mollenauer, 1982). F^+ color centers in calcium-oxide crystals have produced laser output at 357 to 420 nm when pumped by pulsed or continuous-wave ultraviolet lasers (Henderson, 1981), but like many other results, that has been obtained only in experimental demonstrations.

Most color centers which can lase tend to photodegrade under illumination by a pump laser. Only the type II centers and the $Tl^0(I)$ center are stable.

Developers of commercial models concentrate on only a few color-center lasers that are stable enough to meet customer needs. This chapter concentrates on the commercial part of the color-center laser world. Color-center laser technology has been reviewed extensively elsewhere (German, 1989).

Beam Characteristics

Wavelength and Output Power. Figure 25.5 shows the tuning ranges and output powers available at wavelengths longer than 2 μm from a commercial color-center laser without an etalon. Three crystals can be used in the laser, with each providing tuning over part of the range. Insertion of an etalon into the laser cavity to restrict oscillation to a single frequency reduces the output power to 70 percent at the peak wavelength. If a KCl crystal doped to produce $Tl^0(I)$ centers is pumped with a 2-watt (W) YAG laser, the color-center laser can be tuned be-

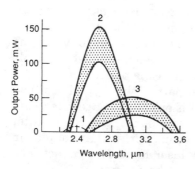

Figure 25.5 Tuning curves for the three color-center crystals used to produce wavelengths longer than 2 μm in a multimode commercial laser, along with output power as a function of wavelength. Crystal 1 (Na–KCl) typically is pumped with 488 and 514 nm from an argon laser; crystals 2 (Li–KCl) and 3 (Li–RbCl) are pumped with the 647-nm red line of a krypton laser. (*Courtesy of Burleigh Instruments Inc.*)

tween 1.43 and 1.58 μm, with peak output over 100 mW at 1.51 μm in either a single mode or two modes.

Efficiency. Power-conversion efficiencies of color centers range from under 0.1 percent off the peak emission wavelength for inefficient materials to 5 percent at the peak of the most efficient types. Efficiency varies considerably between materials and is highest for F_2^+ center materials.

Temporal Characteristics. Commercial color-center lasers normally operate continuously, as do their pump lasers. Modelocked pulses can be generated by synchronous pumping with a modelocked laser; pulses of 10 to 20 picoseconds (ps) are produced at repetition rates of 82 to 164 megahertz (MHz) depending on the pump laser. Color-center lasers themselves can be modelocked, generating pulses as short as 275 femtoseconds (fs) in the laboratory (Islam et al., 1990).

Spectral Bandwidths. Individual longitudinal modes of color-center lasers have linewidths of about 1 MHz, and some color-center lasers operate in a single longitudinal mode. Multimode lasers can operate in as few as two longitudinal modes, with mode spacing of 1.5 GHz. In modelocked operation, the laser operates over a band of equally spaced longitudinal modes, with total linewidth on the order of 20 GHz. Active stabilization of laser frequency is possible and can reduce linewidth to 20 kHz.

Beam Quality, Polarization, and Modes. Output beams are in a single transverse mode, 1.5 to 2 mm in diameter. Divergence is 1.5 to 2.5 mrad. Both figures vary with the square root of wavelength. The laser output is linearly polarized.

Operating Requirements

Input Power. A pump laser with continuous output of one to a few watts is needed, with output wavelength in the visible or near-infrared matched to a color-center absorption band. The standard types used are argon ion, krypton ion, dye, and neodymium-YAG. Equivalent powers are needed for synchronously pumped modelocked operation.

Electrical input to 10 amperes (A) from a 110- or 220-volt (V) source is needed to drive accessories including a vacuum pump, but not counting the pump laser.

Cooling. The color-center laser crystal and its pumping chamber are cooled by a dewar which holds a 2-day supply of liquid nitrogen, to maintain operation at 77 K. When the laser is not operating, the crystal must be stored in a refrigerator.

Required Accessories. A vacuum pump with a cryogenic trap is needed to evacuate both the laser crystal cavity and the tuning arm to low pressure for laser operation. An intracavity etalon is needed for single-frequency operation. Various controls and measurement instruments may also be needed, depending on the application, including wavelength meters and spectrum analyzers.

Operating Conditions and Temperature. Cryogenic operation is needed to avoid damage to color-center crystals. Some brief initial preparation may be required to regenerate color centers after the crystal has been stored. No degeneration occurs at cryogenic temperatures. Precautions also must be taken in shutting down the laser to avoid exposing the crystal to conditions that might cause damage. Details are given in manufacturers' instruction manuals.

Mechanical Considerations. The color-center laser itself measures 72 cm long, 23 cm wide, and 47 cm high. The height is largely due to the liquid-nitrogen dewar; the rest of the body of the laser is only about 15 cm high. The entire assembly weighs about 50 kilograms (kg). (Those figures do not include the vacuum pump or pump laser needed for operation of a color-center laser.)

Safety. Output wavelengths of commercial color-center lasers are in the "eye-safe" region beyond 1.4 μm where absorption within the eye prevents significant fractions of input light from reaching the retina. However, the output beam from a color-center laser includes a small fraction of the pump beam, which is at a shorter wavelength that is not in the eye-safe region. It would be wise to take the same precautions that would be used if the pump laser were in the laboratory by itself, not pumping the color-center laser.

Because of the cryogenic cooling, parts of the color-center laser are cold enough to freeze flesh. The liquid-nitrogen coolant poses similar hazards. However, in commercial models the very cold components normally are inside a narrow-necked dewar.

Special Considerations. The output beam from a color-center laser includes a small part of the pump beam. If the pump beam is visible, it makes the entire output beam visible, a feature that can be useful in alignment. However, to assure spectral purity it may be desirable to filter out the pump wavelength at some point outside the laser cavity. A piece of silicon could block all pump laser wavelengths but transmit all F-center wavelengths.

Reliability and Maintenance

Color-center lasers are built as precision laboratory instruments and should be durable if operated in the manner recommended by the manufacturer. However, the laser crystal is particularly vulnerable to improper operation and handling. Instruction manuals give detailed directions on how to prepare the laser for operation, how to use it, and how to shut it down. It is essential to pay heed to those directions, particularly because the procedures differ markedly from those for other lasers.

Commercial Devices

Standard Configurations.   All commercial color-center lasers are variations on a single basic design, with the crystal chamber housed in a cylindrical dewar. That assembly and the tuning arm are mounted on a rigid metal base about three-quarters of a meter long. The pump laser and vacuum pump must be mounted separately.

Options.   Color-center lasers are offered separately or as part of integrated systems that include vacuum pump and other accessories. Specific pump lasers are recommended by the color-center laser manufacturer, although others with similar characteristics can be used. A sampling of other major options includes

- Synchronously pumped modelocked operation
- An intracavity etalon for single-frequency operation
- Accessories for wavelength tuning
- Measurement accessories

Pricing.   Current (1990) prices of commercial color-center lasers start in the $47,000 range and go upward, depending on the options and accessories chosen.

Suppliers.   Only a single company actively produces and markets color-center lasers, an indication both of the limited market and of the technical sophistication needed to produce a viable product. Experimenters studying the physics of color-center lasers have built their own, but that is not an attractive prospect for the user not deeply involved in color-center physics.

Applications

Virtually all applications of color-center lasers are in spectroscopy or closely related fields which require modest powers and extreme spectral purity. Major applications have been in high-resolution molecular spectroscopy, involving both linear absorption measurements and some nonlinear studies. Intracavity measurements are possible and are attractive where enhanced sensitivity is needed. The narrow linewidth of color-center lasers permits selective stimulation of excited states for chemical kinetics studies. The infrared wavelengths are useful in solid-state spectroscopy, particularly of semiconductors.

The $Tl^0(I)$ center laser emitting in the 1.5-μm region is useful for studies of optical fibers in this region, where the loss of conventional silica glass fibers is lowest. Modelocked models can be used to generate short pulses for measurements of pulse dispersion in fibers, an important issue in the 1.5-μm region.

Some research has been done on the prospect of using highly stabilized color-center lasers as frequency standards. Stabilized bandwidths below 20 kHz have been reported in the laboratory, and even lower bandwidths may be possible.

Some commercial applications for color-center lasers have been investigated, including evaluation of infrared detectors and of the optical characteristics of infrared systems.

Bibliography

B. Fritz and E. Menke: "Laser effect in KCl with F_A (Li) centers," *Solid State Communications 3*:61, 1965 (research report).

Werner Gellerman, Klaus P. Koch, and Fritz Luty: "Recent progress on color-center lasers," *Laser Focus 18*(4):71–75, April 1982 (round-up).

Kenneth B. German: "Color center laser technology," in Peter K. Cheo (ed.), *Handbook of Solid-State Lasers*, Marcel Dekker, New York, 1989, pp. 457–579 (extensive review).

B. Henderson: "Tunable visible lasers using F_+ centers in oxides," *Optics Letters 6*(9): 437–439, 1981 (research report).

Mohammed N. Islam et al.: "Color center lasers passively modelocked by quantum wells," *IEEE Journal of Quantum Electronics 25*(12):2454–2463, 1990 (research report).

Linn F. Mollenauer: "Color center lasers," in Marvin J. Weber (ed.), *CRC Handbook of Laser Science & Technology*, vol. 1, *Lasers & Master*, CRC Press, Boca Raton, Fla., 1982, pp. 171–178 (short review).

Linn F. Mollenauer and D. H. Olson: "Broadly tunable lasers using color center," *Journal of Applied Physics 46*:3109, 1975 (research report).

I. Schneider: "Continuous tuning of a color center laser between 2 and 4 μm," *Optics Letters 7*:271–273, June 1982 (research report).

Fiber Lasers and Amplifiers

Progress on optical fiber lasers and amplifiers during the last few years has been exceptionally rapid even by laser standards. The fiber laser is in some ways a logical miniaturization of solid-state lasers, and the idea of using an optical fiber as a laser medium is not new. The first fiber amplifier was demonstrated in 1963 using a 1-meter (m) neodymium-doped fiber wrapped around a flashlamp (Koester and Snitzer, 1964). However, the concept received little attention until the mid-1980s, when developers of fiber-optic communication systems began seeking new ways to boost signals to extend transmission distances.

Conventional fiber-optic systems use electro-optic repeaters, which detect a weak optical signal, convert it into electronic form, amplify and process the electronic signal, and use it to drive a laser transmitter. The optical-to-electronic conversion could be avoided by simply passing the weak optical signal through a laser amplifier, to boost its strength. One possibility is the use of semiconductor laser chips coated to suppress facet reflection (described in Chaps. 18 and 20). Another is the use of solid-state fiber laser amplifiers.

The basic concept of a fiber laser amplifier is shown in Fig. 26.1. A fiber is made from a solid-state laser material (typically a glass),

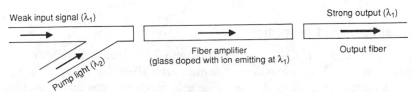

Figure 26.1 Fiber amplifier receives two inputs, a weak signal beam at λ_1 and a strong pump beam at λ_2, which excites ions in the fiber so that they produce stimulated emission at λ_1, amplifying the weak signal.

doped with an ion which emits at the desired wavelength λ_1. It is illuminated from one end by a weak signal at λ_1 and a stronger steady beam at a second wavelength λ_2 which excites the ion in the fiber to its upper laser level. As the weak signal passes through the fiber amplifier, it stimulates emission from excited ions at λ_1. Any pump light remaining after passing through the amplifier may be allowed to propagate along the output fiber, which typically attenuates it more strongly than the signal wavelength.

In practice, doping is confined to the core of a single-mode fiber, creating a strong waveguide effect. This confines the pump light to a small cross-sectional area, making pump intensity high even for modest pump powers. Early demonstrations of fiber lasers used gas or solid-state laser pumps, but for practical applications developers are concentrating on diode laser pumping.

For fiber-optic communications, fiber amplifiers must meet certain criteria:

- Reasonable gain at a broad enough range of wavelengths, typically over 10 nanometers (nm), to allow multiwavelength operation and tolerate variations in laser source wavelength

- Low noise

- Compatible with diode-laser pump sources

- Rapid response to high-speed pulsed signals (i.e., high bandwidth)

Developmental fiber amplifiers have done exceptionally well in meeting these criteria and have demonstrated impressive performance in laboratory experiments. At this writing, results of early field tests also appeared encouraging, and designers of a major transpacific cable have announced plans to use fiber amplifiers in the mid-1990s.

The technology of optical fiber amplifiers is still evolving rapidly as researchers explore their potential. While one can outline the general features of the technology at this early stage, the field is much too young to give definitive details like those for well-established commercial lasers. A few developmental fiber amplifiers are available commercially, but virtually all fiber amplifiers remain in the laboratory.

Introduction and Description

Figure 26.1 shows a fiber laser amplifier which lacks resonator mirrors. It is a traveling-wave device, which amplifies the strength of signals making a single pass along its length. Such devices have received the most attention because they are attractive for signal amplification in fiber-optic systems. It also is possible to build a laser oscillator by placing the fiber in a resonant optical cavity, but such a laser would have limited use as a fiber-optic signal source because it could not be directly current-modulated. (However, it could serve as a narrow-line light source for use with an external modulator.) To date, most development has concentrated on fiber amplifiers, and this chapter thus will concentrate on fiber amplifiers rather than oscillators.

Interest in fiber amplifiers has centered on the two "windows" used for most high-performance optical fiber systems: 1.3 micrometers (μm), where step-index single-mode fibers have pulse dispersion near zero, and 1.55 μm, where silica fibers have lowest attenuation. Developers have identified solid-state laser ions for each window. For 1.3 μm, it is the 1.3-μm line of neodymium, mentioned briefly in Chap. 22. For the longer wavelength, it is the 1.54-μm line of erbium in glass.

The 1.3-μm line is important because most current long-distance fiber-optic systems operate at that wavelength. Unfortunately, the neodymium line suffers from limited gain and is not an ideal match to the zero-dispersion wavelength. The 1.54-μm erbium line is a much better fiber amplifier, but that wavelength is not in common use despite its low loss. The step-index single-mode fibers which are used in most long-distance fiber-optic systems have high pulse dispersion at 1.54 μm, limiting data rates and signal bandwidth. This dispersion problem can be overcome by either using signal sources with very narrow spectral bandwidth or switching to optical fibers of a slightly different design and in which the zero-dispersion wavelength is shifted to 1.55 μm. Both are available commercially, although neither has achieved the widespread acceptance of 1.3-μm fiber-optic hardware. At this writing, erbium-doped fiber amplifiers have a clear technical lead over neodymium fiber amplifiers.

Fiber Amplifier Structures and Hosts

Fiber amplifiers are similar in structure to other optical fibers, with a high-refractive-index core and a lower-index cladding. Total internal reflection confines light in the fiber core. If the core is small enough, about 9 μm in diameter for 1.3-μm light, and slightly larger for 1.55-μm light, it can carry only a single mode of light. Larger-core fibers carry many modes, with the number of possible modes increasing dramatically with core diameter.

In fiber amplifiers, the rare earth dopant normally is confined to the core, as shown in Fig. 26.2. A wide range of dopant concentrations have been tested, with higher concentrations used to make shorter fiber amplifiers. The pump power, doping, and length of the amplifying fiber all are important variables in obtaining optimum performance.

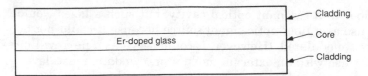

Figure 26.2 Rare earth ions are doped into the fiber core, which has a higher refractive index than the surrounding cladding and confines light by total internal reflection.

Qualitatively, higher doping levels allow shorter fiber lengths to absorb the same pump power. For high efficiency, the fiber should be long enough to absorb most of the pump energy, but not so long that parts of the fiber are unpumped (because unpumped regions would absorb at the laser transition, particularly in erbium-doped fibers). Very low concentrations are used to make fibers with just enough optical gain to offset losses at the amplified signal wavelength, so the fibers effectively have zero loss.

The silica glasses used in conventional optical fibers are the most common hosts for fiber amplifiers. However, fluoride-based glasses, phosphate glasses, and other materials also have been made and tested. Such materials may provide better wavelengths, less excited-state absorption, or other properties that make them attractive in amplifiers.

Erbium-Doped Fiber Amplifiers

Erbium ions in glass form a three-level laser system, with gain over a broad range centered near 1550 nm. Like any three-level laser system, it requires high pump intensity to achieve a population inversion. However, the concentration of pump light in the tiny core of a single-mode fiber makes laser action possible at reasonable pump power levels. Semiconductor diode lasers emit at important pump bands at 980 and 1480 nm, and the performance of erbium-doped glass fiber amplifiers has been very encouraging in system tests. Developmental models are now available.

Energy-Level Structure. Erbium laser action in silica glass takes place on the $^4I_{13/2} \rightarrow {}^4I_{15/2}$ transition of Er^{3+}. It is a three-level laser because the $^4I_{15/2}$ level is the ground state. The erbium ions can be excited to any of several higher levels (including higher sublevels of the $^4I_{13/2}$ state), which decay to the upper laser level. The emission spectrum is about 40 nm wide. The absorption spectrum is shown in Fig. 26.3.

The upper laser level has a 10-ms fluorescence lifetime, long enough to allow a reasonably low threshold even for a three-level laser.

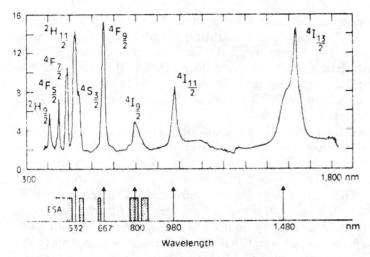

Figure 26.3 Absorption spectrum of erbium-doped silica glass fiber, showing levels to which the ion is excited from the ground state. The line at bottom shows important pump wavelengths; dark bands denote excited state absorption. (*Courtesy of E. Desurvire.*)

Excited-state absorption competes with some possible pump lines, as shown in Fig. 26.3. Pumping is possible at 800 nm, but gains are limited by excited-state absorption. The best results have been obtained by diode laser pumping at 980 and 1480 nm. The latter wavelength is not at an absorption peak, but at wavelengths closer to the 1530-nm peak, the laser becomes a two-level system which cannot produce gain.

The precise energies of each energy level depend on the glass host, so wavelengths differ among glasses. The broad range of possible emission wavelengths is a common feature of glasses, although fluorozirconate glasses offer a slightly broader range near 1.55 μm than does silica glass. Laser action also has been reported on more transitions in fluorozirconate fibers.

Internal Structure. Silica glasses are the best-developed host for erbium fiber amplifiers. Normally erbium doping is confined to the fiber core, which may be a step-index single-mode type or a more complex multilayer or multicore structure. Aluminum may be added to the fiber core to broaden and smooth the emission spectrum, which is important for amplifying multiple signals in the 1.55-μm band. It also allows higher erbium concentrations.

Erbium doping in the core of silica fiber amplifiers varies considerably, from 0.01 to 10,000 parts per million (ppm). Normally, the higher the doping level, the shorter the amplifier. For example, fiber amplifiers containing around 1000 ppm of Er typically are several meters long. A fiber amplifier containing about 100 ppm of Er typically will be over 100 m long. Typical gains of such amplifiers are tens of decibels. Alternatively, a fiber containing 0.01 to 0.1 ppm of Er distributed along its length can provide enough gain to cancel the 0.2 to 0.3 decibels per kilometer (dB/km) loss of optical fibers in the 1.55-μm window (Simpson et al., 1990).

Highly doped fiber amplifiers are discrete devices, intended to produce gain to compensate for the loss of conventional passive optical fibers. Fibers doped with low levels of erbium to balance internal losses are rather different devices designed to serve the same function as a combination of passive fiber segments and "lumped" discrete amplifiers. It may stretch a point to consider such lossless fibers as a special type of amplifier, because they do not have a significant overall gain. However, they do provide amplification, and require pumping to excite erbium atoms as do amplifiers which produce overall gain.

Several groups have investigated fluorozirconate fibers as hosts for Er^{3+} fiber amplifiers. Good results have been obtained pumping at 1480 nm. A 50-cm fiber, doped with 5000 ppm Er^{3+} in its core, generated 20-dB gain, flat to within ±1 dB between 1535 and 1563 nm (Miyajima et al., 1990). Erbium also lases on other lines in fluorozirconate glasses (Smart et al., 1990). However, serious practical problems remain with fluorozirconate glasses.

Pumping.   Erbium-doped fibers normally are pumped from one or both ends with a laser beam coupled into the core of the fiber. As shown in Fig. 26.3, Er doped fibers can be pumped at several wavelengths. Early Er-fiber amplifiers were pumped with 514-nm argon lasers or 647-nm krypton lasers. Pumping in the 800-nm band has also been demonstrated but suffers from excited-state absorption. The preferred choices at this writing are 980 and 1480 nm, both available from semiconductor lasers.

Impressively high gains and powers have been obtained when pumping at 980 or 1480 nm. In recent experiments, 51-milliwatt (mW) output at 1552 nm was generated by pumping a 20-m fiber containing 630 ppm Er^{3+} with 107 mW at 1480 nm, corresponding to 48 percent efficiency (Suzuki et al., 1990). Gain of that amplifier saturated at 28 dB, but higher small-signal gains have been demonstrated at lower powers. Gains over 30 dB have been demonstrated in Er-doped fibers pumped at 980 nm (Shimizu et al., 1990). Recent research has shown that strained-layer lasers can readily be made at 980 nm, and they may offer advantages including low threshold currents and long lifetimes.

Gains to 30 dB have been obtained when pumping with 100 mW at 822 nm (Nakazawa et al., 1990), avoiding excited-state absorption, which has plagued other experiments. That wavelength is attractive because it is readily available from gallium arsenide diode lasers (although the experiments were performed with a titanium-sapphire laser, which is suitable only for use as a laboratory pump source). Developers report that the key requirement for successful 800-nm pumping is low erbium concentrations in the fiber; they used 34-ppm Er in a 100-m fiber.

Another way to avoid excited-state absorption in the 800-nm band is to alter glass composition. Peak small-signal gain of 32 dB and gain of at least 25 dB between 1530 to 1565 have been reported in Er^{3+}-doped silica-core fibers doped with germanium, calcium, and aluminum (Saifi et al., 1991).

Wavelengths and Power Performance. Erbium-doped fiber amplifiers have a wavelength range of about 40 nm, from about 1520 to 1560 nm. A broad range over which amplification is reasonably uniform is desirable for fiber-optic systems which carry signals at multiple wavelengths (wavelength-division multiplexing) in the 1550-nm window. At this writing, the gain is not uniform across the wavelength window for Er-doped silica glass, but the technology is still young. Fluorozirconate glasses offer broader bandwidth and the potential for more uniform gain across the range of amplified wavelengths, but other technical problems remain to be solved.

Experimental erbium-doped fiber amplifiers have achieved gains over 30 dB and output powers of tens of milliwatts. The measured gains normally are small-signal values, for low levels of input power and modest outputs (no more than a few milliwatts, and often under 1 mW). At peak output power, the gain normally is considerably less than the maximum small-signal gain.

Specifications are cautious for the few developmental products. The highest specified fiber-to-fiber small-signal gain is 30 dB, but gain is only 13 dB at rated maximum output of 20 mW. A low-power fiber amplifier delivers up to 1 mW, with small-signal gain of 18 dB and a gain of 12 dB at 1 mW.

Noise Levels and Pulse Characteristics. Several groups have demonstrated multigigabit data transmission using multiple erbium-doped fiber amplifiers, a clear indication of good noise and pulse amplification characteristics. Unlike semiconductor laser amplifiers, fiber amplifiers are not sensitive to polarization of the input beam. Many details remain to be established.

Common Configurations and Commercial Devices. At this writing, Er-doped fiber amplifiers are available only as developmental products. Typically a loop of fiber is mounted in a box, with optical connections to the outside world through standard fiber interfaces, as shown in Fig. 26.4. Laboratory configurations can vary widely. No consensus has yet emerged on the ideal length of fiber amplifiers, which may well depend on the application. For some uses, such as soliton transmission, the best choice may be fibers with low levels of erbium which provide only enough gain to offset internal loss. In other cases, discrete amplifiers may be more desirable.

Most current interest is focused on fiber amplifiers to increase the strength of signals originally generated by semiconductor lasers after they have passed through optical fibers. However, research is also under way on fiber lasers, with their own resonant cavities, which would generate a continuous beam. That continuous output could be modulated externally in communication systems.

Applications. The primary applications envisioned for Er-doped fiber amplifiers are increasing the strength of signals in 1.55-μm communication systems. At this writing, most fiber-optic systems operate at 1.3 μm, but silica fibers have lower attenuation at 1.55 μm, making them more desirable for very long systems, such as transatlantic and transpacific cables, or backbone systems spanning continents.

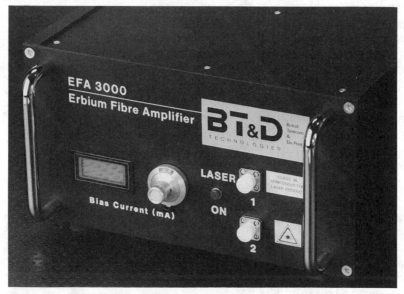

Figure 26.4 A packaged developmental fiber amplifier. (*Courtesy of BT&D Technologies.*)

Erbium-doped fiber amplifiers have been used in demonstrations of in 2.5-gigabit per second transmission over distances to 2200 km without conventional electro-optic repeaters (Saito et al., 1990). Their advantages include operational simplicity, high speed, good gain, and the ability to amplify simultaneously two or more wavelengths which lie within their spectral bandwidth.

Developers of networks which would distribute identical signals to many subscribers through optical fibers also are interested in fiber amplifiers, although not specifically in the 1.55-μm wavelength of Er-doped types. By raising signal strength, fiber amplifiers could let one signal source serve many more subscribers than is now possible because of the high losses inherent in dividing optical signals.

Neodymium-Doped Fiber Amplifiers

Considerable attention has been devoted to neodymium-doped fiber amplifiers, which can operate in the 1.3-μm region where most single-mode fiber-optic systems operate. This system is based on a well-known secondary transition of neodymium which produces 1.3-μm emission from Nd–YAG lasers (see Chap. 22). However, thus far its operational characteristics as a fiber amplifier are inferior to those of the 1.55-μm erbium line.

Energy Levels and Hosts.   The fundamental problem with neodymium-doped fiber amplifiers is excited-state absorption. Figure 26.5 shows the major energy levels involved. Optical pumping populates the $^4F_{3/2}$ upper laser level, which also serves as the upper level for laser tran-

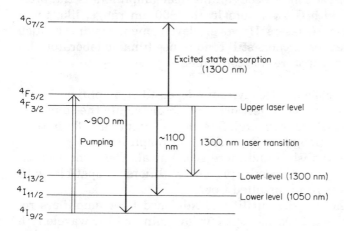

Figure 26.5 Major energy levels in 1300-nm Nd-fiber amplifier, showing excited-state absorption.

sitions at 900 and 1100 nm and is depleted by amplified spontaneous emission at 900 and 1050 nm. Absorption by the upper laser level at 1300 nm in silica glass can raise the Nd ion to the $^4G_{7/2}$ level, suppressing laser action at that wavelength on a transition from the $^4F_{3/2}$ level to the $^4I_{13/2}$ level (Alcock et al., 1986). Although laser amplification can be demonstrated on that transition in silica glass, the wavelength is shifted away from the 1300-nm wavelength used in fiber-optic systems.

Phosphate glasses offer some improvements, but to date the best performance has been in zirconium fluoride glasses, where the excited-state absorption is shifted in wavelength and reduced in strength, giving positive gain between 1310 and 1370 nm. This has allowed saturated output power to exceed 30 mW when pumping with 100 mW at 795 nm (Pedersen and Brierley, 1990). However, even in those experiments the small-signal gain was limited to 6.4 dB at 1337 nm and 4.4 dB at 1320 nm. Theoretical studies indicate that small-signal gain is low on the 1300-nm band even without excited-state absorption (Dakss and Miniscalco, 1990). Such problems could restrict applications of Nd-doped fiber amplifiers.

Internal Structure. Neodymium ions are doped in the core of single-mode fibers to make Nd fiber amplifiers. Researchers are still investigating dopant levels and lengths. High output powers have been reported with 75-cm fluoride fibers in which the cores were doped with 500 ppm of neodymium. Like Er-doped fibers, Nd-doped fibers can be used in oscillators as well as amplifiers.

Pumping. One attraction of neodymium fiber amplifiers is the prospect of pumping with GaAs lasers in the 800-nm range, like those used to pump Nd–YAG lasers. However, development remains at such an early stage that researchers still tend to use tunable laboratory lasers (e.g., titanium-sapphire) for convenience.

Problems and Applications. Neodymium-doped fiber amplifiers would be most attractive for use in networks which distribute the same signals to many subscribers. Although fiber loss is higher at 1.3 μm than at the 1.55-μm wavelength of Er-doped fiber amplifiers, 1.3-μm technology is better established and more economical. The owners of existing fiber-optic systems would prefer amplifiers compatible with their large investment in installed fibers.

Nonetheless, the practical utility of Nd-doped fiber amplifiers remains to be established because of their low gain and the wavelength mismatch with the 1300-nm zero-dispersion point. Erbium-doped fiber amplifiers cannot operate at the same wavelength, but semiconductor laser amplifiers will provide some competition for optical amplification in the 1300-nm window.

Bibliography

I. P. Alcock et al.: "Tunable continuous-wave neodymium-doped monomode fiber laser operating at 0.900–0.945 and 1.070–1.135 μm," *Optics Letters* 11(11):709–711, 1986 (research report).

M. L. Dakss and W. J. Miniscalco: "Fundamental limits on Nd^{3+} doped fiber amplifier performance at 1.3 μm," *IEEE Photonics Technology Letters* 2(9):650–652, 1990 (research report).

E. Desurvire: "Erbium-doped fiber amplifiers for new generations of optical communications systems," *Optics and Photonics News* 2(1):6–11, January 1991 (overview).

C. J. Koester and E. Snitzer: "Amplification in a fiber laser," *Applied Optics* 3(10):1182, 1964 (research report).

Y. Miyajima, T. Sugawa, and T. Komukai: "20 dB gain at 1.55 μm wavelength in 50 cm long Er^{3+}-doped fluoride fiber amplifier," *Electronics Letters* 26(18):1527–1528, August 30, 1990 (research report).

M. Nakazawa, Y. Kimura, and K. Suzuki: "High gain erbium fibre amplifier pumped by 800 nm band," *Electronics Letters* 26(8):548–550, April 12, 1990 (research report).

J. E. Pedersen and M. Brierley: "High saturation output power from a neodymium-doped fluoride fiber amplifier operating in the 1300 nm telecommunications window," *Electronics Letters* 26(12):819–820, June 7, 1990 (research report).

M. A. Saifi et al.: "Er^{3+}-doped GeO_2–CaO–Al_2O_3 silica core fiber amplifier pumped at 813 nm," paper FA6 in *Technical Digest, Optical Fiber Communications Conference*, San Diego, February 1991, by Optical Society of America, Washington, D.C. (research report).

Shigeru Saito et al.: "An over 2,200 km coherent transmission experiment at 2.5 Gbit/s using erbium-doped-fiber amplifiers," Postdeadline Paper PD-2 at *Optical Fiber Communication Conference*, San Francisco, January 22–26, 1990, Optical Society of America, Washington, D.C. (research report).

Makoto Shimizu et al.: "Compact and highly efficient optical fiber amplifier module pumped by a 0.98-μm laser diode," Postdeadline Paper PD-17 at *Optical Fiber Communication Conference*, San Francisco, January 22–26, 1990, Optical Society of America, Washington, D.C. (research report).

J. R. Simpson et al.: "A distributed erbium doped fiber amplifier," Postdeadline Paper PD-19-1 at *Optical Fiber Communication Conference*, San Francisco, January 22–26, 1990, Optical Society of America, Washington, D.C. (research report).

R. G. Smart et al.: "Erbium doped fluorozirconate fibre laser operating at 1.66 and 1.72 μm," *Electronics Letters* 26(10):649–651, May 10, 1990 (research report).

Kazunori Suzuki, Yasuo Kimura, and Masataka Nakazawa: "High power Er^{3+} doped fiber amplifier pumped by 1.48 μm laser diodes," *Japanese Journal of Applied Physics (Letters)* 29(13):L2067–L2069, November 1990 (research report).

Other Solid-State Lasers

Far more solid-state lasers have been demonstrated experimentally than have ever received any kind of commercial development, let alone reached the market status of those described in Chaps. 22 to 26. Technical reviews (e.g., Weber, 1982) list hundreds of solid-state laser lines produced from dopants in crystalline and glass hosts and in materials where the light-emitting ion is part of the crystalline compound. New solid-state laser systems continue to be developed. The most work has gone into developing hosts for the neodymium ion, but many other active species have been studied in a variety of hosts.

Most of these lasers have never gone much farther than experimental demonstration and initial evaluation. Many suffer from serious inherent weaknesses, including low output power, high laser threshold, requirements for cryogenic cooling, difficult crystal growth, poor material quality, or poor heat dissipation. Others simply offer no compelling advantages over better-developed types. Large investments would be needed to develop the newer materials, and they are hard to justify except for the tunable lasers described in Chap. 24 or other types which offer special advantages.

This chapter covers a few alternative solid-state materials which do offer advantages that have pushed them to the threshold of commercial availability, although none have become standard products. (Vibronic lasers are covered in Chap. 24.) They are based on rare earth ions similar to neodymium: erbium-doped glass emitting at 1.54 micrometers (μm) (the same system used in the fiber laser and amplifiers described in Chap. 26); holmium emitting near 2.1 μm; and erbium-doped crystals emitting at several wavelengths, with 2.94 μm the most important at present. Other wavelengths have been available in the past, including lines of erbium at 0.85, 1.22, and 1.73 μm.

Several types of interest have pushed the recent development of new solid-state laser lines. One of the best-funded has been the military search for eye-safe laser range finders and target designators, which can be used safely in training exercises with friendly troops. The 1.06-μm wavelength of Nd–YAG poses a serious eye hazard (as described in Chap. 22), so military developers have sought alternative materials at wavelengths between 1.4 μm (where dangers of retinal damage drop) and 2.2 μm (where the eye becomes sensitive to corneal burns).

Interest in medical applications has pushed development of other solid-state lasers because certain wavelengths are particularly attractive for some procedures. The erbium line near 3 μm, for example, is absorbed by water even more strongly than the 10-μm carbon dioxide line, so it leaves a thinner damaged layer between healthy tissue and the zone removed in surgery. The absorption is so strong that the 3-μm erbium line can be used to cut bone. Other wavelengths are being investigated for other potential applications.

Emerging needs for specific wavelengths for other applications has stimulated researchers to investigate laser lines demonstrated long ago and forgotten, or to seek new lines based on spectroscopic predictions. One example is the development of promethium lasers to fill a need for blue-green lasers for satellite-to-submarine communications (Shinn et al., 1988); the 920-nanometer (nm) output of the Pm^{3+} ion would be frequency-doubled. Another is interest in continuous-wave 1.5-μm lasers as single-frequency light sources for use with external modulators in fiber-optic communications.

Erbium-Glass Lasers

When erbium is doped into certain glasses, it can produce laser output on two closely spaced lines at 1.54 μm. In addition to its use in fiber laser and amplifiers, described in Chap. 26, this system is the leading candidate for eye-safe military laser range finders. Quirks of optical absorption in the eye raise the eye-hazard threshold at 1.54 μm about a factor of 100 above that at other nominally eye-safe wavelengths, so military developers have focused specifically on that wavelength. It also falls in the region of lowest attenuation in silica optical fibers, making it of interest for fiber-optic communications.

Erbium-glass lasers were first tested in the late 1960s but attracted little interest then because no good 1.5-μm detectors were available, and because the glasses then in use limited laser performance. Renewed interest was stimulated by the development of 1.5-μm detectors for fiber optics and of pentaphosphate glasses for laser fusion research, which proved to be excellent hosts for erbium. Ytterbium is added to the glass as a sensitizer, which absorbs pump light from the flashlamp and transfers it to the erbium ions.

Erbium is a three-level laser system in glass and requires excitation of about 60 percent of erbium ions to reach laser threshold. Pumping of bulk lasers has been with xenon flashlamps, and for the laser to operate at room temperature, ytterbium must be added to the glass as a sensitizer, to absorb pump energy and transfer it to the erbium. Optimum erbium concentrations in bulk glass rods are dopings of 0.2 to 0.5% erbium by weight, or 2 to 5 $\times$ 10^{19} Er^{3+} ions per cubic centimeter (cm^{-3}). (Erbium concentrations vary over a wider range in fiber amplifiers.)

One novel and promising scheme for pumping Er-glass lasers is a two-stage approach shown in Fig. 27.1, in which a GaAs semiconductor laser pumps a 1.06-μm Nd–YAG laser and the 1.06-μm output pumps an erbium-glass rod doped with 12% ytterbium and 0.5% Er. Starting with a 1-watt (W) diode array pumping an Nd–YAG rod, this scheme has generated 90-milliwatt (mW) TEM_{00} output at 1535 nm (Anthon et al., 1991).

As a three-level laser system, the Er-glass laser can operate only in pulsed mode at room temperature. Repetition rates of tens of hertz have been demonstrated but are not required for range-finding applications. (Special glasses are required to achieve such high repetition rates.) Pulse energies of 100 millijoules (mJ) are available in standard mode. Erbium-glass also can be Q-switched to generate tens of millijoules (Koechner, 1988, pp. 62–64).

Crystalline Erbium Lasers

The most important line of crystalline erbium lasers lies at 2.94 μm in Er–YAG, which coincides with a strong water-absorption line, making

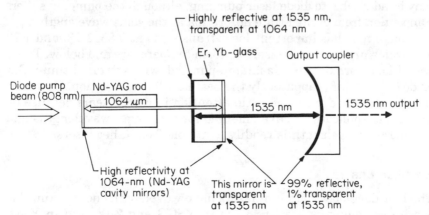

Figure 27.1 Two-stage pumping of an Er-glass laser using a diode laser and Nd–YAG laser. The Nd–YAG and Er-glass cavities overlap, using coatings reflective at one wavelength but transparent at the other. (*Adapted from Anthon et al., 1991.*)

it attractive for laser surgery. The wavelength is slightly different in other hosts. The lowest thresholds have been reported in yttrium scandium gallium garnet (YSGG) codoped with Er and Cr for emission at 2.796 μm (Moulton et al., 1988), but this wavelength may not be as attractive for medicine. The wavelength is too long for use with standard silica optical fibers.

The 2.94-μm line is produced at room temperature from erbium-yttrium aluminum garnet (YAG) doped so heavily with Er that the Er and Y concentrations are similar. When pumped by a flashlamp, it generates 0.1- to 4-millisecond (ms) pulses at repetition rates to 30 hertz (Hz). In laboratory versions, pulse energies can reach 10 joules (J) and average power 20 watts (W). However, the few commercial models available offer pulse energies of no more than a couple of joules in multiple transverse modes, and output in TEM_{00} mode is limited to hundreds of millijoules.

Pump wavelengths for Er–YAG are shorter than 600 nm, limiting pump efficiency. Soviet researchers have demonstrated efficiency to 3 percent, but that is measured as a fraction of pump energy from the flashlamp (Prokhorov and Shcherbakov, 1990). Another limitation on the Er–YAG laser is the very long lifetime of the lower laser level, 2 ms, which is much shorter than the 0.2-ms life of the upper laser level. This makes Q switching difficult, although not impossible; acousto-optic Q switching can generate pulses as short as 50 nanoseconds (ns) (Schnell et al., 1990). Despite these limitations, the medical potential of the wavelength has led to continued development.

Laser pumping of Er–YLF with 0.97-μm output of a tunable titanium-sapphire laser has been demonstrated, generating continuous-wave output at 2.80 μm to tens of milliwatts (Stoneman et al., 1990). That system may be adaptable to diode laser pumping, although the pumping suffers competition from excited-state absorption at the same wavelength.

Erbium also has important laser transitions at 0.85, 1.23, and 1.73 μm, which are four-level lasers at room temperature and below. These transitions normally are flashlamp-pumped, with yttrium lithium fluoride (YLF; $YLiF_4$) apparently the best host for laser action. However, these lasers have attracted little attention in recent years because no important applications have emerged for the longer wavelengths and the shorter wavelength is readily available from other lasers.

Holmium Lasers

The best-developed holmium lasers have wavelengths near 2 μm. The precise wavelengths are 2.088 μm in YSGG and 2.127 μm in YAG. Holmium is added to the crystal along with two other dopants, erbium and thulium, which serve as sensitizers, absorbing the pump light and transferring the energy to Ho^{3+} ions in the crystalline lattice. Without the sensitizers, holmium cannot absorb enough pump energy for use in a practical laser. Diode pumping is possible.

The 2-μm holmium line is attractive for some medical applications, particularly because it can be delivered through optical fibers, but the laser suffers from important practical disadvantages. At cryogenic temperatures, it is a four-level laser system, but at room temperature the lower laser level is thermally populated, making it a three-level laser and hence much less efficient. Holmium can operate continuous-wave at cryogenic temperatures, but operates pulsed only at room temperature. In commercial versions, energies can reach a couple of joules in multimode operation. Energy storage is larger at higher temperatures, so Q switching can generate energetic pulses.

Holmium also has laser lines near 3 μm, and operation at 2.85 and 2.92 μm has been demonstrated in yttrium aluminate ($YAlO_3$) doped with holmium or holmium and neodymium (Bowman et al., 1990). However, it is not clear whether this system will prove practical for medical applications; it is not offered commercially.

Thulium Lasers

Thulium has a transition at 2 μm which can be pumped with a diode laser or lamp to generate high powers for medical applications, laser radars, or laser range finders. Tm–YAG operates untuned at 2.01 μm but can be tuned to 1.95 μm, a wavelength where tissue absorption is significantly higher (Moulton, 1990). Continuous-wave outputs to 1.1 W in TEM_{00} mode have been achieved by pumping with six 1-W diode lasers (Kane and Kubo, 1990). Although this laser appears promising, it is in an early stage of development.

Bibliography

D. W. Anthon, T. J. Pier, and P. A. Leilabady: "1.54 μm high-power diode-pumped erbium laser," paper FB6 in *Technical Digest, Optical Fiber Communications Conference*, San Diego, February 1991, Optical Society of America, Washington, D.C. (research report).

S. R. Bowman et al.: "3 μm laser performance of Ho:$YAlO_3$ and Nd, Ho:$YAlO_3$," *IEEE Journal of Quantum Electronics* 26(3):403–406, 1990 (research report).

Thomas J. Kane and Tracy S. Kubo: "Diode-pumped injection seeded Q-switched Tm: YAG laser," paper SSL1.3/ThL3 in *Leos '90 Conference Digest*, IEEE Lasers and Electro-Optics Society Annual Meeting, Boston, November 4–9, 1990 (research report).

Walter Koechner: *Solid-State Laser Engineering*, 2d ed., Springer-Verlag, Berlin and New York, 1988 (reference on all solid-state lasers).

Peter F. Moulton, Jeffrey G. Manni, and Glen A. Rines: "Spectroscopic and laser characteristics of Er,Cr:YSGG," *IEEE Journal of Quantum Electronics* 24(6):960–973, 1988 (research analysis).

Peter F. Moulton: "Mid-infrared solid-state lasers for medical applications," paper ELT3.1 in *LEOS '90 Conference Digest*, IEEE Lasers and Electro-Optics Society Annual Meeting, Boston, November 4–9 1990 (research report).

Alexander M. Prokhorov, and Ivan A. Shcherbakov: "Soviet developments in solid-state lasers," paper presented at *OPTCON'90* Boston, 1990 (overview of Soviet research).

Stefan Schnell et al.: "Acoustooptic Q switching of erbium lasers," *IEEE Journal of Quantum Electronics* 26(6):1111–1114, 1990 (research report).

Michelle D. Shinn et al., "Spectroscopic and laser properties of Pm^{3+}," *IEEE Journal of Quantum Electronics* 24(6):1100–1108, 1988 (research report).

Robert C. Stoneman, J. G. Lynn, and Leon Esterowitz: "Efficient laser pumped 2.80-μm $Er:LiYF_4$ cw laser," paper JWB1 in *Technical Digest, Conference on Lasers and Electro-Optics*, Anaheim, Cal., May 1990, Optical Society of America, Washington, D.C. (research report).

Marvin J. Weber (ed.): *CRC Handbook of Laser Science & Technology*, vol. 1, *Lasers & Masers*, CRC Press, Boca Raton, Fla., 1982 (reference).

Free-Electron Lasers

The free-electron laser is not yet a commercial product, but it could become important in the future. It promises high power and exceptionally broad tunability, from microwaves to soft x rays. These features are consequences of its unique energy-transfer mechanism—from free electrons traveling through a vacuum to a laser beam. Free electrons lack discrete energy levels, allowing (at least in principle) extremely broad tuning ranges, and intense electron beams can carry extremely high powers.

Fundamental Concepts

In the early 1970s John M. J. Madey, then at Stanford University, proposed the free-electron concept (Madey, 1971). The central idea is to extract light energy from electrons passing through a magnetic field with spatially periodic variations in intensity and direction. His team demonstrated first a free-electron laser amplifier (Elias et al., 1976) and later a free-electron oscillator (Deacon et al., 1977).

High-energy electrons emit light (called *synchrotron radiation*) when a magnetic field bends their path or otherwise accelerates (or decelerates) them. This emission resembles spontaneous emission from atoms or molecules, but free electrons lack discrete energy levels, so synchrotron radiation is not limited to discrete transitions.

In a free-electron laser, energetic electrons pass through a regular array of magnets with alternating polarity, called a "wiggler" or *undulator* magnet, as shown in Fig. 28.1. The regularly spaced magnetic fields bend the electron beam back and forth (thus the name "wiggler") periodically. After the electrons pass through part of the magnet array, they form clumps separated by the period of the magnetic field. Thus each clump experiences the same magnetic field at the same time, separated in space by one period of the magnet. If the

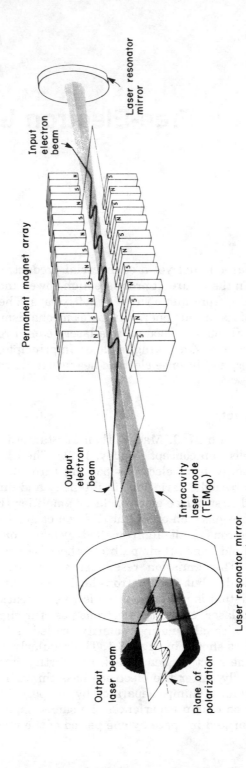

Figure 28.1 Structure of a free-electron laser. (*Courtesy of University of California at Santa Barbara Quantum Institute.*)

Input electron beam

Laser resonator mirror

Permanent magnet array

Output electron beam

Intracavity laser mode (TEM$_{00}$)

Laser resonator mirror and output coupler

Output laser beam

Plane of polarization

electrons have the same energy, they emit synchrotron radiation at the same wavelength, and the light waves from different clumps add together in phase. Although the process is not strictly stimulated emission, the resulting light beam is otherwise similar to conventional laser beams.

Light emitted along the magnet axis can be amplified if its wavelength λ matches a condition for resonance:

$$\lambda = \frac{p}{2[1 - (v^2/c^2)]}$$

where p is the period of the magnetic field (i.e., the distance between pairs of magnets with poles in the same direction), v is electron velocity, and c is the speed of light. Converting velocity to electron energy E gives

$$\lambda = \frac{0.131p}{(0.511 + E)^2}$$

a more useful equation which shows that reducing magnet period or increasing electron energy reduces output wavelength. In practice, the wiggler magnet spacing is usually fixed, and electron energy is varied to tune the wavelength. No single device can be tuned across the entire range of free-electron lasers, but the tuning bandwidth is larger than other lasers. One submillimeter free-electron laser can be tuned from 120 to 800 micrometers (μm), a much greater relative range than the most broadly tunable conventional laser, titanium-doped sapphire (see Chap. 24).

Free-electron lasers also have the potential for extremely high powers because they rely on well-developed technology for producing powerful beams of energetic electrons. Passing electrons repeatedly through the wiggler or recapturing their energy should allow high efficiency. In theory, 20 to 50 percent of the electron-beam energy should be convertible into light; efficiency above 30 percent has been demonstrated at long wavelengths. Power and efficiency are highest at long wavelengths, and free-electron lasers already are the most powerful sources of millimeter and submillimeter waves. Research continues on improving short-wavelength free-electron lasers.

Types of Free-Electron Lasers

Free-electron lasers fall into two major operating regimes, depending on wavelength and the interactions between light waves and electrons. In the Compton regime, wavelengths are less than about 20 μm, electron energies are above 20 mega-electronvolts (MeV), and the interactions are best described as between individual particles. At longer wavelengths, in the Raman or collective regime, the interactions with lower-energy electrons are best described as collective, multiparticle effects.

In the Compton regime, individual electrons have high energy, but total current is low. Energy is transferred between individual electrons and the electromagnetic field, and space-charge effects can be neglected. Gain is low in this mode, and present only for light traveling in the same direction as the electron beam. Compton regime lasers normally operate as oscillators because of the limited gain and low intrinsic efficiency—typically no more than 1 percent. High-reflectivity mirrors are needed to keep cavity losses low.

In the Raman regime, individual electrons have low energy, but total current is high, so electrons need not be considered individually, and space-charge effects are very important. Typical wavelengths are longer than 100 μm, electron energies under 5 MeV, and current densities above 1000 amperes per square centimeter (A/cm^2). Gain and efficiency are much higher in this regime, and in suitable devices gain increases exponentially with wiggler length. Gains of 40 to 50 decibels (dB) have been recorded on one pass. Raman regime lasers are compact, high-gain devices, with loose cavity optics requirements, and can operate as amplifiers as well as oscillators.

In both regimes, free-electron laser gain is higher if the spread of electron energy is narrow—that is, if all electrons have essentially the same energy, a condition called a "cold beam." Gain also depends on electromagnetic effects which influence beam propagation and the bunching of electrons within the beam. Details are quite complex (see, e.g., Brau, 1990; Luchini and Motz, 1990; or Marshall, 1985).

Components of Free-Electron Lasers

There are three critical components in any free-electron laser: the source of the magnetic field, the electron accelerator, and the optical cavity.

Magnetic-Field Sources. The usual picture of a free-electron laser shows an array of magnets stacked with alternating polarity, as shown in Fig. 28.1. Such magnet arrays typically are about 1 to 25 meters (m) long, with spacings on the order of centimeters. Either electromagnets or permanent magnets can be used, and other configurations (such as helical magnets) are possible as long as they produce the required spatial periodicity in the path that the electrons travel.

A strong electromagnetic wave also can provide a magnetic field which varies sinusoidally in intensity along the electron-beam path.

This concept is attractive for short-wavelength free-electron lasers, because it is hard to build magnet arrays with sufficiently short periods. A two-stage free-electron laser could simultaneously produce both long and short wavelengths (Kimel et al., 1986). A conventional wiggler in the first stage would generate a long-wavelength beam, which would provide the wiggler field to generate a much shorter wavelength. With a 10-MeV electron beam generating a 100-μm output, the second stage could produce 119-nanometer (nm) output.

Early free-electron lasers had constant-period wiggler magnets. However, decreasing the magnetic field intensity or period along the wiggler can improve energy extraction efficiency by maintaining the resonance with electrons as they lose energy to the laser beam.

Accelerators. Different types of accelerators can be used.

Radio-frequency linear accelerators, best suited to the Compton regime, use a traveling wave from a pulsed radio-frequency source to accelerate electrons to high energies, but with limited currents. They produce bunches of electrons (called *micropulses*) separated by the period of the driving electromagnetic wave, typically in the gigahertz range. They operate only during *macropulses* of the radio-frequency source lasting microseconds to tens of microseconds. The laser output consists of macropulses containing many micropulses.

Higher beam currents are possible with an induction accelerator, in which a pulse-forming network supplies the accelerating field. Electron pulses lasting tens of nanoseconds are produced at the repetition rate of the pulse-forming network [on the order of 1 hertz (Hz)]. The beam is less concentrated than in a radio-frequency accelerator, and this source is used with Raman regime lasers.

A microtron is a radio-frequency accelerator in which electrons pass repeatedly through the same small accelerator cavity, each time gaining more energy. The electrons follow circular orbits which grow in diameter with each pass through the cavity but remain confined in the accelerator until they reach a certain energy—to a few tens of millions of electron volts—at which point they escape through a shielded channel.

Another alternative for electron energies to 10 MeV is a pulsed electrostatic accelerator, such as a van de Graff generator. Up to 90 percent of the electron-beam energy can be recovered, allowing high efficiency and longer pulses. Some researchers believe very efficient beam recovery could allow continuous submillimeter output.

While most accelerators produce electron beams that make only one pass through the magnet array, a storage ring could make multiple passes. The wiggler magnet would be placed in one arm of the ring, which receives electrons from an external accelerator, confines them in a loop, and accelerates them to high energies. A storage-ring free-electron laser generates pulses as electron clumps pass through the undulator. It operates in the Compton regime, with cavity mirrors placed at the ends of the undulator. Storage rings have produced the shortest wavelengths to date, but their power has been limited.

Optical Cavities and Beam Characteristics

The low gain of short-wavelength Compton regime lasers makes cavity design important. At ultraviolet or visible wavelengths, mirror reflectivity of 99 percent may be needed to sustain oscillation. Multilayer dielectric coatings offer that high reflectivity but are damaged by ultraviolet synchrotron radiation and harmonics of the fundamental wavelength present in a free-electron laser.

It might be possible to avoid this problem by generating only modest powers in an oscillator and boosting output in an amplifier without cavity optics. In initial experiments at 3 μm, gain to 60 percent was measured in an oscillator-amplifier configuration (Vintro et al., 1990).

Applications

Their broad tunability and potential for efficiently generating high powers make free-electron lasers sound attractive for many applications. Unfortunately, their cost is not. Even small electron accelerators are expensive; large ones able to produce high-power beams or the high-energy electrons needed for short wavelengths are even more expensive. Even small free-electron lasers emitting a few hundred watts are likely to cost hundreds of thousands of dollars. Moreover, the technology is far from mature, because the large scale of experiments has led to long lead times.

Current technology works best at submillimeter wavelengths, where free-electron lasers now offer higher narrowband powers than any other source. That spectral region has been little explored, partly because good sources have not been available. Potential applications include spectroscopy, materials research, and plasma heating in fusion devices.

Infrared free-electron lasers could find medical applications (Danly et al., 1987). Laser-tissue interactions depend strongly on wavelength. Free-electron lasers could greatly expand options available beyond about 3 μm, where the only choices now are 10-μm CO_2 lasers, 5-μm CO lasers, and chemical DF lasers at 3.6 to 4.0 μm.

The Strategic Defense Initiative spent hundreds of millions of dollars in the late 1980s trying to develop high-energy free-electron laser weapons with wavelengths of 1 μm or less. Planners envisioned directing the beams from several ground-based multimegawatt free-electron lasers to orbiting relay mirrors which would relay and refocus the beams onto targets in space. However, limited budgets and changing national priorities greatly reduced that effort and delayed plans for a large demonstration laser at the White Sands (N.M.) Missile Range.

Bibliography

Charles Brau: *Free-electron Lasers*, Academic Press, Boston, 1990 (overview).

Bruce G. Danly, R. J. Temkin, and George Bekefi: "Free-electron lasers and their application to biomedicine," *IEEE Journal of Quantum Electronics QE-23*(10):1739–1750, 1987 (review paper).

David A. G. Deacon et al.: "First operation of a free-electron laser," *Physical Review Letters 38*:892, 1977 (research report).

L. R. Elias et al.: "Observation of stimulated emission of radiation by relativistic electrons in a spatially periodic transverse magnetic field," *Physical Review Letters 36*:717, 1976 (research report).

Luis Elias: "Free-electron laser research at the University of California, Santa Barbara," *IEEE Journal of Quantum Electronics QE-23*(9):1470–1475, 1987 (research summary).

Isidoro Kimel, Luis R. Elias, and Gerald Ramian: "The UCSB Two-Stage FEL experiment," *Nuclear Instruments and Methods in Physics Research A250*:320–327, 1986 (research report).

P. Luchini and H. Motz, *Undulators and Free-Electron Lasers*, Oxford Science Publications, Oxford, 1990 (monograph).

John M. J. Madey: "Stimulated emission of bremsstrahlung in a periodic magnetic field," *Journal of Applied Physics 42*:1906, 1971 (first proposal for a free-electron laser).

Thomas C. Marshall: *Free-electron Lasers*, Macmillan, New York, 1985 (overview).

John A. Pasour: "Free-electron lasers," *IEEE Circuits and Devices Magazine*, March 1987, pp. 55–64 (review article).

L. Vintro et al.: "Observation of gain in a free-electron laser master oscillator power amplifier," *Physical Review Letters 64*(14):1662–1665, April 2, 1990 (research report).

X-Ray Lasers

The x-ray laser is a seemingly logical but in practice quite difficult extension of conventional laser technology. Visible and near-ultraviolet lasers operate on electronic transitions in the outer or valence shells of atoms. Electronic transitions from outer shells to inner shells involve much more energy, and thus produce x rays, which have much shorter wavelengths. In theory, a population inversion on such an inner-shell transition should, under the right conditions, produce an x-ray laser.

The problem is that the right conditions are extremely difficult to produce. X-ray transitions are so energetic that it takes extremely high peak powers to produce population inversions. Moreover, x-ray transitions have very short excited-state lifetimes, so the energy must be applied and extracted very quickly.

Two fundamentally different approaches were demonstrated successfully in the 1980s by the Lawrence Livermore National Laboratory. One used the most intense energy source ever produced by the human race—a nuclear explosion—and details remain highly classified. The other used short, intense pulses from high-energy lasers built for fusion research, and is unclassified.

X-ray laser research began in the early 1960s but soon foundered because of the difficulty of producing the conditions required for lasing. By the late 1970s many unsuccessful experiments had been tried (Waynant and Elton, 1976), and most government support had dried up. Ironically, the first sign of progress appeared soon afterward, when British researchers saw evidence of laser gain in highly ionized carbon (Jacoby et al., 1981).

Bomb-Driven X-Ray Lasers

Military research on bomb-driven x-ray lasers began in the 1970s but attracted little attention until early 1981, when *Aviation Week &*

Space Technology reported that the Livermore had demonstrated one (Robinson, 1981). Livermore officials later admitted that they had demonstrated a bomb-driven x-ray laser, but never have confirmed other information reported in the article, and most details remain classified. If the reported 1.4-nanometer (nm) wavelength was accurate, it remains the shortest wavelength at which lasing has been reported.

Careful analysis of the sparse public information (see, e.g., Hecht, 1984, pp. 126–136) indicates that the intense thermal x rays from an underground nuclear blast powered the x-ray laser. It vaporized the laser medium, probably a rod or foil, stripping away many of its electrons to produce a plasma. Before the plasma could disperse, some electrons fell back to the inner shells, creating a population inversion. Spontaneous emission of a few x rays stimulated emission from other ions which were (briefly) in the excited state. X-ray transitions have very high gain, so no cavity mirrors were needed.

Beam size and shape depended on plasma geometry. A cylindrical plasma would emit the most x rays in both directions along its length, where the most amplification is possible. X-ray laser pulses must be very short because the excited-state lifetime is on the order of picoseconds and the plasma and nuclear fireball disperse rapidly.

Edward Teller strongly argued for the bomb-driven x-ray laser, and it became a controversial cornerstone of Ronald Reagan's Strategic Defense Initiative. However, serious technical issues caused further controversy in the mid- to late 1980s, and the bomb-driven x-ray laser has since been downgraded in importance (Morrison, 1990).

Laboratory X-Ray Lasers

Development of "laboratory" x-ray lasers followed a qualitatively similar approach—hitting the laser material with an intense burst of energy. However, the energy for the first demonstration was a 0.5-nanosecond (ns) pulse of light with peak power in the terawatt (10^{12} W) range from a massive neodymium-glass laser, built for fusion experiments. It was focused onto a line on a thin foil, which it ionized, producing a plasma and a population inversion on an x-ray transition. The first solid evidence for x-ray laser action was on two lines of highly ionized selenium (Se^{24+}) at 20.6 and 20.9 nm (Matthews et al., 1985).

Collisions in the long, thin laser-produced plasma actually generate the population inversion, exciting the Se^{24+} ion to a high energy level. Spontaneous emission of an x-ray photon stimulates emission from other excited Se^{24+} ions, producing a very short pulse of amplified spontaneous emission. (No mirrors were used in the original experiments.)

Several additional x-ray laser lines have been demonstrated since the early Livermore experiments. Other researchers have varied the basic technique, but the fundamental approach remains excitation of a linear plasma with a high-intensity laser pulse. Directionality depends on the shape of the illuminated plasma because the laser lacks a true resonant cavity.

Wavelengths range from about 4 to over 30 nm, with progress toward shorter wavelengths. In 1990 Livermore observed gain at 4.316 nm in tungsten missing 46 electrons (W^{46+}) (MacGowan et al., 1990), the first public report of gain below the 4.36-nm K-edge absorption barrier of carbon. That wavelength range could be useful in studying living cells.

Livermore's building-sized fusion laser is hardly a practical pump source for most experiments. Other groups are trying to make more compact and efficient x-ray lasers, probably at longer wavelengths. Efforts include both seeking transitions which could be pumped with less laser power, and developing more practical pump lasers able to deliver the ultrashort high-power pulses needed to excite an x-ray laser.

At this writing, it is not clear what shape a "practical" laboratory x-ray laser will take, or when one might be available. One expert projects wavelengths of 5 to 20 nm, with peak powers of 1 to 10 megawatts (MW) in a pulse of 0.1 to 1 ns, generated with an efficiency of 10^{-6}, implying a pump laser peak power of 10^{12} to 10^{13} W (Elton, 1990, p. 241).

Applications

As with chemical and free-electron lasers, the primary impetus for most x-ray laser research was interest in high-energy laser weapons. The potential small size and high power make the bomb-driven x-ray laser attractive as a space-based weapon. Chemical lasers require bulky fuel tanks, and free-electron lasers require even bulkier sources of electrical power, but some advocates said a single small nuclear explosive could power up to a few dozen x-ray lasers. However, physical reality appears to have shot down such visions.

Laboratory x-ray lasers could find important research applications. One intriguing possibility is x-ray holograms of living cells to record three-dimensional images (Solem and Baldwin, 1982). Wavelengths in the 4-nm range should be strongly scattered by microstructures within the cells. (The intense x rays would destroy the cells, but the recording would preserve an image.)

Monochromatic, coherent x rays also could be useful in materials research. Much brighter than current synchrotron sources, they could perform now-impossible measurements and be used in electron spectroscopy and physical research. Tightly focused x-ray beams could be used for micromachining and photolithography, with much higher resolution than ultraviolet lasers. Such higher resolution would allow fabrication of finer diffraction gratings and smaller components on semiconductor chips.

Bibliography

Raymond C. Elton: *X-ray Lasers*, Academic Press, Boston, 1990 (comprehensive reference).

Jeff Hecht: *Beam Weapons: The Next Arms Race*, Plenum, New York, 1984 (overview).

D. Jacoby et al.: "Observation of gain in a possible extreme ultraviolet lasing system," *Optics Communications* 37(3):193–196, 1981 (research report).

B. J. MacGowan, et al.: "Demonstration of x-ray amplifiers near the carbon K edge," *Physical Review Letters* 65:420–423, 1990 (research report).

Dennis L. Matthews et al.: *Physical Review Letters 54*:110, 1985 (research report).

David Morrison: "The rise and stall of the X-ray laser," *Laser & Optronics 9*(8):23–24, August 1990 (news analysis).

Clarence A. Robinson, Jr.: "Advance made on high-energy laser," *Aviation Week and Space Technology*, February 23, 1981, pp. 25–27 (news report leaking classified information).

Johndale C. Solem and George C. Baldwin: "Microholography of living organisms," *Science 218*:229–234, October 15, 1982 (review article).

Ronald W. Waynant and Raymond C. Elton: "Review of short-wavelength laser research," *Proceedings of the IEEE 64*(7):1059–1092, 1976 (review article).

Types of Laser

Commercial Laser Types, Organized by Wavelength

Wave-length, μm	Type	Chapter	Output type and power
0.152	Molecular fluorine (F_2)	13	Pulsed, to a few watts average
0.192	ArF excimer	13	Pulsed, to tens of watts average
0.2–0.35	Doubled dye	17	Pulsed
0.222	KrCl excimer	13	Pulsed, to a few watts average
0.235–0.3	Tripled Ti-sapphire	24	Pulsed
0.248	KrF excimer	13	Pulsed, to over 100 W average
0.266	Quadrupled Nd	22	Pulsed, watts
0.275–0.306	Argon ion	8	Continuous-wave (CW), 1-W range
0.308	XeCl excimer	13	Pulsed, to tens of watts
0.32–1.0	Pulsed dye	17	Pulsed, to tens of watts
0.325	He–Cd	9	CW, to tens of milliwatts
0.33–0.36	Ar or Kr ion	8	CW, to several watts
0.33–0.38	Neon	8	CW, 1-W range
0.337	Nitrogen	14	Pulsed, under 1 W average
0.347	Doubled ruby	23	Pulsed, under 1 W average
0.35–0.47	Doubled Ti-sapphire	24	Pulsed
0.351	XeF excimer	13	Pulsed, to tens of watts
0.355	Tripled Nd	22	Pulsed, to tens of watts
0.36–0.4	Doubled alexandrite	24	Pulsed, watts
0.37–1.0	CW dye	17	CW, to a few watts
0.442	He–Cd	9	CW, to over 0.1 W
0.45–0.52	Ar ion	8	CW, to tens of watts
0.48–0.54	Xenon ion	16	Pulsed, low average power
0.51	Copper vapor	12	Pulsed, tens of watts
0.523	Doubled Nd–YLF	22	Pulsed, watts
0.532	Doubled Nd–YAG	22	Pulsed to 50 W or CW to watts
0.534, 0.538	He–Cd	9	CW, milliwatts, in white-light laser
0.5435	He–Ne	7	CW, 1-mW range
0.578	Copper vapor	12	Pulsed, tens of watts
0.594	He–Ne	7	CW, to several milliwatts

Commercial Laser Types, Organized by Wavelength (*Continued*)

Wave-length, μm	Type	Chapter	Output type and power
0.612	He–Ne	7	CW, to several milliwatts
0.628	Gold vapor	12	Pulsed
0.6328	He–Ne	7	CW, to about 50 mW
0.635–0.66	InGaAlP diode	19	CW, milliwatts
0.636	He–Cd	9	CW, milliwatts, in white-light laser
0.647	Krypton ion	8	CW, to several watts
0.67	GaInP diode	19	CW, over 10 mW
0.68–1.13	Ti-sapphire	24	CW, watts
0.694	Ruby	23	Pulsed, to a few watts
0.72–0.8	Alexandrite	24	Pulsed, to tens of watts (CW in lab.)
0.73	He–Ne	7	CW, 1-mW range
0.75–0.9	GaAlAs diode	19	CW, to many watts in arrays
0.98	InGaAs diode	19	CW, to 50 mW
1.047 or 1.053	Nd–YLF	22	CW or pulsed, to tens of watts
1.061	Nd-glass	22	Pulsed, to 100 W
1.064	Nd–YAG	22	CW or pulsed, to kilowatts
1.15	He–Ne	7	CW, milliwatts
1.2–1.6	InGaAsP diode	20	CW, to 100 mW
1.3–1.4	Overtone HF	11	CW or pulsed, to tens of watts
1.313	Nd–YLF	22	CW or pulsed, to 0.1 W
1.315	Iodine	16	Pulsed, to several watts average
1.32	Nd–YAG	22	Pulsed or CW, to a few watts
1.4–1.6	Color center	25	CW, under 1 W
1.523	He–Ne	7	CW, milliwatts
1.54	Erbium-glass (bulk)	27	Pulsed, to 1 W
1.54	Erbium-fiber (amplifier)	26	CW, milliwatts
1.75–2.5	Cobalt-MgF_2	24	Pulsed, 1-W range
2–4	Xe–He	16	CW, milliwatts
2.1	Holmium	27	Pulsed, watts
2.3–3.3	Color center	25	CW, under 1 W
2.6–3.0	HF chemical	11	CW or pulsed, to hundreds of watts
2.94	Erbium-YAG	27	Pulsed, to tens of watts
3.3–29	Lead salt diode	21	CW, milliwatt range
3.39	He–Ne	7	CW, to tens of milliwatts
3.6–4.0	DF chemical	11	CW or pulsed, to hundreds of watts
5–6	Carbon monoxide	16	CW, to tens of watts
9–11	Carbon dioxide	10	CW or pulsed, to tens of kilowatts
10–11	Nitrous oxide (N_2O)	16	CW
40–1000	Far-infrared gas	15	CW, generally under 1 W

Index